Fundamentals of Power Electronics

Fundamentals of Power Electronics

Editor

Tudor Volkov

scitus academics

Fundamentals of Power Electronics
Edited by **Tudor Volkov**

ISBN: 978-1-68117-122-7
Library of Congress Control Number: 2015950367

SCITUS Academics LLC,
www.scitusacademics.com
Box No. 4766, 616 Corporate Way,
Suite 2, Valley Cottage,
NY 10989

Notice

Printed in United States of America on Acid Free Paper

Table of Content

Preface

Power electronics is the application of solid-state electronics to the control and conversion of electric power. It also refers to a subject of research in electronic and electrical engineering which deals with the design, control, computation and integration of nonlinear, time-varying energy-processing electronic systems with fast dynamics.

The power electronics field has evolved beyond the status of comprising one or two special-topics courses. In the breadth versus depth tradeoff, it no longer makes sense for one textbook to attempt to cover all of these courses; indeed, each course should ideally employ a dedicated textbook.

In the power electronics literature, much has been made of the incorporation of other disciplines such as circuits, electronic devices, control systems, magnetic, and power applications, into the power electronics field. Yet the field has evolved, and now is more than a mere collection of circuits and applications linked to the fundamentals of other disciplines. There is a set of fundamentals that are unique to the field of power electronics. It is important to identify these fundamentals, and to explicitly academic conferences, and other affairs around these fundamentals. This book is organized around the fundamental principles, while the applications and circuits are introduced along the way as examples.

CHAPTER 1

Operation and Control of Wind Farms in Non-Interconnected Power Systems

Inanes D Margaris[1,2], Anca D Hansen[1], Nicolaos A Cutululis[1], Poul Sørensen[2] and Nikos D. Hatziargyriou[1]

[1]National Technical University of Athens (NTUA), Greece,

[2]Risø DTU National Laboratory for Sustainable Energy, Denmark

[3]Public Power Corporation (PPC) S.A., Greece

INTRODUCTION

Autonomous power systems are characterized by the absence of interconnections with neighbouring systems due to geographical, economic and political reasons. These systems face particular problems associated with safety and reliability during the design and operation procedure associated with safety and reliability. Typical problems include large variations in frequency because of the low inertia and the large fluctuations voltage due to the low short circuit ratio. The quality of the provided electricity to consumers is threatened. At the same time, special features of non interconnected systems, such as concentration of production in a limited number of power stations, the large size of the units in relation to the load, the need for larger spinning reserve due to the absence of interconnections, and the small stability margins raise the impact on safety and cost of operation.

Under these conditions, the effective handling of transient phenomena arising due to serious disorders is particularly critical. The systems should respond adequately to dynamic events and ensure static anddynamic safety. The most common faults that may cause undesired events are the loss of transmission lines, the sudden loss of load, and short circuits – especially three phase errors – and loss of production units. Based on collected operational data, incidents of loss of unit during operation are quite common and cause serious problems, therefore require special treatment. In several cases, such events have led in the past in smaller or even general black-outs.

These problems are becoming more intense due to the increasing penetration of wind power in the last decade. Since renewable energy sources and particularly wind energy have stochastic behavior, the power output is not guaranteed. This is the main factor that imposes restrictions on the expansion because in general, distributed energy sources do not contribute to the control and regulation of the system in the same way as conventional units. Another important point, which differentiates the turbines compared with conventional synchronous generators used in electric systems, is associated with the technology of converting mechanical energy into electrical. The wind turbines are in large proportion equipped with asynchronous generators (possibly in conjunction with electronic power converters) and therefore have substantial differences in the dynamic response over conventional units. For these reasons, limits are always imposed in the instantaneous penetration of wind power. These limits vary across the power systems, depending on the specific circumstances prevailing in each autonomous system, both in terms of conventional units (e.g. production technology, control capabilities, etc.) and wind farms (size and technology of the wind turbines, dispersion of wind turbines on the island, etc.). It is often the case that the limit set by the system operator for the instantaneous penetration of wind power is around 30% -40% of the load. In order to allow both the evaluation of the dynamic behaviour of autonomous systems after severe disturbances (e.g. ability of the system to restore frequency back to the desired limits after a major disturbance, such as loss of production and / or lines) as well as the definition of safe penetration limits, it is essential to conduct numerous studies. These include transient stability, load - frequency regulation, etc. The development of appropriate models for dynamic simulations in non interconnected systems is critical.

POWER SYSTEM MODEL

Thermal Power Plant Models

The conventional generating capacity comprises usually diesel, gas and steam plants with different ratings and control attributes. Each thermal plant contains several control blocks, which are essential for power system of dynamic simulations, e.g. voltage controller, primary controller (governor), primary mover unit and the synchronous generator. In many cases, due to lack of accurate data, simplified models for the conventional units are used in simulations. In this study, the exact models for each unit were used to ensure optimal representation of the interaction between wind farms and the power system.

The following three different models, already existing as built-in standard models in PowerFactory, (DIGSILENT, 2006), are used for the governors: GAST2A model for the gas turbines, DEGOV1 model for thediesel generators and IEEEG1 general model for the steam plants. A detailed description of the GAST2A built-in model in PSS/E for the governor used in the gas plant is described in (Mantzaris et al., 2008), while details on the corresponding standard IEEEG1 model for the governor in the steam plant can be found in (DIGSILENT, 2006). The parameters of these models, validated both in Matlab and PSS/E software packages, are presented in (Mantzaris et al., 2008). For the Automatic Voltage Regulators (AVR), the built-in SEXS model of PowerFactory is used with adjusted parameters for each unit.

Dynamic Load Models

The electrical loads of the systems include typically various kinds of electrical devices. An appropriate approach for the dynamic modeling of the loads connected to Medium Voltage (MV) feeders is to assume constant impedance of the loads during dynamic simulations, (Cutsem&Vournas, 1998):

$$P = P_0 (V / V_0)^2$$

$$Q = Q_0 (V / V_0)^2$$

where P, P_0 and Q, Q_0 are the active and reactive power consumed by the load for voltage equal to reference voltage V, V_0 respectively.

Protection System

The protection system was also modeled in the simulation platform. The settings for both under/over voltage and under/over frequency protection system are crucial for the operation and dynamic response of the system during transient instances. As mentioned in the Introduction, non interconnected system, like the one used in this report as a study case, face the problem of significant variations in voltage and frequency. The relays, which act on either the production (protection of the conventional units or protection of the wind turbines), or the demand side (relays attached on the

Medium Voltage feeders) decide the disconnection of equipment or loads, when the limits set by the system operator (or the production unit user) are violated. Regarding the loads, this leads to the so called load shedding, which often determines also the dynamic security margins for the system. It is often the case, in isolated systems, with low inertia, that during frequency variations, large proportion of the load is disconnected to avoid further frequency drop and possible frequency instability, i.e. due to sudden loss of a production unit.

The voltage and frequency protection system was modeled specifying the lower (or upper) limit of the value and the time duration, during which the variable measured, is out of the accepted range. One kind of under/over frequency protection operating in modern power systems is the so called ROCOF protection (Rate of Change of Frequency). The relays controlled by this system, open when the frequency changes at a rate faster than the specified one for a specific time. Thus, a part of the substation loads is disconnected. However, in many non-interconnected systems, especially those designed many decades ago, the under/over frequency protection system controlling the relays at substation loads measures the actual frequency and not the rate of change. Thus, if the frequency drops lower than a specified limit for specific time duration, the relay is ordered to open.

As a case study the small size island system of Rhodes is used. Rhodes power system for the reference year 2012 includes a 150 kV transmission system, two power plants, distributed in the north and in the south, as shown in Figure 1, and five wind farms. A significant proportion of the generation comes from wind turbines and diesel units. In 2012, the total installed wind power capacity and the maximum annual power demand are assumed to be about 48 MW and 233 MW, respectively (seeTable 1).

The present Rhodes power system model is based on dynamic models of conventional generating units, loads and wind turbines. In order to be able to perform power system simulation studies for 2012, the present system model has to be modified with additional generating units and wind farms, which are expected to be online by the year of study, 2012, (Margaris et al. 2009). The protection system, mainly under/over frequency and voltage protection relay is also included in the dynamic power system model. In the reference year study 2012, five wind farms with different technologies will be connected online in Rhodes power system. Table 2 depicts the wind turbine technology and the size of each wind farm.

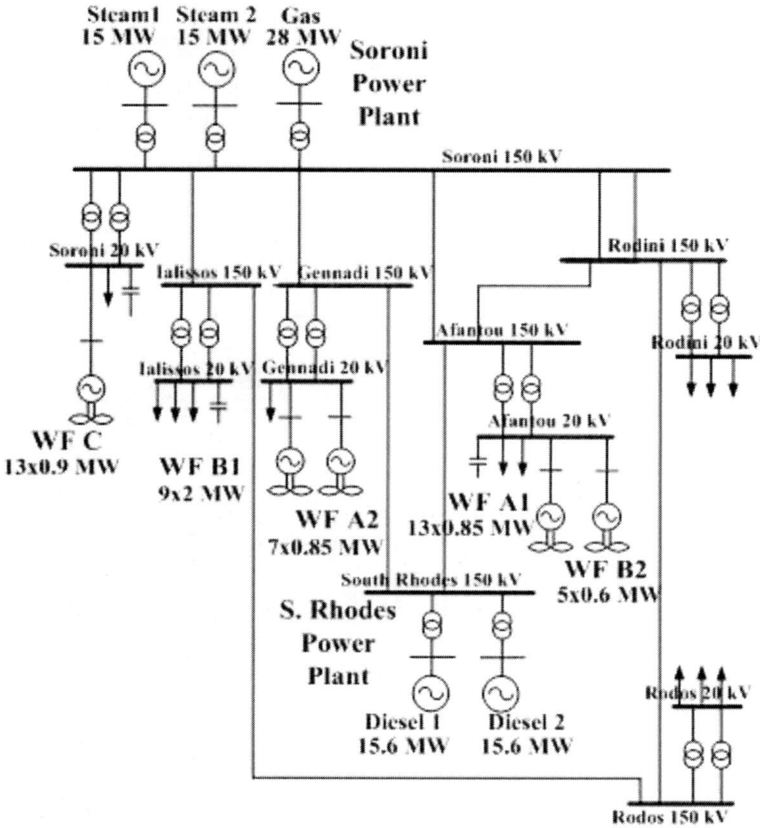

Figure 1. Rhodes power system.

The basic characteristics of Rhodes power system in 2012 are summarized in Table 1:

Table 1. Basic Characteristics of Rhodes Power System (2012).

Rhodes power system	
Max Power Demand (MW)	233.1
Rated Thermal Power (MW)	322.9
Rated Wind Power Generation (MW)	48.8

Table 2. Wind Farms in Rhodes Power System (2012).

	Wind Turbine Technology	Installed Capacity (MW)
Wind Farm A1	DFIG	11.05
Wind Farm A2	DFIG	5.95
Wind Farm B1	PMSG	18
Wind Farm B2	PMSG	3
Wind Farm C	ASIG	11.7

Load Scenarios

Regarding the first step of the approach, the operating scenarios have to be carefully defined. These scenarios are based on collected operational data of the power system and correspond to the possible severe condition of operation. In this way, it is ensured that their analysis covers the intermediate modes of operation in terms of security. Three reference scenarios were defined as follows:

- The Peak Load Demand scenario – SCENa
- The Maximum Wind Power Production scenario (in absolute values of power) – SCENb
- The Maximum Wind Power Penetration scenario (in percentage of the load demand) – SCENc

The first scenario is the base case scenario and is used to evaluate the operational mode of the system in terms of security without significant wind power production, because annual peak load occurs in a hot summer day with typically very low wind. The second scenario is used to investigate security with large wind power production levels. In this case the levels of wind penetration are quite high going beyond 20% of the total load demand. The third scenario examines a penetration level above the 30% margin, which has been used until now for wind energy as a rule of thumb in autonomous island systems.

Static Security Analysis

Under the different scenarios, the secure operation of the system for steady operation has to be ensured, based on the N and N-1 criteria. Among the security requirements which have to be fulfilled by the power system are the following:

- The loading of the transmission lines should be within the accepted limits
- Bus voltages should be in the range of ±5%around the nominal voltage for normal operation (N)
- Bus voltages should be in the range of ±10%around the nominal voltage for emergency operation (N-1)

WIND POWER FLUCTUATIONS

This part addresses different grid integration issues of large wind farms in non-interconnected power systems with respect to secure operation during variable wind and load profiles. Today, the power systems all over the world need a dramatic and continuous restructuring, as different renewable energy technologies are going to replace some of conventional units in the near future. This means, that there is urgent need for accurate modeling of various different generation technologies and novel wind turbine control strategies to fulfill requirements set by the TSOs, in order to ensure the dynamic security of such power systems.

Especially referring to wind power, the fluctuating nature of wind power imposes serious challenges to system operators. Power system inertia, protection relays settings, voltage and frequency stability in autonomous power systems have to be carefully and thoroughly analyzed before the penetration margin levels are expanded.

In most of the cases, operation experience defines the accepted penetration levels keeping the margin at 25-30% of peak annual load. However, higher or lower values can actually be accepted depending on the combination of power generator technologies online, (Margaris et al. 2009) – as it is the case of the specific power system under study here.

Modeling considerations of the power system under study are presented, especially speed governors and automatic voltage regulators of the conventional units. Defining accurate models of these system components is of vital importance for the overall system performance.

Modeling and control issues for three different wind turbine technologies are described. The study case of Rhodes island is presented through simulations for various load and wind scenarios. The frequency fluctuations are calculated using wind time series based on measured and validated results, (Sørensen et al., 2007).

Wind Farms Modelling

Modern power systems include a variety of wind turbine technologies. The different response of each kind in dynamic phenomena requires separate and detailed modeling of each one. Three wind turbine technologies are considered here, namely Doubly Fed Induction Generator (DFIG), Permanent Magnet Synchronous Generator (PMSG) and Active Stall Induction Generator (ASIG) based wind turbines.

In order to investigate the interaction between these large wind farms and the power system, an aggregated method for modeling the wind farms is used, (Akhmatov et al., 2003; Poeller& Achilles, 2003). Such modeling approach is commonly used for power system studies, as it reduces substantially both the complexity of the system and the computation time, without compromising the accuracy of the simulation results.

Models for all these different wind turbine technologies are implemented, including the main components of each wind turbine configuration:

- Drive train and aerodynamics
- Pitch angle control system
- Control system
- Protection system

The system configuration, as well as some modeling and control issues for each wind turbine technology are described in the following section.

System Configuration of Variable Speed DFIG Wind Turbine

The DFIG wind turbine configuration stands nowadays as the mainstream configuration for large wind turbines, (A.D. Hansen & L.H. Hansen, 2007). To ensure a realistic response of a DFIG wind turbine, the main electrical components as well as the mechanical parts and the controllers have to be considered in the model. The model used in this study for the wind farms with DFIG wind turbines is described in detail in (A.D. Hansen et al., 2006). The DFIG system is essentially a wound rotor induction generator with slip rings, with the stator directly connected to the grid and with the rotor interfaced through a back-to-back partial-scale power converter (A.D. Hansen & G. Michalke, 2007). The converter consists of two conventional voltage source converters (rotor-side converter RSC and grid-side converter GSC), and a DC-bus, as illustrated in Figure 2.

A two-mass model is used to represent the drive train to illustrate the dynamic impact of wind turbines on the grid properly. A large mass for the wind turbine rotor and a small mass for the generator are thus connected by a flexible shaft characterized by stiffness and damping for the low-speed shaft. A simplified aerodynamic model, based on a two-dimensional aerodynamic torque coefficientCqtable, (A.D. Hansen & G. Michalke, 2007), is typically sufficient for such studies.

The control system consists of a pitch control system and an electrical control system of the converters. The pitch control system is realized by a PI controller with antiwind-up, using a servomechanism model with limitation of both the pitch angle and its rate-of-change. As the pitch angle controls directly the generator speed to its reference signal, this control is able to prevent over-speeding both in normal operations and during grid faults, by limiting the mechanical power extracted from the wind and thus restoring the balance between electrical and mechanical power.

The efficiency of the pitch control system is crucial for studies like in this work, where the response of the wind farms under variable and extreme wind conditions is vital from the power system perspective.

The electrical control system, depicted in Figure 2, is essential for a good behavior of a DFIG wind turbine during both normal and grid fault operations. Decoupled control of active and reactive power is applied through vector control techniques, (A.D. Hansen & G. Michalke, 2007), allowing for changes in the active and reactive power in the range of milliseconds. The RSC controls mainly the active and reactive power delivered to the grid, while the GSC ensures nominal voltage at the common DC-bus at unity power factor operation of the converter. As illustrated in Figure 2 and described in details in (Gail et al., 2006; M.H. Hansen et al., 2005), the control of the converters is based on cascade control loops: a very fast inner current controller regulates the currents to the reference values that are specified by the outer slower power controller.

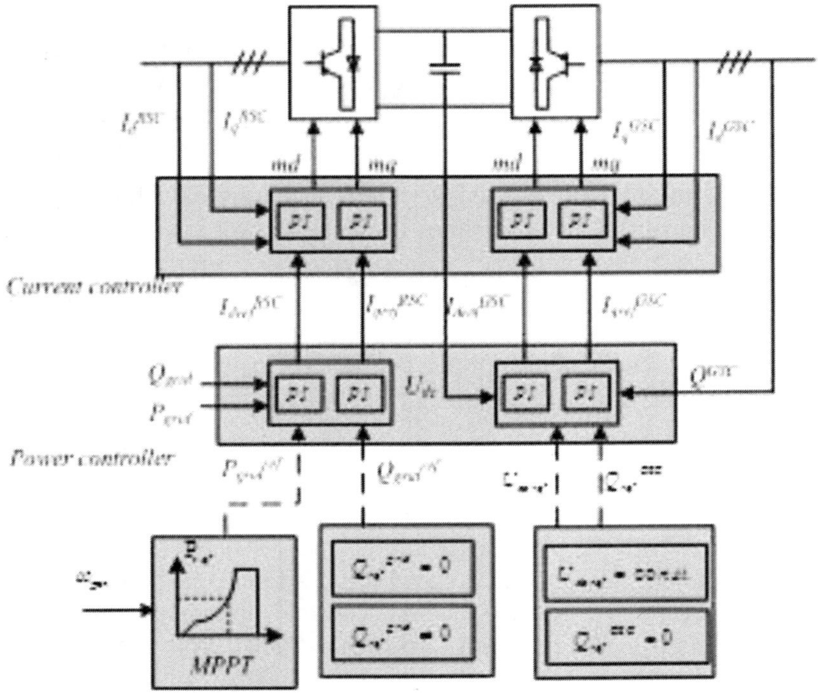

Figure 2. Electrical control scheme for the DFIG wind turbine.

System Configuration of Variable Speed PMSG Wind Turbine

Similar to the DFIG wind turbine configuration, the PMSG model consists both of a wind turbine mechanical level (i.e. aerodynamics, gearless drive-train and pitch angle control) and a wind turbine electrical level (i.e. multi-pole PMSG with a full-scale frequency converter and its control).

The synchronous generator is connected to the grid through a full-scale frequency converter system that controls the speed of the generator and the power flow to the grid. The full-scale frequency converter system consists of a back-to-back voltage source converter (generator-side converter and the grid-side converter connected through a DC link), controlled by IGBT switches. The rating of the converter system in this topology corresponds to the rated power of the generator plus losses. The presence of such a converter enables the PMSG to keep its terminal voltage on a desired level and to adjust its electric frequency according to the optimized mechanical frequency of the aerodynamic rotor, independently of the fixed electric frequency and the voltage of the AC grid.

The wind turbine mechanical level of PMSG (aerodynamics, drive train and the pitch control) is modelled similarly to that for the DFIG wind turbine. The generic control strategy of the frequency converter for the

PMSG wind turbine, as illustrated in Figure 3, is described in details in (A.D. Hansen & G. Michalke, 2008).

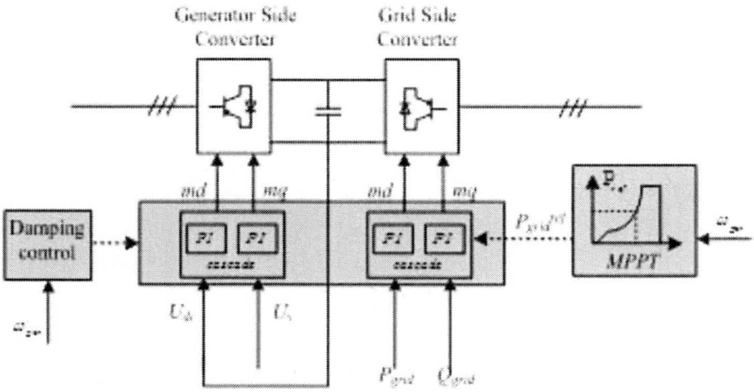

Figure 3. Power converter control strategy for the variable speed wind turbine with multi-pole PMSG.

The damping controller ensures stable operation for the wind turbine, by damping the torsional oscillations excited in the drive-train and reflected in the generator speedωgen. The generator-side converter controller keeps DC-bus voltage U_{dc} constant and controls the generator stator voltage $U_s$to its rated value in the stator voltage reference frame. The advantage of controlling the generator stator voltage Us to its rated value is that the generator and the power converter always operate at the rated voltage, for which they are designed and optimized. The grid-side converter controller controls independently the active Pgridand the reactive Qgrid power to the grid in the grid voltage reference frame. The controllers of the converter are also based on control loops in cascade, similarly to the DFIG control scheme.

System Configuration of Fixed Speed Asig Wind Turbine

The sub-models for aerodynamics, mechanical components and the squirrel cage induction generator in the ASIG wind turbine configuration are as described in (A.D. Hansen et al., 2003). The drive train is again represented by a two-mass model, similar to the other configurations. The turbine's power is controlled directly by the pitch controller through the pitch angle. The operation of the wind turbine is based on two control modes:

- Power limitation - where the turbines power is limited to the rated power above rated wind speed. The stall effect is thus controlled by adjusting accordingly the pitch angle.
- Power optimization - where the aerodynamic efficiency and thus the power output is maximized for wind speeds below rated wind speed. The pitch angle as function of the wind speed is shown in Figure 4.

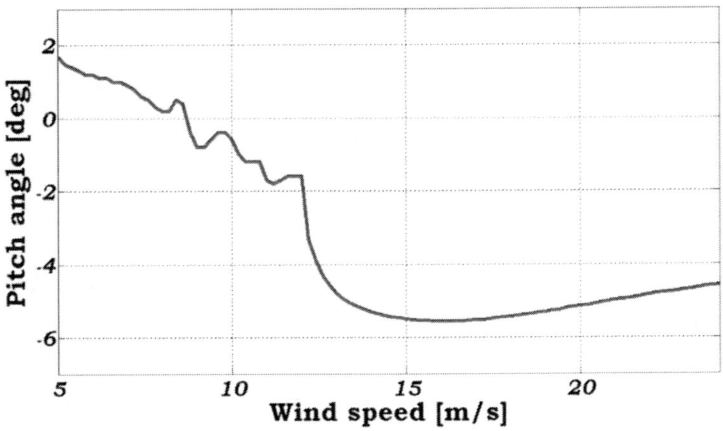

Figure 4. Pitch angle as a function of the wind speed.

Notice that all three wind turbine configurations are using a gain scheduling procedure, (Ackermann, 2005), in their pitch control system in order to compensate for the existing non-linear aerodynamic characteristics. The fundamental principle of the gain scheduling feature is that the proportional controller of the pitch angle controller is varied so that the total gain of the system remains unchanged for all the operating points of the wind turbine.

Simulation Results

In the following, the emphasis is made on the secure operation of the system under variable load and wind profiles. The goal of these simulations is to illustrate and evaluate the interaction between the five wind farms with Rhodes power system during different load scenarios and winds. The first set of simulations focuses on the dynamic response of the three types of wind farms configurations, considered in this article and on the impact on the system during deterministic wind speed steps. The second set of simulations is carried out in order to illustrate and evaluate the fluctuations that occur in the generator speed and the power of wind turbines in the presence of a turbulent wind. The attention is also drawn to the impact of these fluctuations on the power system of Rhodes and to the wind farm controller capability to ensure safe operation of the wind farms during different variable load and wind profiles.

Turbulent Wind Speed in Wind Farms

In the following, simulations with turbulent wind speed time series are presented and discussed. The goal of these simulations is to evaluate the fluctuations in the system frequency due to wind speed fluctuations during different load scenarios. It is assumed that all five farms are running with turbulent wind speed.

The wind turbine time series, which are inputs to the wind farms, are generated based on the wind speed fluctuation model developed by Sorensen et. al., (Sørensen et al., 2007). This wind speed fluctuation model has been validated against wind speed measurements from large wind farms.

For each wind farm site and each wind and load scenario, different wind time series are generated for 10 minutes. The correlation between the wind speeds of the wind turbines in a wind farm is taken into account in the wind speed fluctuation model. The correlation aspect is even more important for such isolated power system as it is the case of Rhodes island. As the distances between the wind farms are quite small, the wind speeds seen by the wind farms can be highly correlated.

Figure illustrates the wind time series, which has been applied to the wind farms, during the load scenario with maximum load demand, i.e. SCENa. Notice, that all wind farms are running in power optimization mode, as their wind speeds are below the rated wind speed.

In this scenario, the wind farms produce in total 7.5 MW, accounting only for 3% of the total demand. The system frequency fluctuations due to the wind speed fluctuations are negligible, as illustrated inFigure 6, thus the wind power impact on the power system is in this case very small. As expected, although visible, the fluctuations are of quite small magnitude and they are therefore not leading to problems in the system security.

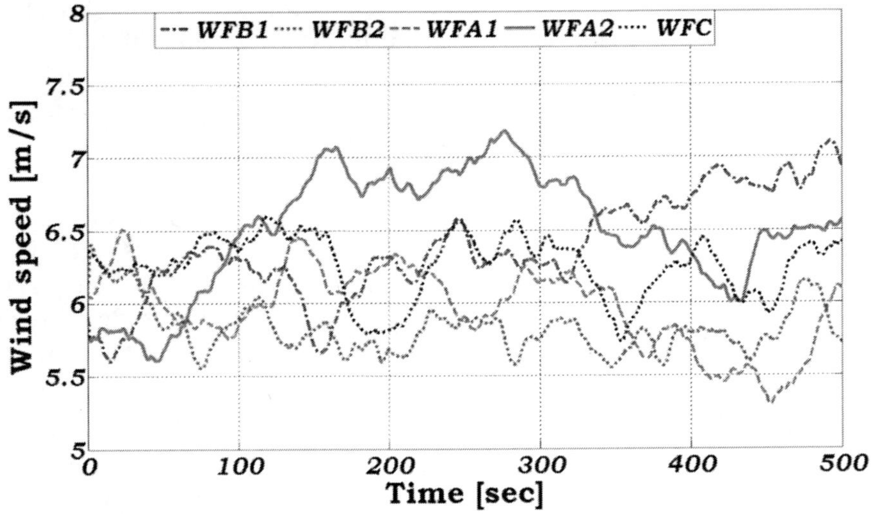

Figure 5. Wind time series applied to the wind farms – SCENa.

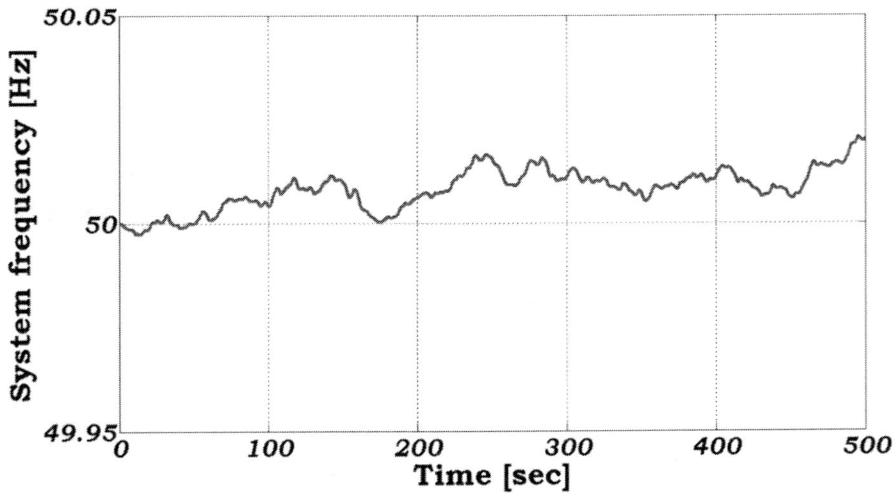

Figure 6. System frequency for wind time series –SCENa.

Figure 7 illustrates the wind time series, which has been applied to the wind farms, during the load scenario with maximum wind power production, i.e. SCENb.

Notice that in this SCENb, the wind turbines operate in a narrow range around their rated capacity and it can thus be presumed that the wind power fluctuations' impact on the Rhodes power system to be more significant compared to the SCENa.

Due to the highly correlated the wind speeds, the form that frequency fluctuations appear to have inFigure 8, is not directly related to a specific wind time series, but rather to the combination of the active power fluctuations delivered by all five wind farms.

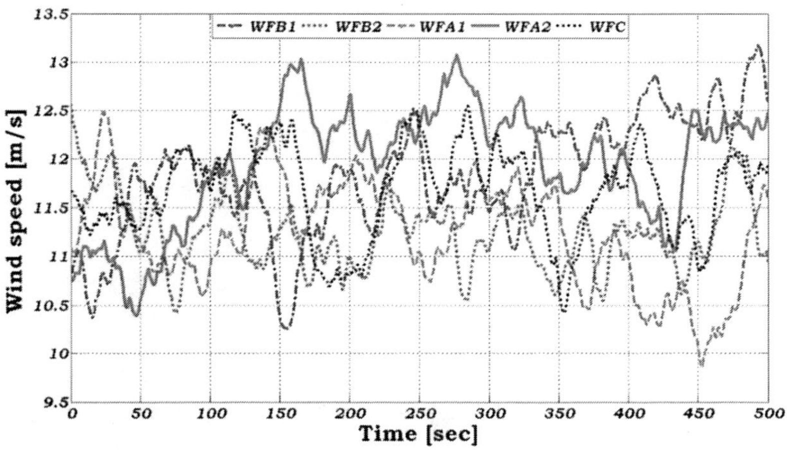

Figure 7. Wind time series applied to the wind farms –SCENb.

In this scenario, the wind farms produce in total 45.2 MW, which corresponds to 26% of the total demand. Notice that the system frequency, illustrated in Figure 8, has very fast dynamic deviations in the range 49.96 Hz – 50.1 Hz, which is considered safe from the power system security point.

Even though the wind power penetration is significant in SCENb, the frequency fluctuations are not considered large enough to impose security questions. The high correlation among the wind in all the wind farms is ensuring that the system frequency is not strongly dependent on a sudden drop or increase of the wind in only one wind farms. One of the most important factors influencing the frequency deviations – besides the parameters of the conventional units and their emergency rate of power undertake – is of course the size of each wind farm and the type of generator used.

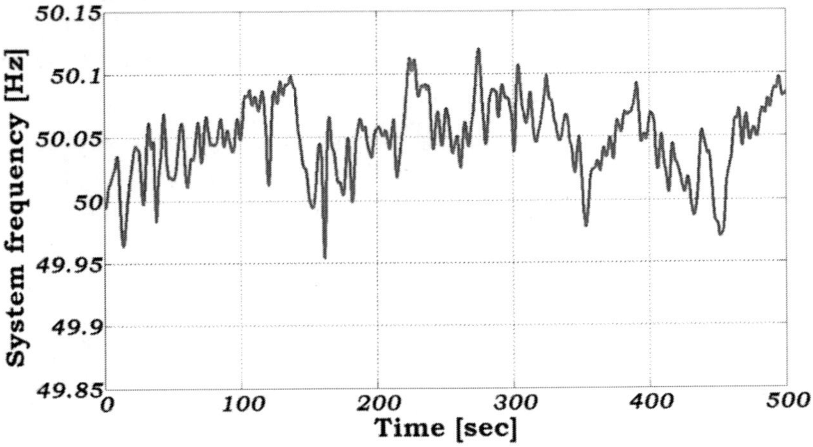

Figure 8. System frequency for wind time series –SCENb.

The following graphs show the performance during the scenario SCENb of two wind turbine technologies, i.e. ASIG wind turbine and PMSG wind turbine, which are picked-up to illustrate the behavior of a generic fixed and variable speed wind turbine, respectively. Beside the wind speed, typical quantities, as generated power to the grid, generator speed and the pitch angle are illustrated.

Figure 9 illustrates the wind speed time series applied as input to an ASIG wind turbine inside wind farm WFC. Notice that the wind turbine is simulated at an average wind speed of about 11.5 m/s. This operational point corresponds to a transition operational regime for the wind turbine, between power optimization and power limitation regime.

The strong correlation between the wind speed fluctuations, active power output and generator speed is obvious in Figures 9 -11. As expected for a ASIG wind turbine, the 3p fluctuation in the wind is visible both in the active power output and the generator speed. It should be noticed that, in this scenario SCENb, the ASIG wind farm produces around 6 % of the total demand.

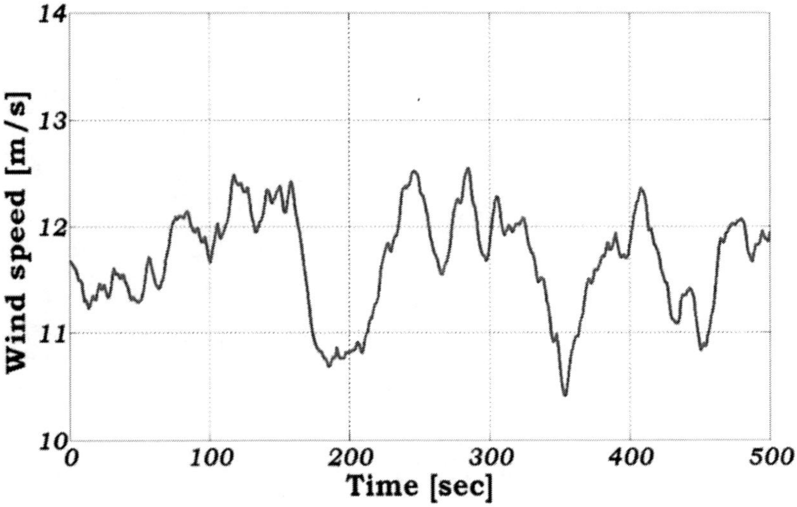

Figure 9. Wind time series in ASIG wind farm WFC –SCENb.

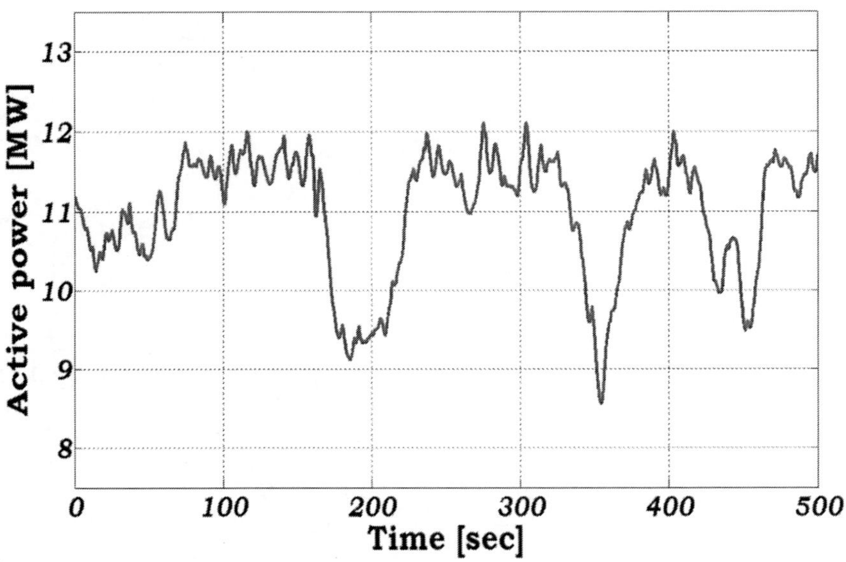

Figure 10. Wind farm active power output for ASIG wind farm WFC – SCENb.

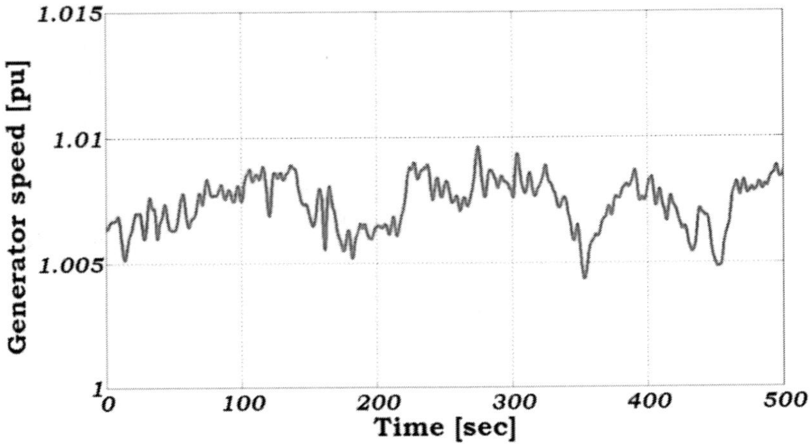

Figure 11. Generator speed for ASIG wind turbine in wind farm WFC –SCENb.

Figure 12 shows the pitch angle of ASIG wind turbine during SCENb. When the wind speed is less than the rated wind, the wind turbine produces maximum possible power. In this case, the pitch angle is equal with its optimal value, i.e. -1.7 deg. When the wind speed is above rated wind speed value, the pitch angle corresponds to the values which keep the turbine power to the rated power. Notice that the pitch system responds with delay due to the pitch rate limiter existing in the actuator. Despite the very fast wind speed deviations, the pitch control system manages to ensure smooth transition between the optimization and limitation range whenever the wind speed crosses the rated value.

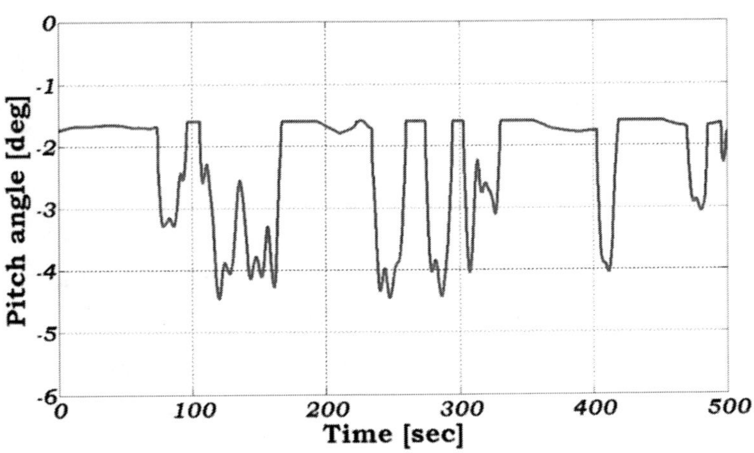

Figure 12. Pitch angle for ASIG wind turbine in wind farm WFC - SCENb

Figures 13-16 illustrate how the wind farm WFB2 equipped with PMSG wind turbines behaves during the scenario SCENb. Figure shows the wind speed time series, which has been applied as input to a PMSG wind turbine inside wind farm WFB2.

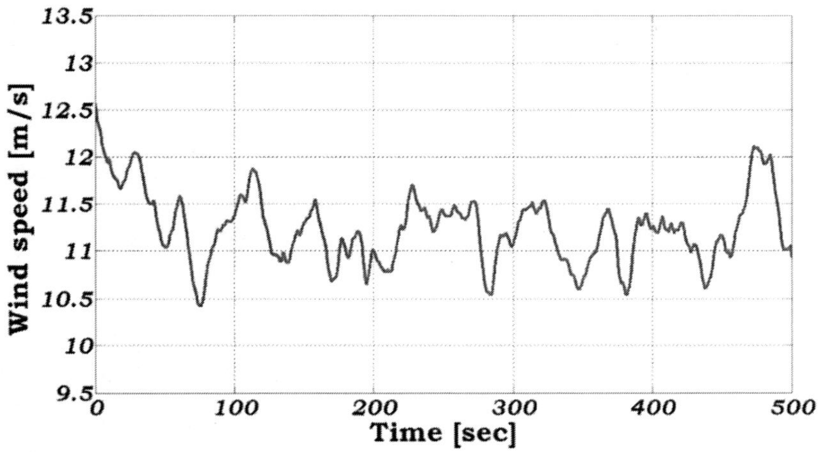

Figure 13. Wind time series in PMSG wind farm WFB2 –SCENb.

Similar to the ASIG wind turbine, discussed previously, the operation point of the PMSG wind turbine corresponds again to a transition operational regime for the wind turbine, between power optimization and power limitation regime.

This wind farm has rated capacity 3 MW, which stands for 1% of the total demand. As illustrated inFigure14, whenever the wind speed goes above the rated value, the pitch control manages to keep the active power output constant and equal to nominal power.

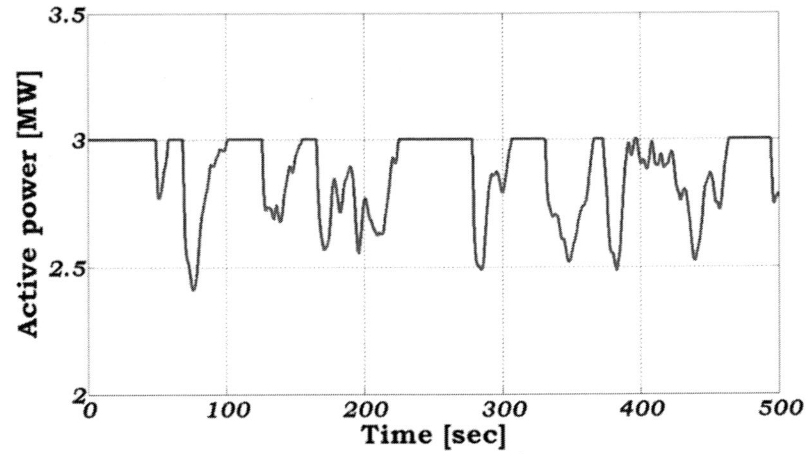

Figure 14. Wind farm active power output for PMSG wind farm WFB2 – SCENb.

Notice that the generator speed of the PMSG wind turbine is tracking the slow variation in the wind speed each time the turbine is running in the operation mode, while in power limitation mode, the speed controller tries to control the generator speed to its rated value.

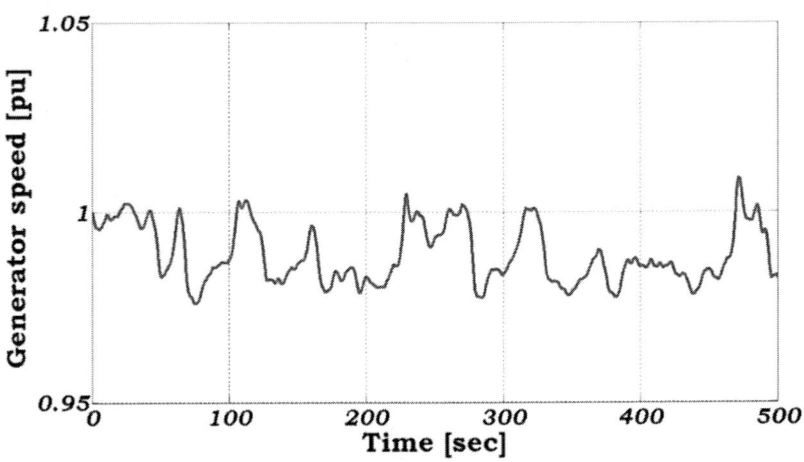

Figure 15. Generator speed for PMSG wind turbine in wind farm WFB2 – SCENb.

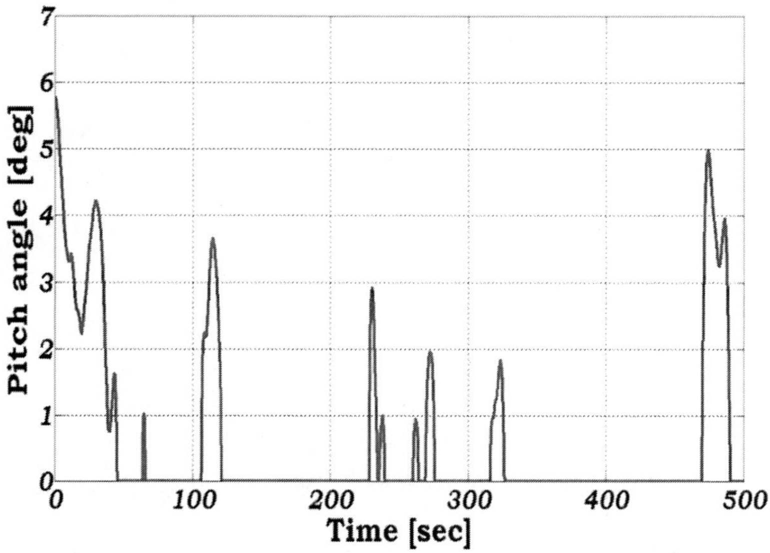

Figure 16. Pitch angle for PMSG wind turbine in wind farm WFB2 – SCENb.

The damping controller manages to damp the torsional oscillations in the drive train and ensures safe operation of the wind turbine. Compared to the results from the ASIG wind farm above, the fast deviations in the wind are filtered out from the electrical power. The generator is decoupled from the grid through the converter at its terminal and any rapid fluctuations in the wind are not influencing the power delivered to the grid. Figure 16 gives the pitch angle, which is increased from its optimal value each time the turbine is running in power limitation mode.

In this scenario, the wind power penetration is maximum, reaching 32% of the total demand. The wind farms however are not producing as much as in SCENb. This scenario is considered the worst case scenario, as the fluctuations in the wind may have serious impact on the power system operation.

Figure 17 illustrates the wind time series, which has been applied to the wind farms, during the load scenario with maximum wind power penetration, i.e. SCENc.

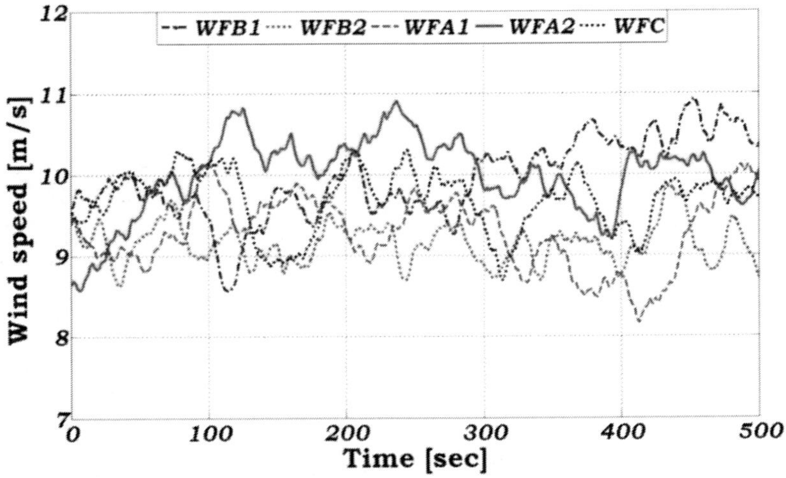

Figure 17. Wind time series applied to the wind farms –SCENc.

The range of operation for the wind turbines in this case covers the optimization area, where the control of the wind turbine has to ensure optimum power output.

illustrated in Figure 18, due to the wind fluctuations, the system frequency varies between 50 Hz and 50.25 Hz, which is considered safe for the system operation. Although the wind speeds in each wind farm may have sudden changes, as it is illustrated in Figure 17, the overall response of the system is satisfactory, and the power outputs from the windfarms seem to counteract each other in the frequency impact.

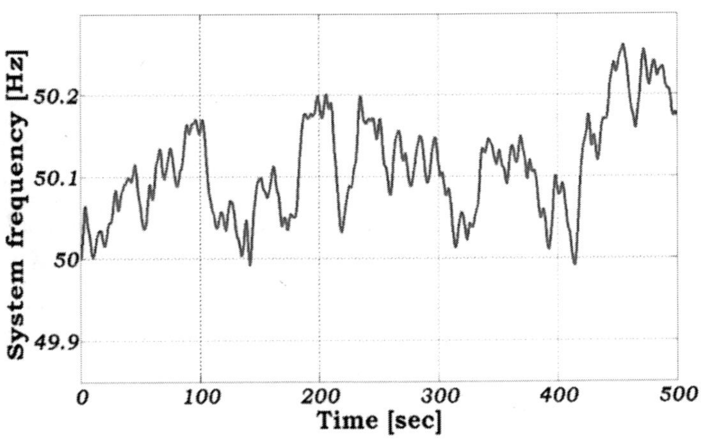

Figure 18. System frequency for wind time series – SCENc.

The system frequency deviates more in this case than in SCENb (see Figure 8), which is due to the increased wind power penetration levels in this scenario. Nevertheless, the emergency rate of power undertaken by the conventional units is sufficient to overcome the rapid active power fluctuations produced by the wind farms on the island.

In the following, the attention is directed in details towards the performance of two wind turbine technologies during SCENc, this time on a ASIG wind turbine and a DFIG wind turbine. Again typical quantities, as generated power to the grid, generator speed and the pitch angle are illustrated.

Figure 19 illustrates the wind speed time series, which has been applied as input to the ASIG wind farm WFC. In scenario SCENc, this wind farm produces around 7 MW i.e. 8 % of the total demand.

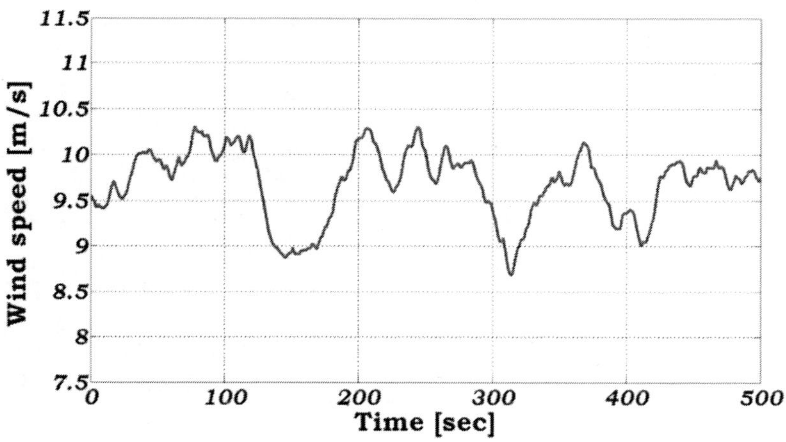

Figure 19. Wind time series in ASIG wind farm WFC –SCENc.

Notice in Figures 19-21 that the fluctuations in wind speed, active power output, generator speed are strongly correlated in this case, and the fast dynamic deviations are also here significant. Figure 22shows the pitch angle, which is reduced accordingly although restricted by the pitch rate limiter when the wind speed changes at very high rate.

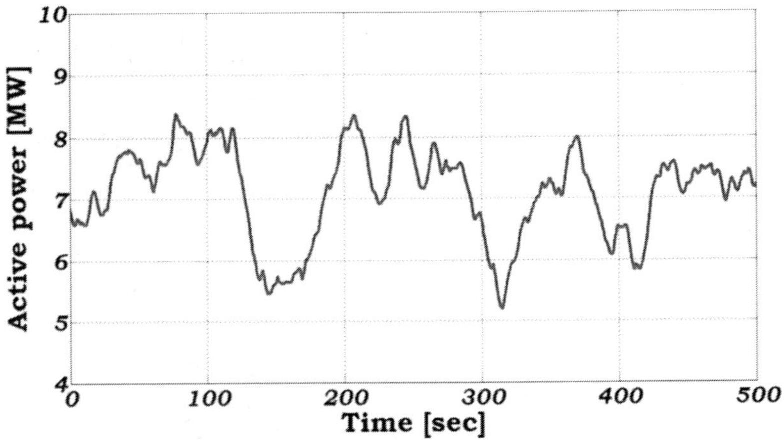

Figure 20. Wind farm active power output for ASIG wind farm WFC – SCENc.

The dynamic performance of the DFIG wind farm WFA1 during SCENc is in the following addressed in order to illustrate the efficiency of the designed models. This farm produces almost 6.5 MW i.e. 7.5% of the total demand. Figure 23 shows the wind speed time series, which has been applied as input to the DFIG wind farm during SCENc.

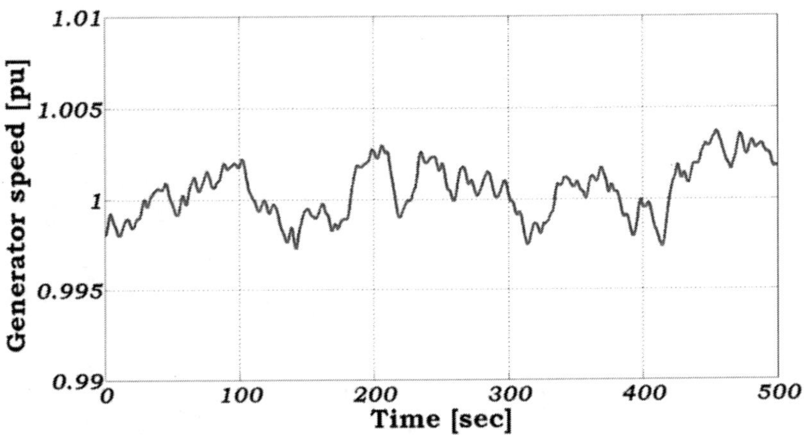

Figure 21. Generator speed for ASIG wind turbine in wind farm.

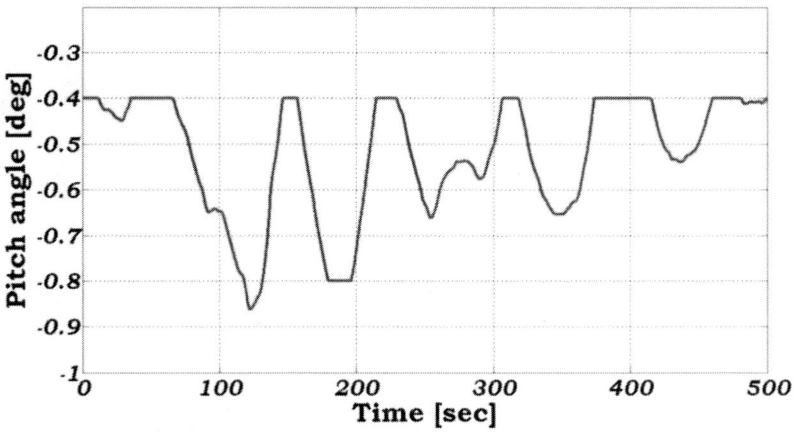

Figure 22. Pitch angle for ASIG wind turbine in wind farm WFC –SCENc.

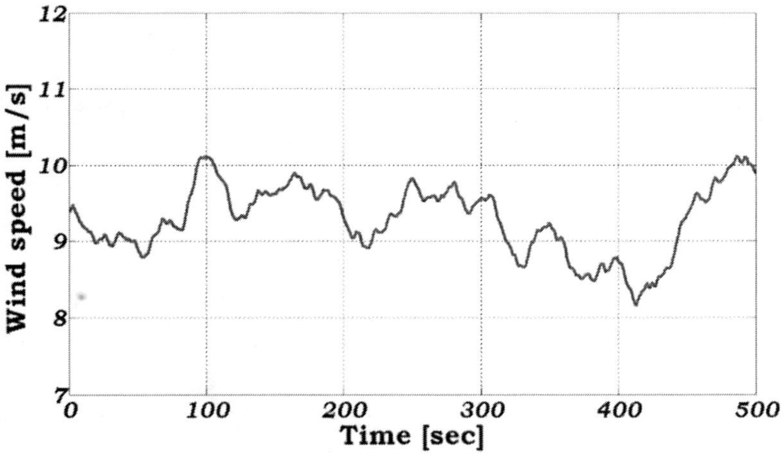

Figure 23. Wind time series in DFIG wind farm WFA1 –SCENc.

The Maximum Power Tracking (MPT) strategy, implemented in the RSC (rotor-side converter) of the DFIG system ensures optimum operation of the wind turbine maximizing the aerodynamic coefficient Cp in every wind speed. In this operational area for the DFIG wind turbines, small changes in the rotor speed result in significant changes in the active power output of the system. This is due to the fact, that the designed MPT characteristic is very stiff in this area.

Notice that, contrary to an ASWT wind farm, where the correlation between the system frequency and the wind turbine response is very strong, in the case of DFIG configuration the generator is partially decoupled from the system frequency, as seen in Figure 18s 18, 21 and 25.

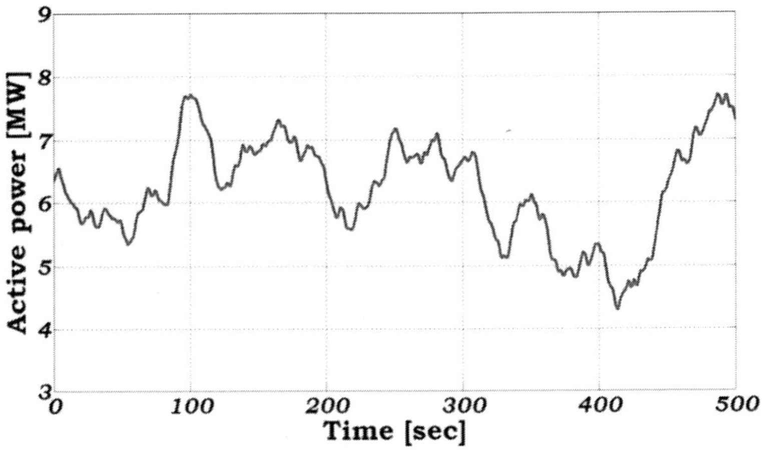

Figure 24. Wind farm active power output for DFIG wind farm WFA1 –SCENc.

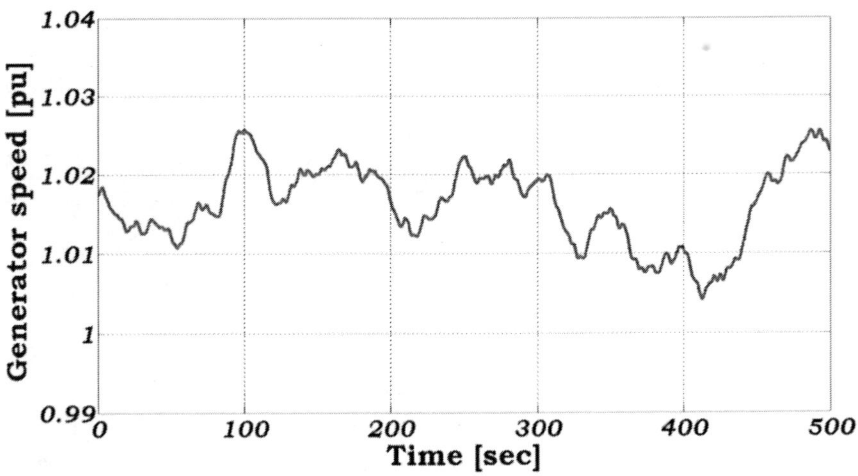

Figure 25. Generator speed for DFIG wind turbine in wind farm WFA1 – SCENc.

The generator speed is continuously adapted to the wind speed in order to extract maximum power out of wind. As wind farm is running in optimization mode, the pitch angle is passive, being kept constant to its optimal value.

FREQUENCY CONTROL OF WIND POWER

Increasing wind power penetration especially in non-interconnected systems is changing gradually the way grid frequency control is achieved. The technical requirements set by the networks operators include various aspects, such as fault ride-through of wind turbines during faults, voltage-reactive power control and overall control of the wind farms as conventional power plants. A key aspect of the operation of wind farms in autonomous power systems is the frequency control. This session presents the results from a study on the Rhodes power system already presented above, focusing on frequency control. The Rhodes power system has been used to address all the main issues related to system secure operation under different system conditions. The response of the wind farms in frequency disturbances is analyzed and the different characteristics of each wind turbine type related to frequency are described. Three different wind turbine configurations have been used – Active Stall Induction generator (ASIG), Doubly Fed Induction generator (DFIG) and Permanent Magnet Synchronous generator (PMSG). An auxiliary control has been designed for the DFIG type to enhance the capability to support the frequency control. The load shedding following severe frequency disturbances is calculated and the under/over-frequency protection relay settings are discussed under the novel system conditions. Results for different system conditions and control methods are presented and discussed focusing on the ability of modern wind turbine technologies to assist in frequency control in isolated power systems during severe disturbances in the production-consumption balance.

As wind power penetration increases in modern power systems, a variety of technical and regulatory issues regarding the interaction between large wind farms and power system is under constant discussion. The system operators are setting onerous requirements that that wind farms have to fulfill. Among these, voltage and frequency control play an important role. Frequency control has started to appear as emerging need under increasing wind power penetration conditions and due to the extended replacement of conventional generators by large wind farms in power supply. The impact of wind farms in frequency phenomena is even more vital in non-interconnected power systems, where the power system inertia is limited.

It is often the case, that when auxiliary services of wind turbines, like frequency control, are investigated, simple models are used for either the power system or the wind turbines. In this article, detailed model for all different components of the system were used to evaluate the system response in serious events with maximum accuracy. The dynamic security of power systems has to be carefully examined, before wind power penetration limits are expanded. The response of conventional units, the load dependency on frequency and voltage and the wind turbines' response during events that affect system frequency are some of the key aspects that have to be modeled in detail for this kind of investigations.

In this article the main issues of frequency control in isolated power systems, with high wind power penetration are investigated. Rhodes power system, includes three different types of conventional generators – namely gas, diesel and steam units – and three different types of wind turbines – Active Stall Induction generator (ASIG), Doubly Fed Induction generator (DFIG) and Permanent Magnet Synchronous generator (PMSG) wind turbines. This variety of components gives the chance for a wide range investigation of frequency issues for modern power systems.

Definitions regarding the frequency control in autonomous power systems are given and the protection relay system settings related to under/over frequency deviations are discussed. The response of different wind turbine technologies during frequency events is explained. Three different primary frequency control schemes implemented in the developed model are analyzed. Finally, the basic characteristics of the power system are given followed by a brief description of the available wind turbine technologies available on line in the system for the reference year 2012. Results from various simulations are discussed and the capability of modern wind farms to provide auxiliary frequency control is demonstrated.Frequency definitions and protection system.

In this section, some basic definitions on frequency are given to introduce the main issues regarding frequency control with emphasis on isolated power systems.

In the power system, frequency is the variable indicating balance or imbalance between production and consumption. During normal operation, the frequency should be around the nominal value. The deadband which is considered as safe operation in most European grid codes is the zone 50±0.1Hz, although the limits vary between the different system operators in Europe, mainly due to the different characteristics of each grid. The range 49÷50.3 Hz is in general the dynamic security zone that in most of the cases is not allowed to be violated at any means, (Lalor et al., 2005). However, these safety margins for frequency deviations are often expanded in autonomous power systems, where system inertia is low, to avoid constant load shedding whenever the balance between production and consumption is lost.

In case of sudden generation loss or large load connection, the frequency of the frequency starts to drop. The two main system functions that ensure return of an unbalanced system to nominal frequency are:

- Primary Control: During the first 30-40 sec after the event leading to frequency deviation, the rotational energy stored in large synchronous machines is utilized to keep he balance between production and consumption through deceleration of the rotors. The generation of these units (often referred to as primary control units) is thus increased until the power balance is restored and the frequency is stabilized.
- Secondary control: After the primary response of the generators, a slow supplementary control function is activated in order to bring frequency back to its nominal value. The generators connected to the system are ordered to change their production accordingly either through an Automatic Generation Control scheme, either through manual request by the system operator – which is often the case in isolated systems like Rhodes.

These two main frequency control functions are illustrated in Figure 26 for a sudden drop in system frequency.

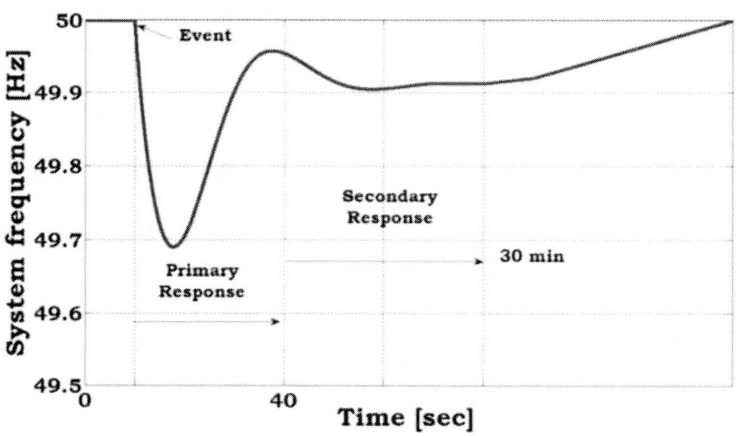

Figure 26. Definitions of frequency control in power systems

The rate, at which the frequency changes following an event i.e. drops in Figure 34, is depending on the so called inertia of the system thus the total angular momentum of the system which is the sum of the angular momenta of the rotating masses in the system i.e. generators and spinning loads. The frequency control implemented in this study tries to improve the system response in terms of initial Rate of Change of Frequency and minimum frequency (frequency nadir). Therefore, the discussion and also the results will emphasize the capability of wind farms to provide with primary frequency support in the first seconds after the event which causes the frequency deviation.

Response of Wind Turbines to Frequency Events

The replacement of conventional synchronous generators by wind farms in modern power systems with increased wind power penetration, changes the way traditional frequency control was treated in power systems. Wind turbines connected to the grid, depending on their configuration, have a different response to frequency deviations. In this Session, the relation between each wind turbine configuration and its response during frequency deviations is discussed and explained.

Response of ASIG Wind Turbines in Frequency Events

One of the most common wind turbine configurations in modern power systems is the standard fixed speed wind turbine based on induction generator connected to the rotor through a transmission shaft and a gearbox. The Active Stall Induction Generator wind turbine model developed to simulate the fixed speed wind turbines in this study is described in previous sections.

As described in (Morren et al., 2006), the induction machine based wind turbine inertia response is slower and lower than the conventional synchronous generator's response. This difference is mainly because of the reduced coupling of the rotational speed of the WT and the system frequency and of the lower inertia constant of the WT compared to a standard conventional generator connected to the grid.

However, in the case of a frequency drop, like the one illustrated in Figure 34, the inertia response of the ASIG wind turbine is not negligible due to usual low nominal slips. The rotor of the ASIG is decelerating following the system frequency drop. The kinetic energy which was accumulated in the rotating mass is transformed into electrical energy delivered to the grid. The amount of the available kinetic energy is determined from the total angular momentum of the WT – thus the sum of the angular momenta of the electrical generator, the rotating blades and the gearbox – and the rotational speed. There are some studies estimating this available energy (Ullah et al., 2008; Ramtharan et al., 2007) through rough estimations.

A comparison between the inertia response provided by the three different wind turbine configurations studied in this article is given in Figure 27 for loss of the largest infeed in thesystem of Rhodes. When the frequency starts to drop, the ASIG provides with significant active power surge to the grid, thus, reducing the initial rate of change of frequency. The response of the DFIG and PMSG wind turbine types is explained further below. It is obvious that, fixed speed wind turbines have an intrinsic behavior that provides auxiliary service to the system during frequency imbalances, although they can not contribute to other services, i.e. voltage – reactive power control, in the same way the variable speed wind turbines can.

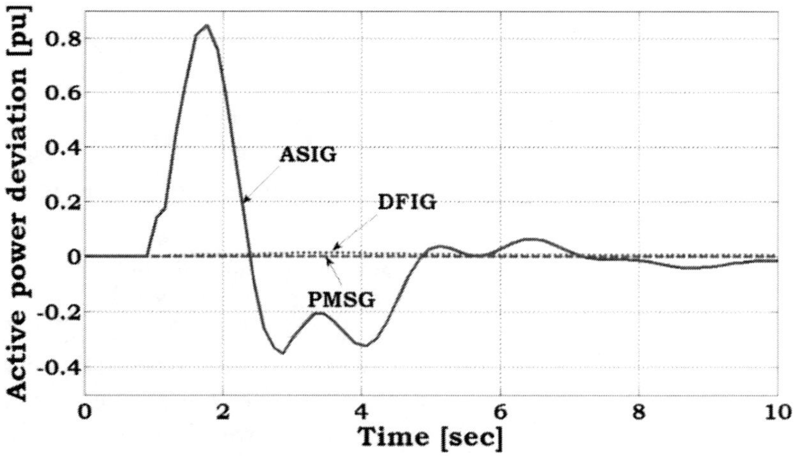

Figure 27. Change in active power production during a frequency drop for the three main wind turbine configurations.

Response of DFIG Wind Turbines in Frequency Events

The DFIG wind turbine configuration, which is the most common configuration for large wind turbines, is mainly based on an induction generator and a frequency converter connected to the rotor circuit via slip rings. Details on the model developed in this study, including control aspects, can be found in section 3.1.1.

As described in (Ullah et al., 2008), the response of a DFIG wind turbine is slightly different than the one described above for synchronous and induction generators. The inertial response of the DFIG type is mostly based on the applied control scheme acting on the converter connecting the rotor to the grid (Morren et al., 2006). The overall response can be explained as the result of two opposite torques acting on the rotor during a frequency change, i.e. a frequency drop: a decelerating torque, proportional to the rate of change of the rotor speed dωdt and therefore to the frequencydfdt, which makes the rotor speed follow the frequency drop – an accelerating torque, which is produced by the difference in the electromagnetic torque, controlled by the speed controller of the machine, and the aerodynamic torque acting on the rotor of the turbine. This last component tends to cancel the decreasing effect that would eventually make the DFIG have a similar response to a simple induction generator connected to the grid, (Ekanayake et al., 2003).

Response of PMSG Wind Turbines In Frequency Events

A multi-pole PMSG wind turbine is connected via a full-scale frequency converter to the grid. The converter decouples the generator from the grid; the generator and the turbine system are not directly subjected to grid faults in contrast to the direct grid connected wind turbine generators. Therefore, the power output from the WTG does not change and no inertial response is obtained during a frequency event. The rotor speed of the multi-pole synchronous generator is not connected with system frequency at any means.

Large wind turbines nowadays substitute conventional generators in modern power systems under increasing wind power penetration conditions. The effect on the power system inertia and the availability of inertia response from wind turbines have become key issues for the secure integration of wind energy into the electrical grids, especially in autonomous power systems like Rhodes. In power systems, like the one studied in this article, regular load shedding occurs due to large frequency deviations. Although sufficient spinning reserve is ensured to overcome any frequency problems, increasing wind power penetration is challenging the system security.

Supplementary control attributes have been proposed in the literature in order to achieve active frequency control by the wind turbines, (Morren et al., 2006; Ullah et al., 2008; Ekanayake et al., 2003; Ekanayake& Jenkins, 2004; Holdsworth et al., 2004; Suwannarat et al., 2007). In most of these publications, simple models for either the power system or the wind turbines are used based mainly on the assumption that the aerodynamic torque acting on the rotor during the frequency event does not vary significantly. In this section, the three different frequency control methods, which were applied in the DFIG wind turbine models used in the Rhodes power system model, are described:

- Inertia Control
- Droop Control
- Combined Control

Results from frequency events in the Rhodes power system when these control methods are used in the wind farms equipped with DFIG wind turbines are given in Section VI. The general control scheme is illustrated in Figure 28.

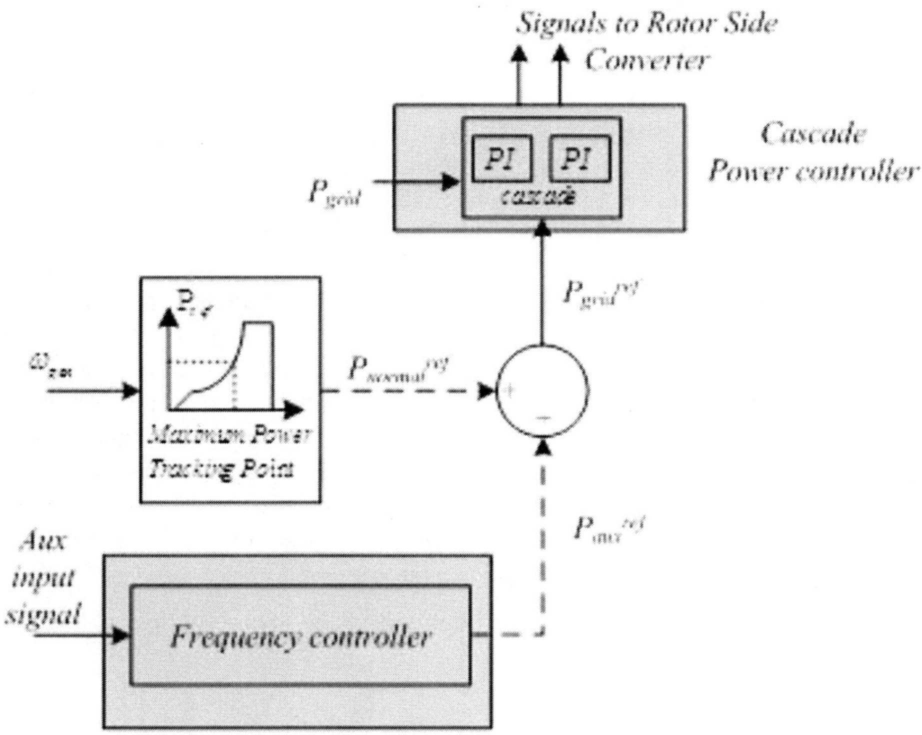

Figure 28. General frequency control scheme for DFIG wind turbines.

In the first method, the inertial response of the DFIG is restored through an additional loop in the power reference block providing the active power reference signal to the Rotor Side Converter. Details for the basic control structure of the DFIG model designed in this study for normal operation can be found in section 3.1.1. Figure29 shows the inertia control loop.

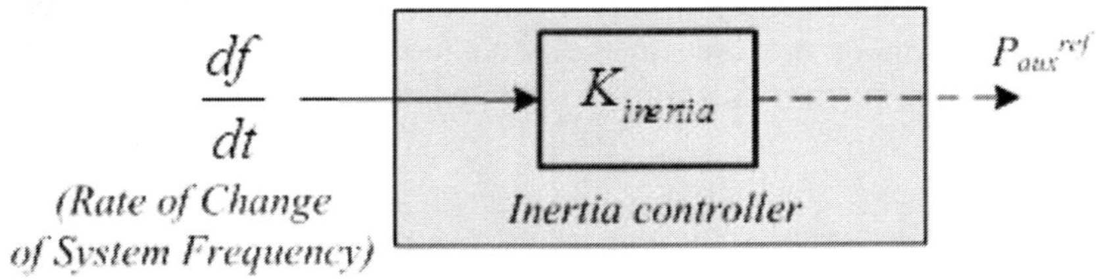

Figure 29. Inertia controller for DFIG wind turbine.

This feature is often referred to as "virtual inertia" effect, thus the control aim is to control the DFIG wind turbine to adjust its power output when subjected to frequency deviations. The rate of change of frequency defines the additional power reference signal, which is added to the normal power reference provided by the Maximum Power Tracking Controller. This auxiliary signal introduces a term proportional to2Hdfdt, where H is the total inertia constant of the wind turbine, (Suwannarat et al., 2007). This inertia constant is expressed in seconds and represents the time that the wind turbine is able to provide with rated active power when decelerating the rotor from the nominal speed down to zero using only the available kinetic energy in the rotor mass, (Ullah et al., 2008). Thus, in physical terms the values of the proportional parameter Kinertia shown in Figure 29 are restricted by this amount of energy. However, the wind turbine could be ordered to provide with even more active power, given that the stability of the system is ensured.

The second control method applied in the DFIG models is the Droop control, illustrated in Figure 30.

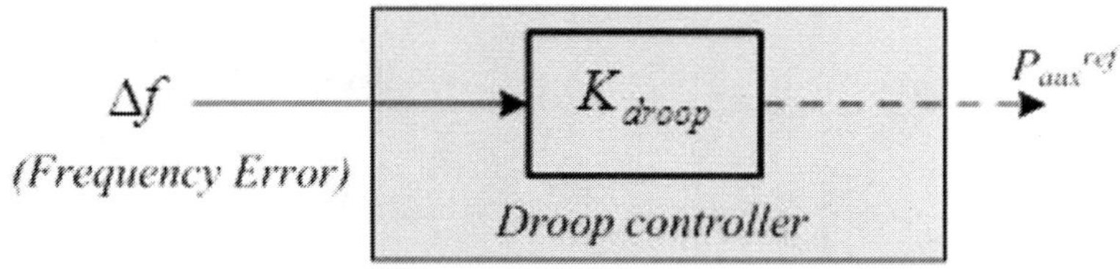

Figure 30. Droop controller for DFIG wind turbine.

In this case, the additional reference signal is equal to:

$$P_{aux}^{ref} = K_{droop}\left(f - f_o\right)$$

where f_o is the nominal system frequency, 50 Hz for the Rhodes power system. This control method is based on the primary frequency control applied to conventional generators. Typical values for the droop parameter of large conventional units are 3%-5%, depending on the type of unit.

This control loop aims to decrease the accelerating torque acting on the generator rotor during a frequency drop, as described above for DFIG wind turbines, (Ekanayake et al., 2003). The droop control can be assumed to be implemented in the wind farm controller level instead of individual wind turbine controller. This means that the overall wind farm controller provides the auxiliary signal Pauxref which is distributed to the individual wind turbine controllers. In that case, the communication delays should be taken into consideration, as the rate at which the wind farm changes its output during the first milliseconds following the frequency event is crucial for the overall system response. Results from both control levels, thus Droop controller on individual wind turbine controller and Droop controller on wind farm controller, are shown and compared.

The last method tested in this study, is actually a combination of the two first control methods. Based on the analysis made in (Ekanayake et al., 2003) and referred in section 2.2 for DFIG wind turbines, the sum of Droop and Inertia control should manage to counteract the opposite torques acting on the generator rotor during frequency phenomena. The Combined control scheme is given in Figure 31.

As a first approach, this last method of Combined Control seems to be optimum for the DFIG wind turbines. In most of the publications available, Droop Controller and Inertia controller have been treated independently. Discussion on the results from each control method proposed here is made in section 4.2.

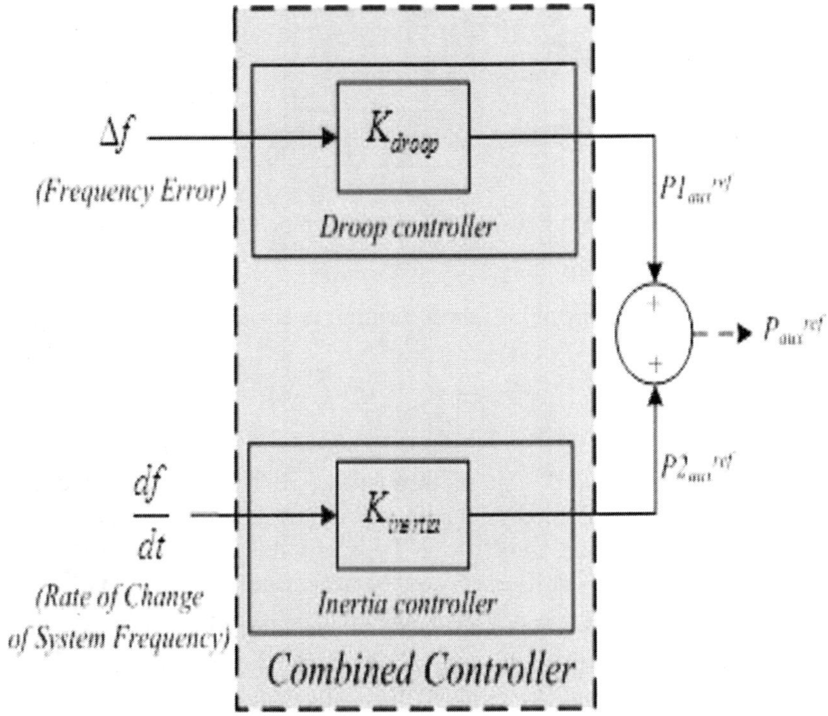

Figure 31. Combined controller for DFIG wind turbine.

Results

In this section, results from the Rhodes power system are presented. The frequency control capability of DFIG wind turbines is investigated for the load scenarios defined above. The standard event often used to check the dynamic security of power systems, thus the loss of the largest conventional generator, is simulated and the frequency response of the system under the different frequency control methods is illustrated. The emphasis on these results is given on the first seconds of the primary control operation of the system and the load shedding following the event is computed, based on the action of the under-frequency protection relay settings.

In SCENb, the wind farms online produce close or equal to their rated capacity. The total wind power production is 45.21 MW which stands for 27% of the total demand. The wind speeds in both wind farms equipped with auxiliary frequency control capability the wind is considered constant during the event studied, close to 11.5 m/s. The largest conventional unit produces 20.1 MW when ordered to trip, leading to loss of 11% of the total production. Figure 32 shows the response of three wind farms during this fault, when no auxiliary control is activated in the wind turbines. The wind farm with ASIG wind turbines increases its active power output during the first seconds following the frequency drop, contributing to the systeminertia. On the other hand, the wind farm with DFIG wind turbines has almost negligible power contribution, while the PMSG wind farm does not change its active power output during the frequency drop. These results confirm the analysis made in Section III regarding the natural response of each wind turbine type. The change in active power output for each wind farm inFigure 32 is given in p.u. of the rated capacity of each wind farm.

Figure 33 shows the frequency response for all different control methods for frequency control in the wind farms with DFIG wind turbines described in Section IV. In SCENb, these two wind farms produce in total 15 MW – 9% of the total demand. In the same figure, the results for Droop control implemented in the wind farm level or the wind turbine level are included.

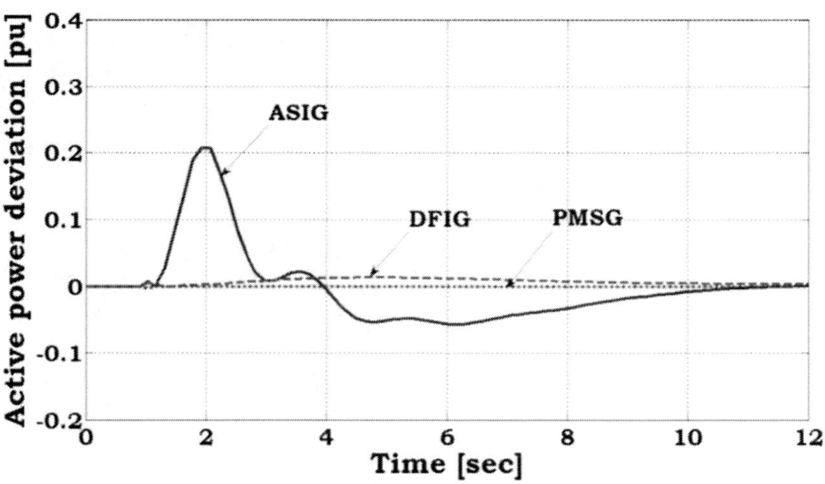

Figure 32. Change in power in different wind turbine configurations during frequency drop –SCENb.

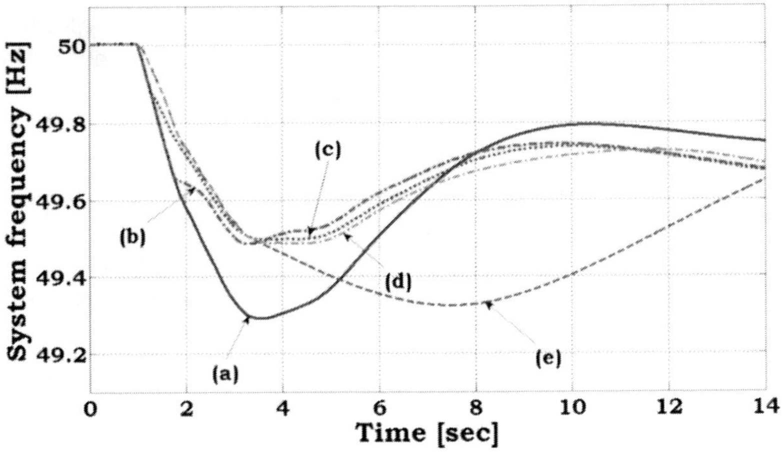

Figure 33. System frequency for largest unit loss when frequency control is applied by DFIGs – (a) No auxiliary control, (b) Droop control on WF level, (c) Droop control on WT level, (d) Combined control, (e) Inertia control –SCENb.

When no aux control (case (a)) is used in the DFIG wind farms, the frequency reaches its minimum (see also Table 2 below) with the highest initial rate of change of frequency. In this scenario, even in the case with no auxiliary control provided from wind farms there is no load shedding as the system inertia is high enough to ensure moderate frequency drops. The Droop control implemented in the wind farm control level (case (b)) does not manage to improve the maximum rate of change of frequency, although the minimum frequency is higher compared to the case (a). The best case, in terms of minimum frequency, is as expected case (c), where Droop control is implemented in the wind turbine control level. On the other hand, the Inertia control (case (e)) achieves the slowest rate of change of frequency although the frequency minimum is the lowest among the different frequency control methods proposed here. The optimum performance seems to be achieved through the Combined control scheme (d) where both minimum frequency and maximum rate of change of frequency are improved. This last control method seems to combine the pos from the Droop and Inertia control schemes. The droop control implemented in the wind turbine control level (case (c)) has slightly higher minimum frequency but the difference is negligible (0.01 Hz).

Figure 34 shows frequency drop during the first 2 seconds after the loss of the conventional unit to clarify the effect of each control method on the initial rate of change of frequency.

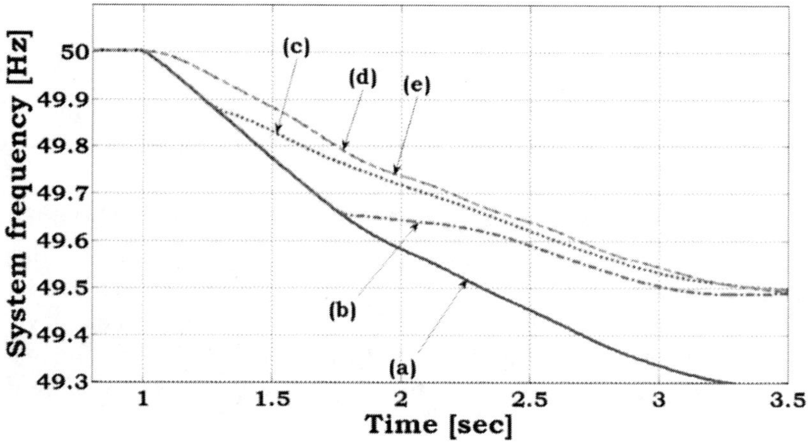

Figure 34. System frequency for largest unit loss when frequency control is applied by DFIGs – Zoom in the first seconds after the event (a) No auxiliary control, (b) Droop control on WF level, (c) Droop control on WT level, (d) Combined control, (e) Inertia control –SCENb.

The results for this scenario are summarized in Table 3, where the minimum frequency, the maximum rate of change of frequency and the load shedding are computed for all the cases demonstrated above.

Table 3. Results for SCENb– loss of largest infeed.

Frequency Control Scheme	Minimum Frequency (Hz)	Maximum Rate of change of frequency (Hz/sec)	Load Shedding (MW)	
(a)	No auxiliary control	49.29	-0.48	0
(b)	Droop control on WF level	49.49	-0.48	0
(c)	Droop control on WT level	49.41	-0.48	0
(d)	Combined control	49.49	-0.36	0
(e)	Inertia control	49.32	-0.36	0

The contribution of the wind farms during the frequency drop is obvious from the results presented above. From the wind turbine side now, the results for the rotor speed and the active power output of wind farm A1 (see Table 2) equipped with DFIG wind turbines are illustrated here for all the cases of frequency control.

As already discussed in section 2.2, the rotor speed of the DFIG wind turbines is not affected if no auxiliary frequency control is applied. Therefore, the inertia response of the DFIG is negligible (seeFigure 35). In all the other cases, the rotor decelerates while the system frequency drops. Thus, the kinetic energy accumulated in the rotor mass is converted to electrical energy and delivered to the grid – giving the power surge during the primary frequency control period shown in Figure 36.

When Inertia control is used (case (e)), the rotor speed goes back to its pre fault value, as the auxiliary power reference signal is calculated based on the derivative of the frequencydfdt. When the frequency stabilizes after the primary frequency control period, although not nominal yet as the secondary control has not been activated in this time frame, the derivative goes to zero and the auxiliary power reference signal goes also to zero. However, in the rest of the frequency control schemes, when droop control is used, the rotor speed is decreasing reaching another steady state value, as the difference from the nominal value remains even after the stabilization of the frequency. This means, that the wind turbine is no longer operating in the maximum power tracking curve as the normal control demands, (Ekanayake& Jenkins, 2004). The tertiary response which re-establishes the rotor speed of the wind turbine may take place several seconds after the event, when the power system has overcome the imbalance and the stress to stabilize the frequency. This procedure could also be implemented as part of an Automatic Generation controller operating in the whole power system, including the wind farms as active components, (Margaris et al., 2010).

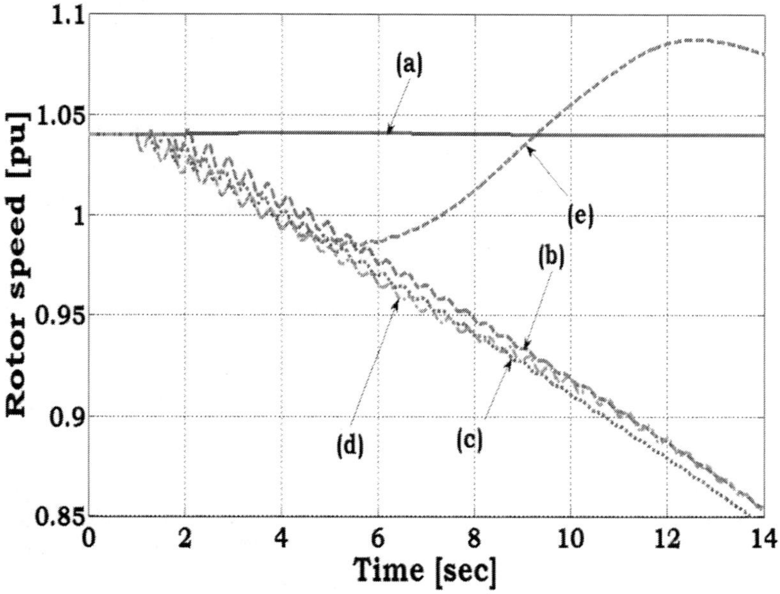

Figure 35. Rotor speed deviations after largest unit loss for different frequency control methods applied in DFIGs - (a) No auxiliary control, (b) Droop control on WF level, (c) Droop control on WT level, (d) Combined control, (e) Inertia control –SCENb.

The active power output of the wind farm A1 is given in Figure 36. In cases (d) and (e), where Inertia control and Combined control are used respectively, the wind farm increases its active power at a high rate, thus leading to lower rate of change of frequency as described in Table 3 above. In case (a) of course, when no auxiliary control is provided the active power change is negligible. In cased (b), where the Droop controller is assumed to work in the wind farm control level, the power surge is delayed compared to case (a).

In SCENc, the wind power penetration is maximum. The total wind power production is 28.2 MW in total 83 MW of demand (34 %). Although, the wind farms produce less than in the Maximum Wind Power Production scenario (SCENb) studied above, the impact of wind power in the power system operation is considered far more significant. The system inertia is decreased in this case, making the frequency control task in the system more complex. The wind speeds in this scenario is almost 9.3 m/s for the wind farms A1 and A2 (see Table 2).

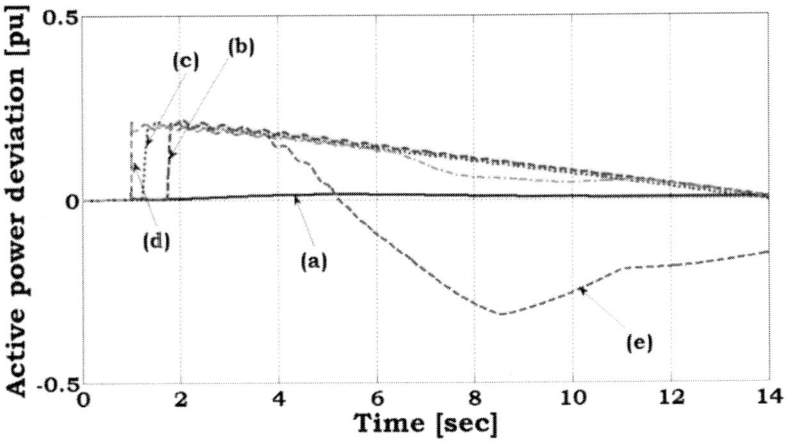

Figure 36. Change in active power output after largest unit loss for different frequency control methods applied in DFIGs - (a) No auxiliary control, (b) Droop control on WF level, (c) Droop control on WT level, (d) Combined control, (e) Inertia control –SCENb.

The largest conventional unit in the system produces 21 MW before the protection system acts to take it out of operation – this means production loss equal to 25 % of the total demand. The fault is severe and the power system stability is checked for all the frequency control schemes designed in this study.

Figure 37 illustrates the response of the three different wind turbine configurations available in the Rhodes power system, during the event, when no auxiliary control is activated in the DFIG wind farms. The comments made in section 2.2 are confirmed by the results (see alsoFigure 32).

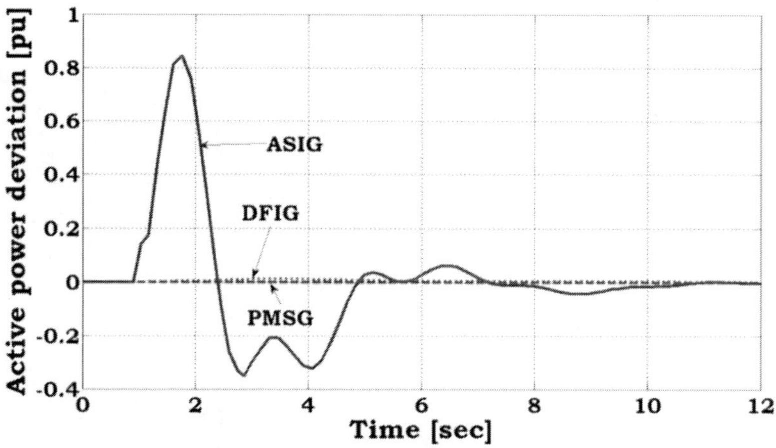

Figure 37. Change in power in different wind turbine configurations during frequency drop – SCENc.

Comparing to Figure 32, which demonstrates the response for SCENb, the contribution of the ASIG in SCENc is higher. The change in the active power production of the wind farm with ASIG wind turbines is higher than 0.8 pu compared to almost 0.2 p.u. in SCENb. This can be explained comparing the frequency response in both cases. In SCENb (see Figure 33 – case (a)), the frequency does not decrease as much as in SCENc (see Figure 38 – case (b)), therefore, the rotor of the ASIG wind turbines decelerates more in the first scenario, leading to higher active power contribution. InFigure 37 the response of wind farms equipped with DFIG or PMSG wind turbines is almost negligible, as explained in section 2.2.

In Figures 38 and 39 the system frequency for all the different frequency control schemes implemented in the wind farms A1 and A2 (see Table 2) is shown. In this scenario, the minimum frequency following the event is very low compared to SCENb, whereas the maximum rate of change of frequency is significantly higher.

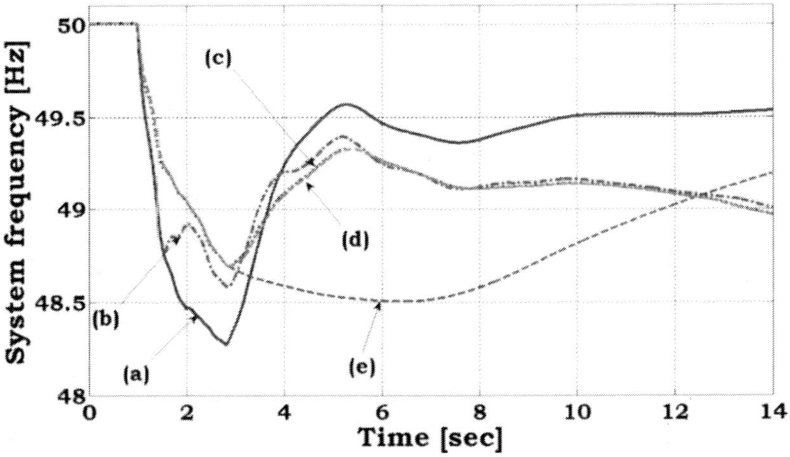

Figure 38. System frequency for largest unit loss when frequency control is applied by DFIGs – (a) No auxiliary control, (b) Droop control on WF level, (c) Droop control on WT level, (d) Combined control, (e) Inertia control –SCENc.

In case (a), when the wind farms do not have auxiliary frequency control, the frequency drops below 48.5 Hz which is the upper zone of the under-frequency protection relay settings acting on the loads. This drop leads to disconnection of 15.1 MW of load – 18 % of the total demand. This load shedding is not considered accepted in terms of dynamic security terms, (Margaris et al. 2009). However, in all the other cases, where the frequency control is activated in the DFIG wind farms, the load shedding is avoided totally. The maximum frequency drop appears in case (e), where the inertia controller is used. The optimum frequency drop in terms of minimum frequency is achieved in cases (c) and (d), thus when either Droop control is implemented on the wind turbine control level (case (c)) or when the Combined control is used (case (d)).

In this scenario, the effect of auxiliary frequency control on the maximum rate of change of frequency is very crucial. As illustrated in Figure 39, where the initial drop of the frequency for all cases is zoomed in, and as summarized also in Table 4, this rate is very high compared to SCENb (see alsoTable 3). The inertia of the system in this case is lower because the number of the conventional generators connected to the system in SCENc, and which are the ones determining the system inertia in large percentage, are reduced. The rate of change of frequency is close to 2.8 Hz/sec (in absolute value) in cases (a) and (b), although in the last case the minimum frequency does not drop below 48.5 Hz. Inertia control manages to reduce the rate to less than 1.8 Hz/sec, which is the highest rate among all the cases. Here also, as explained for SCENb above, the Combined controlseems to be the best compromise in terms of minimum frequency and maximum rate of change.

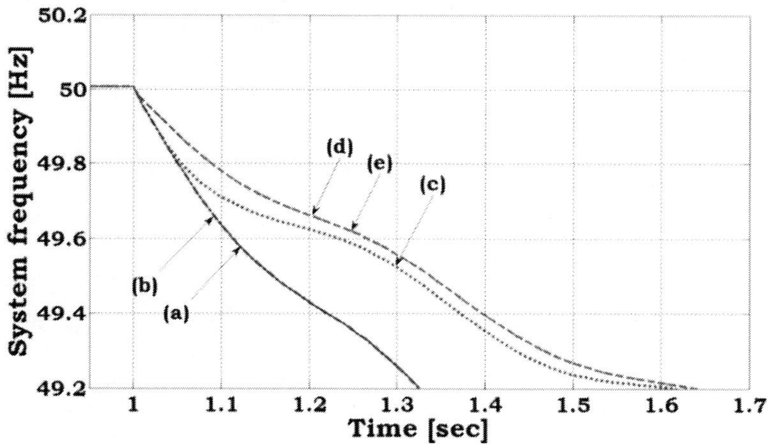

Figure 39. System frequency for largest unit loss when frequency control is applied by DFIGs – Zoom in the first seconds after the event (a) No auxiliary control, (b) Droop control on WF level, (c) Droop control on WT level, (d) Combined control, (e) Inertia control –SCENc.

Table 4. Results for SCENc – loss of largest infeed.

Frequency Control Scheme	Minimum Frequency (Hz)	Maximum Rate of change of frequency (Hz/sec)	Load Shedding (MW)
(a) No auxiliary control	48.28	-5	15.1
(b) Droop control on WF level	48.58	-5	0
(c) Droop control on WT level	48.69	-5	0
(d) Combined control	48.69	-3.8	0
(e) Inertia control	48.5	-3.8	0

Figure 40 and Figure 41 show respectively the rotor speed deviation and the change in active power output for wind farm A1, during the frequency drop. The comments made in SCENb are also valid here, regarding the differences among the various frequency control schemes. However, looking at the active power produced by the wind farm A1 in SCENc the contribution of the wind farms is more significant during the primary frequency response. The rotor deceleration is higher, thus more kinetic energy from the rotor mass is delivered to the grid as active power.

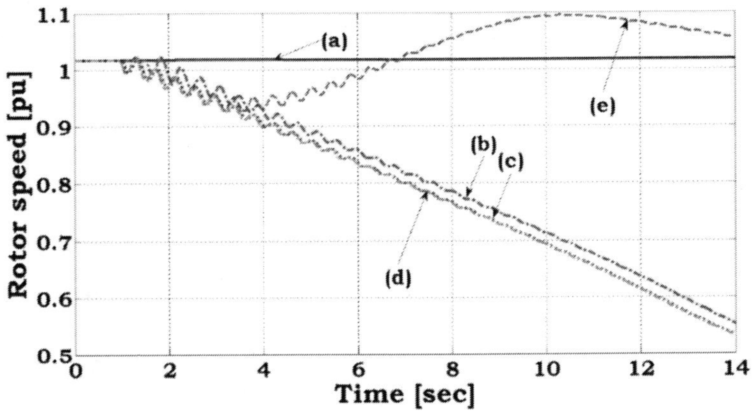

Figure 40. Rotor speed deviations after largest unit loss for different frequency control methods applied in DFIGs - (a) No auxiliary control, (b) Droop control on WF level, (c) Droop control on WT level, (d) Combined control, (e) Inertia control – SCENc.

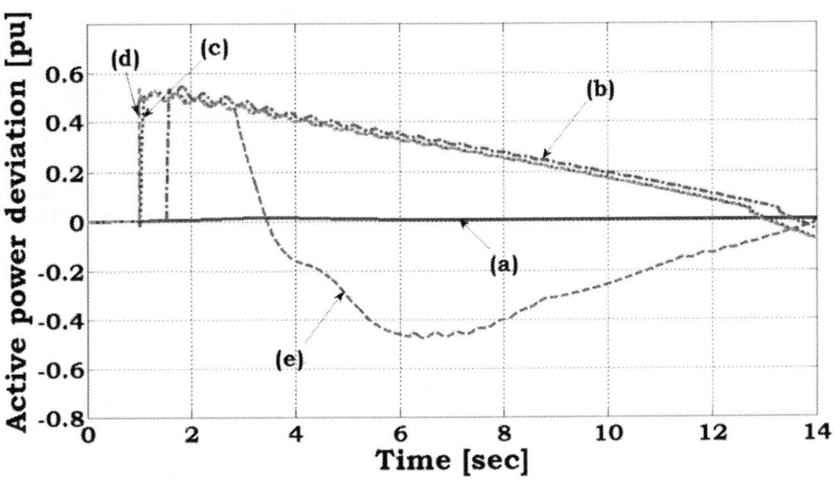

Figure 41. Change in active power output after largest unit loss for different frequency control methods applied in DFIGs - (a) No auxiliary control, (b) Droop control on WF level, (c) Droop control on WT level, (d) Combined control, (e) Inertia control –SCENc.

In the last part of the results section, comparative results will be shown for SCENb between the cases, where Combined frequency control is implemented only in one wind farm and the later, where the control is incorporated in both wind farms with DFIG wind turbines. All the results presented previously are extracted when both wind farms A1 and A2 have this frequency control capability.

As shown in Figure 42 and also summarized in Table 5 below, in case (d1) only wind farm A1 provides auxiliary frequency control leading to bigger frequency drop and higher rate ofchange of frequency compared to case (d2) where also wind farm A2 is equipped with this control capability.

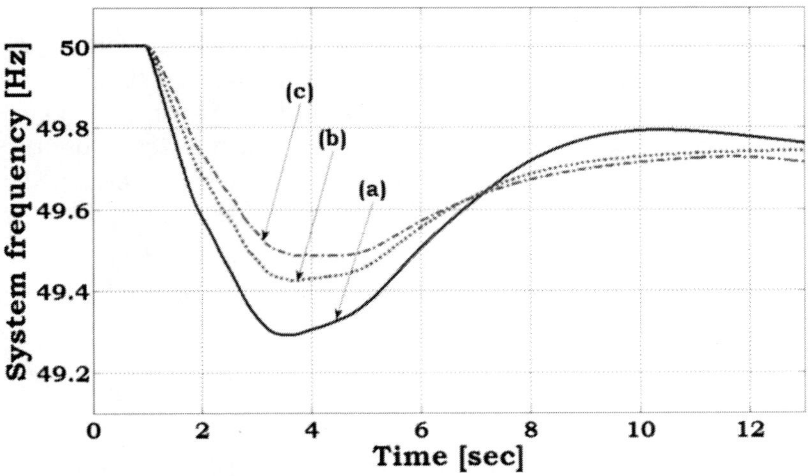

Figure 42. System frequency for largest unit loss for different frequency control methods applied in DFIGs - (a) No auxiliary control, (b) Combined control provided by only one wind farm (c) Combined control provided by two wind farms.

Table 5. Results for SCENb – loss of largest infeed.

Frequency Control Scheme	Minimum Frequency (Hz)	Maximum Rate of change of frequency (Hz/sec)	Load Shedding (MW)
(a) No auxiliary control	49.29	-0.48	0
(b) Combined control through one wind farm	49.43	-0.4	0
(c) Combined control through two wind famrs	49.49	-0.36	0

Thus, the inertia of the system, including the “virtual” inertia provided by the DFIG wind turbines, is obviously reduced when the proportion of the wind turbines providing auxiliary control is smaller. It is noted here, that in this comparison the Combined control method was chosen, as it is concluded by the previous results that this scheme achieves the best performance among the other proposed ones.

The conclusions from all the results presented in this section will be summarized in the next session, where also discussion on the expansion of the maximum allowed penetration in autonomous systems, like Rhodes, is made.

CONCLUSIONS

This chapter presented a detailed investigation of Rhodes power system dynamic security in the case of reference year 2012, when large wind farms are expected to be connected to the island. The conventional units, including Automatic Voltage Regulators and Speed Governors, were modeled in detail. The response of the governors during dynamic phenomena in the system is essential for the overall response. Especially the emergency rate of power undertake by the all different kinds of units i.e. steam, gas and diesel plants, is a major factor defining the overall system operation. The presented models includes also the models of the five wind farms which are connected in the system, comprising detailed models for three different wind turbine technologies – namely Doubly Fed Induction Generator (DFIG), Permanent Magnet Synchronous Generator (PMSG) and Active Stall Induction Generator (ASIG) based wind turbines.

Different load scenarios, relevant for the island operation, are analyzed, such as maximum wind power penetration scenario, maximum wind power production scenario and maximum load demand scenario.

In many non-interconnected systems the penetration of wind power has started to reach the boundaries of 30%. Serious concerns are expressed by the operators of these systems and wind fluctuations are one of the aspects that has to be evaluated before expanding the penetration levels, (Margaris et al., 2011). From the results, it is concluded that even in the worst case scenario, the Maximum Wind Power Penetration, the frequency fluctuations resulting from the wind speed fluctuations are not considered to pose security questions for the power system. However, the impact of wind variations is obvious in the system frequency and the correlation between wind speeds and system frequency has to be always investigated before reviewing the penetration levels.

Increased wind power penetration does not only limit the ability of thermal plants to undertake power, but also limits inertia. Systems inertia in cases of increased wind penetration is crucial for system stability as it defines the rate of frequency drop. The designed models for the conventional units in Rhodes power system, were sufficient to undertake power despite the rapid wind changes and the resulting active power fluctuations delivered by the wind farms. This study has proven that, when it comes to wind power fluctuations, the penetration levels of wind power can be expanded beyond 30% of the load.

Another issue, which should be remarked, is that the increased wind power penetration does not only limit the ability of thermal plants to undertake power, but also the inertia of the system. Systems inertia in cases with increased wind penetration is crucial for system stability, as it defines the rate of frequency drop.

The second part of the investigations provided a survey on frequency control issues of the autonomous power system with high wind power penetration, (Margaris et al., 2011). Frequency response during severe faults, i.e. the sudden loss of the biggest conventional generator, was simulated for two different load scenarios. These scenarios correspond to different system inertia but also to different operating points of the wind turbines. The wind turbines have different response when the system frequency varies, depending on the specific electrical configuration characteristics. Although the fixed speed ones contribute to the system inertia during frequency phenomena, this is not the case for DFIG and PMSG wind turbines, where the rotor is partially or totally detached form the system frequency respectively. The under-frequency protection system acting on the loads connected to MV substations can measure either the rate of change of frequency or the actual frequency to act on the relays. Modeling the protection system in the power system provides more accurate results regarding the load shedding, a variable that defines the dynamic security level of a system. As wind power penetration is increasing in modern power systems, the wind turbines have to contribute to the frequency stability of the system, acting similar to conventional power plants. In this article, three different frequency control schemes were investigated to enhance the primary frequency support of DFIG wind turbines. Simulation ended up to the following conclusions:

Non interconnected systems face the problem of decreased system inertia, especially when wind farms tend to substitute conventional units under increasing wind power penetration conditions. The DFIG wind turbines can be equipped with auxiliary Inertia control, providing with valuable inertia response during the first seconds following the frequency event. This control, reestablishes in the DFIG wind turbines the intrinsic characteristic of the fixed speed ASIG wind turbines, thus, using the kinetic energy accumulated in the rotor mass to support the grid during frequency variations.

Additional control methods can also be used in DFIG wind turbines, like Droop control used in conventional generators providing primary frequency support. The case, where the Droop control is implemented in the wind farm control level, instead of the turbine level, was also investigated to check the effect of communication delay in the response of the wind farm. Although when the Inertia control is used, the rate of change of frequency is significantly reduced, the Droop control seems to benefit more power system when looking at the minimum frequency after the event.

A Combined control, where both Inertia and Droop control schemes are used, was incorporated in the model. This last method manages to combine the positive effects of both methods, improving not only the initial rate of change of frequency after the fault but also ensuring lower frequency drop.

In some cases, when wind power production is higher than 30 % of the total demand, the auxiliary frequency control implemented in two wind farms in the Rhodes power system, manages to avoid load shedding totally. Therefore, the rule of thumb of 30 % penetration, which is often used in autonomous power systems, can be further expanded as long as auxiliary frequency control is provided by wind farms.

The benefits of the primary frequency support from modern wind turbines increase as the number of the turbines with this capability rises. This means that, if all new wind farms installed in autonomous power systems are equipped with primary frequency control capability, the frequency stability can be ensured even for penetration levels that today are hard to consider.

From the wind turbine side, in some cases the turbine may be forced to operate away from the maximum power-tracking curve, which means economic cost for the wind farm. So, before the system operators set any requirements for frequency control, the economic costs of frequency control for the wind turbine owner should be addressed.

The review of the frequency protection system settings can be done, as long as the frequency stability of the system is ensured. In many cases, the protection settings are quite sensitive and large amounts of load are cut off. The review of the protection system in modern power systems has to follow the progress made in the wind farms' capability to support the grid during disturbances.

Although, technology such as flywheels can support system inertia in autonomous power systems, advanced frequency control implemented in wind turbines will make it possible to achieve the penetration levels for wind power that today seem hard to reach.

There are some measures that can further enhance the dynamic security of systems like Rhodes:

The review of the frequency protection system settings can be done, if the dynamic security is ensured through FRT capability of wind farms online. The protection settings are quite sensitive and large amounts of load can be cut off just after the fault incident.

Systems like SVC or STATCOM for fast voltage control systems can guaranty the uninterrupted operation of wind parks, especially those consisting of fixed speed wind turbines. For instance, the substation, where most of these wind parks are connected, could be considered as the most appropriate for this installation. Under these conditions, wind power penetration could increase beyond 30% of the load, keeping the dynamic security of the system in the acceptable levels.

REFERENCES

1. T. . Ackermann, Ed, 2005 Wind Power in Power Systems, John Wiley and Sons, 978-0-47085-508-9
2. V. Akhmatov, 2003 Analysis of dynamic behavior of electric power systems with large amount of wind power, PhD thesis, 2003, Ørsted DTU.
3. T. V. Cutsem, C. Vournas, 1998 Voltage Stability of Electric Power Systems, Springer, 978-0-79238-139-6
4. DIgSILENT 2006 DIgSILENT technical documentation-PowerFactory.
5. J. Ekanayake, L. Holdsworth, N. Jenkins, 2003 Control of DFIG wind turbines. Power Engineer, 17 1 (2003), 2832 .
6. J. Ekanayake, N. Jenkins, 2004 Comparison of the response of doubly fed and fixed-speed induction generator wind turbines to changes in network frequency. IEEE Trans. Energy Conversion, 19 4 (2004), 800802 .
7. G. Gail, A. D. Hansen, T. Hartkopf, 2006 Controller design and analysis of a variable speed wind turbine with doubly-fed induction generator, EWEC, Athens, 2006.
8. D. Hansen, C. Jauch, P. Sørensen, F. Iov, F. Blaabjerg, 2003 Dynamic wind turbine models in power system simulation tool DIgSILENT. Risø-R-1400(EN), 2003.
9. D. Hansen, P. Sørensen, F. Iov, F. Blaabjerg, 2006 Centralised power control of wind farm with doubly-fed induction generators. Renewable Energy, 31 (2006), 935951 .
10. Hansen A.D., and Hansen L.H. 2007 Market penetration of different wind turbine concepts over the years, EWEC, Milano, 2007.
11. D. Hansen, G. Michalke, 2007 Fault ride-through capability of DFIG wind turbines. Renewable Energy, 32 (2007), 15941610 .
12. D. Hansen, G. Michalke, 2008 Modelling and control of variable speed multi-pole PMSG wind turbine. Wind Energy, 11 5 (2008), 537554 .
13. M. H. Hansen, A. D. Hansen, T. J. Larsen, S. Øye, P. Sørensen, 2005 Control design for a pitch-regulated variable speed wind turbine, Risø-R-1500 (EN), 2005.
14. L. Holdsworth, J. B. Ekanayake, N. Jenkins, 2004 Power system frequency response from fixed speed and doubly fed induction generator-based wind turbines. Wind Energy, 7 DOI:10.1002/we.105, 2135 .
15. Jauch, A. D. Hansen, P. Sørensen, F. Blaabjerg, 2004 Simulation Model of an Active-stall Fixed-speed Wind Turbine Controller. Wind Engineering, 28 2 (2004), 177195 .
16. G. Lalor, A. Mullane, M. J. O'Malley, 2005 Frequency Control and Wind Turbine Technologies. IEEE Transactions on Power Systems, 20 4 (2005).

17. J. Mantzaris, M. Karystianos, C. Vournas, 2008 Comparison of Gas Turbine and Combined Cycle Models for System Stability Studies, 6th Mediterranean. Conf. MedPower, Thessaloniki, Greece, 2008.
18. D. Margaris, J. C. Mantzaris, M. E. Karystianos, A. I. Tsouchnikas, C. D. Vournas, N. D. Hatziargyriou, I. C. Vitellas, 2009 Methods for evaluating penetration levels of wind generation in autonomous systems, IEEE PowerTech Conf., Bucharest, June 2009.
19. D. Margaris, A. D. Hansen, N. A. Cutululis, P. Sørensen, N. Hatziargyriou, 2010 Impact of wind power in autonomous power systems- Power fluctuations- Modeling and Control issues. Wind Energy, online: DOI: 10.1002/we.417. 2010.
20. D. Margaris, S. A. Papathanassiou, N. Hatziargyriou, A. D. Hansen, P. Sørensen, 2011 Frequency control in autonomous power systems with high wind power penetration. under review in IEEE Transactions.
21. Morren, J. Pierik, S. W. H. de Haan, 2006 Inertial response of variable wind turbines. Electric Power Systems Research. Elsevier, 76, 980987 .
22. M. Poeller, S. Achilles, 2003 Aggregated wind park models for analyzing power system dynamics, Fourth international workshop on large-scale integration of wind power and transmission networks, Billund, Denmark, October 2003.
23. G. Ramtharan, J. B. Ekanayake, N. Jenkins, 2007 Frequency support from doubly fed induction generator wind turbines. IET Renew. Power Gener., 1 1 (2007), 39 .
24. P. Sørensen, N. A. Cutululis, A. Vigueras-Rodríguez, H. Madsen, P. Pinson, L. E. Jensen, J. Hjerrild, M. Donovan, 2007 Modelling of power fluctuations from large offshore wind farms. Wind Energy, 11 (2007), 2943 .
25. Suwannarat, B. Bak-Jensen, Z. Chen, H. Nielsen, J. Hjerrild, P. Sørensen, A. D. Hansen, 2007 Power system operation with large scale wind power integration, PowerTech 2007, Lausanne Switzerland, July 2007.
26. N. R. Ullah, T. Thiringer, T. Karlsson, 2008 Temporary Primary Frequency Control Support by Variable Speed Wind Turbines- Potential and Applications. IEEE Transactions on Power Systems, 23 2 (2008).

CITATION

IoannisMargaris, Anca Hansen, PoulSørensen, NicolaosCutululis and Nikos Hatziargyriou (2011). Operation and Control of Wind Farms in Non-Interconnected Power Systems, Wind Farm - Impact in Power System and Alternatives to Improve the Integration, Dr. GastÃ³n Orlando Suvire (Ed.), ISBN: 978-953-307-467-2, InTech, DOI: 10.5772/16532.

CHAPTER 2

A Virtual Synchronous Machine Implementation for Distributed Control of Power Converters in SmartGrids

Salvatore D'Arco[1], Jon Are Suula[2], Olav B. Fosso[2]

[1]SINTEF Energy Research, 7465 Trondheim, Norway

[2]Department of Electric Power Engineering, Norwegian University of Science and Technology, 7495 Trondheim, Norway

ABSTRACT

The ongoing evolution of the power system towards a "SmartGrid" implies a dominant role of power electronic converters, but poses strict requirements on their control strategies to preserve stability and controllability. In this perspective, the definition of decentralized control schemes for power converters that can provide grid support and allow for seamless transition between grid-connected or islanded operation is critical. Since these features can already be provided by synchronous generators, the concept of Virtual Synchronous Machines (VSMs) can be a suitable approach for controlling power electronics converters. This paper starts with a discussion of the general features offered by the VSM concept in the context of SmartGrids. A specific VSM implementation is then presented in detail together with its mathematical model. The intended emulation of the synchronous machine characteristics is illustrated by numerical simulations. Finally, stability is assessed by analysing the eigenvalues of a small-signal model and their parametric sensitivities.

INTRODUCTION

The increasing penetration of power generation from renewable energy sources and the transition from a centralized power production model to distributed generation are expected to pose serious challenges to the development and operation of future power systems. This tendency is a strong motivation behind the paradigm shift from the traditional power system architecture towards an approach ensuring more flexibility and coordination between the generation units and loads that is promised by "SmartGrids"[1]. At the same time, the share of the electric power transferred through the power system which is processed by at least one power electronic conversion stage in the path from primary energy conversion to final consumption is continuously increasing. Already in 2007, it was estimated that this share would reach 80% around 2015 [2], and even if the development has been slightly slower, such a high share of power electronic conversion is expected to be exceeded during the coming years. Thus, power electronic control will have a crucial role in the emerging SmartGrid scenario, as the presence of power converters in the power system and their impact on global stability and controllability continues to increase.

Although the ongoing SmartGrid developments point towards an increasing level of communication and integration between various elements of the power system, distributed architectures with local primary control of converters combined with centralized secondary control seem to be an appropriate approach for optimizing steady-state operation while ensuring immediate response to transient events. Thus, converter units should be able to react autonomously to abrupt changes in the power system operating conditions, while complying on a longer time scale with the set-points and service requirements requested by the system operator through external communication.

In classical power systems, the Synchronous Machine (SM) with speed governor and excitation control offers favourable features to support the system operation within a distributed control scheme. Indeed, SMs contribute to the system damping through their inertia, participate in the primary frequency regulation through the droop response of the speed controller, and provide local control of voltage or reactive power flow. These capabilities, and especially the inertial and damping response common to all SMs, are not inherently offered by the power electronics interfaces commonly adopted for the integration of renewable energy sources. A distributed model for production and local control is alsoopening the possibility of islanded operation, which is inherently feasible with one or more controllable SMs in the islanded area. Such islanding operation is usually more complex to achieve with power converter interfaces designed for integration with a large-scale power system.

Power from many traditional large-scale generation facilities is currently being replaced by distributed generation capacity from wind power and photovoltaics. The traditional control structures implemented in the power converters for these applications rely on the synchronization to a stable grid frequency supported by large rotating inertias and are not inherently suitable in a SmartGrid context. Thus, from an implementation perspective, significant research efforts are still devoted towards development of control schemes for power electronic converters explicitly conceived to address the conditions emerging in future SmartGrids. Given the inherent benefits of the SMs outlined above, a captivating approach is the control of power electronic converters to replicate the most essential properties of the SM and by that gain equivalent features from a functional point of view. Thus, several alternatives for providing auxiliary services like reactive power control, damping of oscillations and emulation of rotating inertia with power electronic converters have been proposed [3], [4], [5], [6], [7] and [8]. Some of these control strategies are explicitly designed to mimic the dynamic response of the traditional SM, and can therefore be classified in broad terms as Virtual Synchronous Machines (VSM).

During the last decade, several concepts for VSMs have been presented with different names and different practical implementations [4], [8], [9], [10], [11] and [12]. The first review studies providing an overview of implementations have been recently published in[10] and [13], with an attempt to define a classification framework presented in [10]. The review in [10] also highlights how some implementations offer only partially the benefits of the SMs while only a few can ensure features as island operation of single or multiple units.

Most previous studies of VSM-based control strategies have presented particular implementation schemes which have been verified by time-domain simulations and/or laboratory experiments. A first study that included detailed modelling and small-signal stability of a particular VSM implementation was presented in [14]. However, this model was mainly developed for tuning of the converter control loops and did not consider the primary power-frequency control or the dynamics of the grid frequency detection needed to ensure an implementation of the VSM damping effect that adapts to variations in the grid frequency. A VSM system model addressing also these issues was recently presented in [15].

This paper includes a comprehensive treatment of a particular VSM implementation, starting from a discussion of the comparative advantages offered by the VSM concept in the context of SmartGrids in abstract terms. Then a brief overview of the VSM development status is offered, with the purpose of identifying general preferences for selecting specific implementations for future SmartGrid applications. The selected implementation is based on an internal representation of the SM inertia and damping behaviour through a reduced order swing equation, together with cascaded voltage and current controllers for operating a Voltage Source Converter (VSC), based on the general scheme from [15]. The paper derives step-by-step a detailed nonlinear mathematical model for this VSM implementation, and a corresponding small signal model in order to apply linear analysis techniques to the system in the perspective of stability assessment and controller tuning. The effect of system parameters on the poles of the linearized system model is also analyzed by calculating the parametric sensitivities of the system eigenvalues. The features and performance of the investigated VSM and its linearized small-signal model is verified with reference to a few selected cases by numerical simulations.

APPLICATION OF VIRTUAL SYNCHRONOUS MACHINES IN THE SMARTGRID CONTEXT

Power generation from distributed renewable energy sources like wind and photovoltaic power plants is usually connected to the power grid through actively controlled power electronic converters, and similar interfaces are applied for energy storage systems and an increasing share of controllable loads. The conventional scheme for such grid connected power converters is based on current controlled Voltage Source Converters (VSCs), which are synchronized to the measured grid voltage through a Phase Locked Loop (PLL) [16]. This approach usually requires a relatively strong grid with the presence of units that can maintain and stabilize the grid frequency and voltage. Even if auxiliary services like frequency and voltage support can be provided by current controlled VSCs, this functionality must be added through additional outer loop controllers which are not inherently applicable for operation in islanded mode [17]. Although this approach can be suitable for a relatively low penetration of grid connected converters, it does not seem sustainable for operation in a long term SmartGrid perspective with the expected dominant presence of power electronic conversion units and a high degree of flexibility in the network configurations.

Challenges for Power Converter Control in Future SmartGrids

In the last decade, several alternative concepts and approaches for control and operation of power converters distributed in the power system have emerged. A noticeable example from the overall system operation point-of-view is the concept of Virtual Power Plants (VPPs) that aims to aggregate generation resources, energy storages and loads into clusters that can be controlled by the distribution system operator in a similar way as traditional power plants [18] and [19]. Such VPPs should coordinate the controllable units while ensuring supply to the uncontrolled loads in the system, but must also be able to supply auxiliary services like control of voltage or reactive power flow and support the frequency regulation of the system. In small isolated power systems, or in case parts of the distribution system should be able to operate in islanded mode, the generation units aggregated together in one VPP must also ensure a sufficient system inertia to keep the system stable while maintaining the power balance without large frequency deviations. The VPP concepts currently under development are capable of providing frequency-activated power support to the system within a few seconds, and can therefore ensure the steady-state power balance [20]. However, a faster response is required for ensuring an inertia-based power-frequency balance that will be able to keep the system stable in transient conditions. As mentioned, such a power-frequency response is a natural feature of traditional SMs which is not inherently present in the current controlled VSCs usually applied for integrating renewable sources to the grid. SMs offer also additional advantages as automatic synchronization and power sharing in response to changes in the operating conditions.

From these considerations it appears that a SmartGrid can represent a challenging environment for power converter control schemes, especially due to the possible large penetration of power electronics and the variability of the operating conditions. Indeed, a large penetration of converter-interfaced units will correspond to a lower level of physical inertia than in traditional power system dominated by large synchronous machines, and in certain conditions it can even be necessary to operate pure power electronic, inertia-less, systems. Considering that a flexible SmartGrid framework can result in more frequent reconfigurations of the power system, with corresponding variations in equivalent grid impedance and the possibility to operate parts of the system as groups of electrical islands, the variability of the conditions can introduce a further dimension of complexity. This can also lead to larger and more frequent transients and corresponding requirements for the control systems to maintain individual stable operation as well as contributing to the system stability in a wide range of operating conditions. The design of power electronic control schemes should cope with these challenges and preferably mitigate their effects.

One of the general characteristics of the emerging SmartGrid scenario is the presence of a communication infrastructure that can increase the volume of signal interaction between controllable units in the power system and facilitate their coordination. This can lead to a wide range of options for centralized control schemes, like the mentioned concept of VPPs, where references are determined by a centralized control unit and distributed to the individual converters. However, it should be noted that reducing the necessity of communication between the units, especially during transients, can increase the robustness of the system and reduce the risks in the event of temporary unavailability of the communication infrastructure. Thus, distributed control concepts where individual units can autonomously define their transient response based on local measurements are still relevant. Moreover, decentralized schemes where only steady-state references or set-points are distributed from a centralized system controller can lower the requirements in terms of bandwidth and latencies for the communication infrastructure, resulting in lower installation and operation costs. Multiple examples of possible solutions for decentralized control of power electronic converter can be found in the large literature on isolated MicroGrids [17], [21] and [22] although an exhaustive analysis is beyond the scope of this paper. However, not all of these schemes are suitable for SmartGrid applications where the converter are expected to operate most of the time in grid connected mode. It should also be mentioned that most control schemes that allow for both grid connected and stand-alone operation while also maintaining some decentralized control features, tend to be fairly complicated since transitions between these two modes usually require a reconfiguration of the control structure.

General Characteristics of VSMs

In the emerging SmartGrid context, the VSM concept can offer a basis for realizing flexible decentralized converter control schemes that can operate both in grid connected and islanded conditions, and that can almost seamlessly switch between the corresponding operating modes. Furthermore the inherent inertial characteristic of the VSM can provide services as frequency support and transient power sharing as primary control actions. These are indeed based only on local measurement and do not depend on external communications as in typical alternative schemes. Still, there is no conflict between this local controllability and the ability to operate in a hierarchical structure while following external references and set-points provided by a centralized controller for optimizing the system operation. Moreover, a further advantage of the VSM approach lies in its conceptual simplicity, due to the immediate and intuitive physical interpretation of its behaviour with analogy to the corresponding behaviour of a physical machine.

The dominant behaviour of SMs in terms of inertia response and damping can be modelled by the traditional swing equation [23]. Considering these general characteristics, several control strategies have been developed for allowing power electronic converters to provide synthetic or virtual inertia to the power system, and have been proposed for a variety of applications like for instance wind turbines, energy storage systems and HVDC transmission schemes [3], [4], [5], [6], [7], [8], [9], [10], [11], [12], [13],[14], [24], [25] and [26]. Some of these control methods provide a synthetic inertial response to variations in the grid frequency and only a few aims to explicitly replicate the features of the traditional SMs. However, emulation of the inertia and damping effects requires an energy buffer with sufficient capacity to represent the energy storage effect of the emulated rotating inertia available. Thus, the amount of virtual inertia that can be added to the system by a single VSM unit will be limited by the DC-side configuration and by the current rating of the converter.

The current state of the art on the VSM concept has been presented in [10] and [13]. These reviews identify that some of the implementations proposed as VSMs do not exploit the full potential of the concept because they still rely on a PLL for detecting the grid voltage phase angle, the grid frequency and its derivative, thus requiring the presence of rotating inertia in the grid. Other proposed implementations of the VSM concept are based on the simulation of an internal mathematical SM model inside the control system providing a voltage reference output for the PWM [9]. However, direct open loop PWM signal generation from these voltage references prevents the possibility to explicitly embed the limitations and controlled saturations of voltages and currents that are normally required as protective functions for safe operation of power electronic converters. These protective functions can be easily included in a cascaded control scheme [10], [17], [27] and [28] where the output from the VSM inertia emulation is used as reference for a voltage control loop cascaded with an internal current control loop. Numerically, this approach is sufficiently robust for practical implementations and will in the following be assumed as the reference VSM scheme, elaborated from previous studies in [14] and [15]. It can also be noted that the implementation of a VSM based on the swing equation providing references for operation of the converter, under certain conditions has been shown to be equivalent to the frequency-droop-based control strategies first developed for Uninterruptable Power Supply (UPS) systems and MicroGrids, as demonstrated in [10] and [29]. However, the interpretation of the parameters in a VSM approach seems to be simpler and more intuitive than the equivalent parameters in the commonly applied MicroGrid schemes and, thus, preferable.

MATHEMATICAL MODEL OF THE VSM REFERENCE IMPLEMENTATION

This section describes the control scheme for the selected VSM reference implementation, considering each functional block, and derives the corresponding mathematical model.

Control System Overview

An overview of the studied VSM configuration is shown in Fig. 1, where a VSC is connected to a grid through an LC filter. In the following, the switching effects of the VSC are neglected and an ideal average model is assumed for modelling the converter. Furthermore, no application-specific constraints of the DC-side of the VSC are considered and, thus, modelling and control of the energy source or storage on the DC side of the converter is not further discussed.

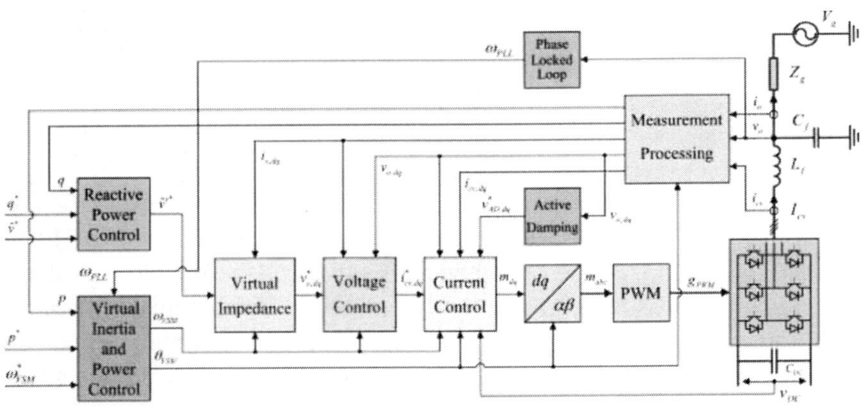

Figure 1. Overview of investigated system configuration and control structure for the Virtual Synchronous Machine.

The VSM-based power control with virtual inertia provides frequency and phase angle references ω_{VSM} and θ_{VSM} to the internal control loops for operating the VSC, while a reactive power controller provides the voltage amplitude reference $\hat{v}^{r*}$. Thus, the VSM inertia emulation and the reactive power controller appear as outer loops providing the references for the cascaded voltage and current controllers. A PLL detects the actual grid frequency, but this frequency is only used for implementing the damping term in the swing equation. Thus, the operation of the inner loop controllers does not rely on the PLL as in conventional VSC control systems, but only on the power-balance-based synchronization mechanism of the VSM inertia.

Modelling Conventions

In Fig. 1, upper case symbols represent physical values of the electrical circuit. The control system implementation and the modelling of the system are based on per unit quantities, denoted by lower case letters where the base values are defined from the apparent power rating and the rated peak value of the phase voltage [30].

The modelling, analysis and control of the electrical system is implemented in Synchronous Reference Frames (SRFs). The transformation from the stationary reference frame into the SRFs are based on the amplitude-invariant Park transformation, with the *d*-axis aligned with a voltage vector and the *q*-axis leading the *d*-axis by 90°[30]. Thus, the magnitude of current and voltage vectors at rated conditions is 1.0 pu. Whenever possible, SRF equations are presented in complex space vector notation as:

$$\mathbf{x} = x_d + j \cdot x_q \tag{1}$$

Thus, active and reactive powers can be expressed on complex or scalar form as:

$$\begin{aligned} p &= Re(\mathbf{v} \cdot \mathbf{i}) = v_d \cdot i_d + v_q \cdot i_q \\ q &= Im(\mathbf{v} \cdot \breve{\mathbf{i}}) = -v_d \cdot i_q + v_q \cdot i_d \end{aligned} \tag{2}$$

The current directions indicated in Fig. 1 result in positive values for active and reactive powers flowing from the converter into the grid.

System Modelling

In the following sub-sections, the implementation of each functional block of the VSM-based control and the mathematical models of all system elements from Fig. 1 are presented as a basis for developing a non-linear model of the system. This system model will also be used to establish a linearized small-signal state-space representation.

VSM Inertia Emulation and Active Power Droop Control

The emulation of a rotating inertia and the power-balance based synchronization mechanisms of this virtual inertia is the main difference between the investigated VSM control structure and conventional control systems for VSCs. The VSM implementation investigated in this case is based on a conventional swing equation representing the inertia and damping of a traditional SM [10] and [14]. The swing equation used for the implementation is linearized with respect to the speed so that the acceleration of the inertia is determined by the power balance according to:

$$\frac{d\omega_{VSM}}{dt} = \frac{p^{r*}}{T_a} - \frac{p}{T_a} - \frac{p_d}{T_a} \tag{3}$$

In this equation, $pr*$ is the virtual mechanical input power, p is the measured electrical power flowing from the VSM into the grid, and pd is the damping power, while the mechanical time constant is defined as Ta (corresponding to $2H$ in a traditional SM). The per unit mechanical speed ω_{VSM} of the virtual inertia is then given by the integral of the power balance while the corresponding phase angle θ_{VSM} is given by the integral of the speed. A block diagram showing the implementation of the VSM swing equation is shown on the right in Fig. 2.

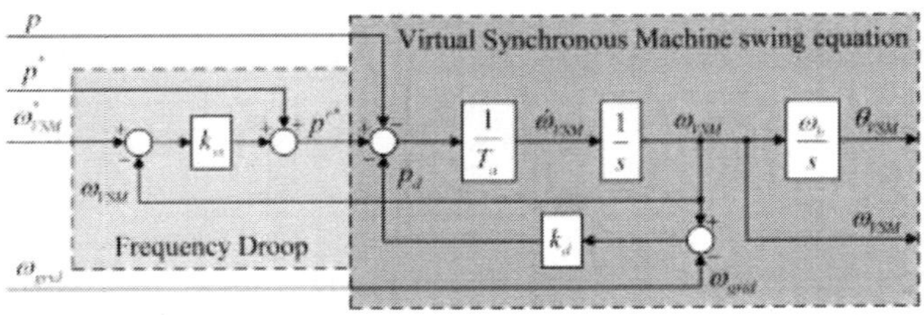

Figure 2. Virtual Synchronous Machine inertia emulation with power-frequency droop.

The VSM damping power pd, representing the damping effect of a traditional SM, is defined by the damping constant kd and the difference between the VSM speed and the actual grid frequency. Thus, an estimate of the actual grid frequency is needed for the VSM implementation. As indicated in the figure, the frequency estimate is in this case labelled as ωPLL and is provided by a PLL.

An external frequency droop, equivalent to the steady-state characteristics of the speed governor for a traditional synchronous machine, is included in the power control of the VSM as shown in the left part of Fig. 2. This power-frequency droop is characterized by the droop constant $k\omega$ acting on the difference between a frequency reference ω^*_{VSM} and the actual VSM speed ω_{VSM}. Thus, the virtual mechanical input power pr^* to the VSM swing equation is given by the sum of the external power reference set-point, p^*, and the frequency droop effect, as shown on the left of Fig. 2.

For modelling the VSM in a SRF, the phase angle of the VSM in grid connected mode should be constant under steady-state conditions and should correspond to the phase displacement between the virtual position of the VSM internal voltage and the position of the grid voltage vector. Since only the deviation of the VSM speed from the actual grid frequency should be modelled to achieve this, a new set of variables representing the speed deviation $\delta\omega VSM$ and the corresponding phase angle difference $\delta\theta VSM$ is introduced. Thus, the power balance of the VSM inertia can be expressed by (4), while the VSM phase displacement is defined by (5):

$$\frac{d\delta\omega_{VSM}}{dt} = \frac{p^*}{T_a} - \frac{p}{T_a} - \frac{k_d(\omega_{VSM} - \omega_{PLL})}{T_a} - \frac{k_\omega(\omega_{VSM} - \omega^*)}{T_a} \tag{4}$$

$$\frac{d\delta\theta_{VSM}}{dt} = \delta\omega_{VSM} \cdot \omega_b \tag{5}$$

Since the VSM speed in steady state will become equal to the grid frequency ωg, the frequency deviation $\delta\omega_{VSM}$ will return to zero under stable grid connected operation.

The actual per unit speed of the VSM shown in the block diagram of Fig. 2 can be expressed from the speed deviation $\delta\omega_{VSM}$ resulting from (4) and the grid frequency ωg as given by (6). The corresponding VSM phase angle θ_{VSM} is then defined by (7)

$$\omega_{VSM} = \delta\omega_{VSM} + \omega g \tag{6}$$

$$\frac{d\theta_{VSM}}{dt} = \omega_{VSM} \cdot \omega_b \tag{7}$$

The phase angle θVSM will then become a saw-tooth signal between 0 and 2π, which is the phase angle that will be used for the transformation between the rotating reference frame defined by the VSM inertia and the three-phase signals, as indicated in Fig. 1.

Reactive Power Droop Controller

The droop-based reactive power controller applied in this case is similar to the controllers commonly applied in microgrid systems [17] and [27]. The voltage amplitude reference $\hat{v}^{r*}$ used for the inner loop voltage and current control is then calculated by (8) where $\hat{v}^{*}$ is the external voltage amplitude reference and $q*$ is the reactive power reference. The gainkq is the reactive power droop gain acting on the difference between the reactive power reference and the filtered reactive power measurement qm. The state of the corresponding first order low pass filter applied in this case is defined by (9), where ωf is the cut-off frequency. A block diagram of the resulting control structure is shown in Fig. 3:

$$\hat{v}^{r*} = \hat{v}^{*} + k_q(q^{*} - q_m) \tag{8}$$

$$\frac{dq_m}{dt} = -\omega_f \cdot q_m + \omega_f \cdot q \tag{9}$$

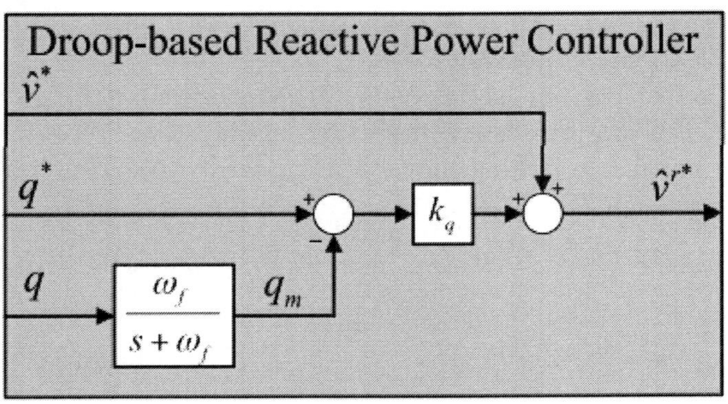

Figure 3. Reactive power droop controller.

Reference Frame Orientations

The synchronization of the VSM control system to the grid is based on the phase angle orientation of the virtual rotor of the VSM, and the phase angle θ_{VSM} is used in the transformations between the stationary reference frame and the VSM-oriented SRF. Thus, the power balance of the VSM swing equation will ensure the synchronization to the grid voltage without the need for a traditional PLL. Since the VSM-oriented SRF in steady state rotates with the same frequency as the grid voltage, this phase angle will be continuously increasing between 0 and 2π, as indicated in the vector diagram shown in Fig. 4. According to its definition, the phase angle $\delta\theta_{VSM}$ is instead representing the phase difference between the VSM SRF orientation and the rotating grid voltage vector, as also indicated in Fig. 4.

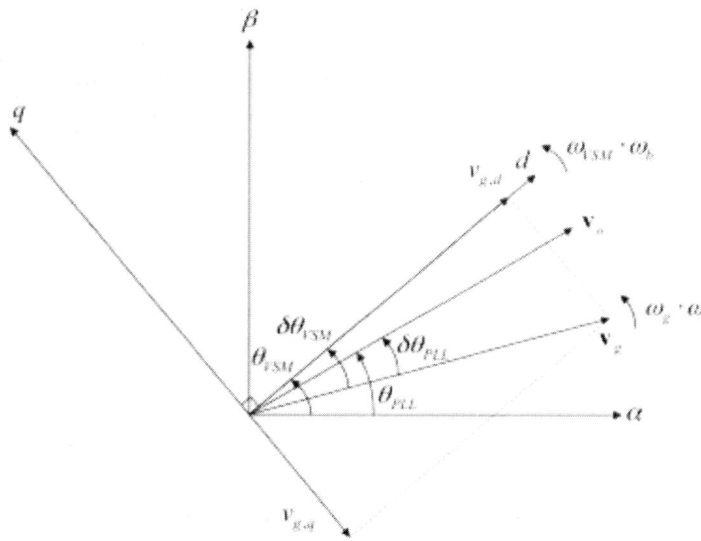

Figure 4. Vector diagram defining the SRF and voltage vector orientations.

The VSM-oriented SRF is used for both control and modelling of the system, and therefore, also the model of the electrical system will be represented in this reference frame. This has significant advantages for the modelling of the system, since multiple reference frame transformations between a local SRF for controller implementation and a global SRF for electrical system modelling can be avoided. Considering the amplitude of the equivalent grid voltage $\hat{v}_g$ to be known, the voltage vector **v**g in the VSM-oriented SRF can then be expressed by (10):

$$\mathbf{v}_g = \hat{v}_g e^{-j\delta\theta_{VSM}} \tag{10}$$

By the power-balance-based synchronization effect of the VSM swing equation, the control system defines its own reference frame orientation with respect to the grid voltage. In principle, no additional reference frames are needed to model the system from Fig. 1. However, since an estimate for the grid frequency is used to implement the VSM damping effect, a PLL operating on the measured voltage $\mathbf{v}_o$ at the filter capacitors is implemented as part of the control system. Thus, this PLL will establish its own SRF aligned with the voltage vector $\mathbf{v}_o$. The phase angle displacement of this PLL with respect to the grid voltage can then be defined as $\delta\theta_{PLL}$ in a similar way as for the phase angle displacement of the VSM. The detailed implementation of the PLL will be presented in the following sub-section, but the definition of its steady state phase displacement $\delta\theta_{PLL}$ with respect to the grid voltage, and the corresponding phase angle θ_{PLL} between the rotating PLL-oriented SRF and the stationary reference frame, is shown in Fig. 4.

According to the definitions indicated in Fig. 4, the phase angle between the VSM and PLL oriented SRFs will be defined by the difference between the VSM and PLL angles. For modelling of the PLL in its own reference frame, the voltage $\mathbf{v}_o$ at the filter capacitors can be transformed from the VSM-oriented reference frame to the PLL-oriented reference frame by:

$$\mathbf{v}_o^{PLL} = \mathbf{v}_o^{VSM} e^{-j(\delta\theta_{PLL} - \delta\theta_{VSM})} \tag{11}$$

Phase Locked Loop

The Phase Locked Loop (PLL) applied in this case for tracking of the actual grid frequency is based on [31] and [32] and its structure is shown in Fig. 5. This PLL is using first order low-pass filters on the estimated *d*- and *q*-axis voltage components and an inverse tangent function to calculate the phase angle error of the PLL. This phase angle error e_{PLL} is the input to a PI controller tracking the frequency of the measured voltage. For the practical implementation, the estimated frequency ω_{PLL} is then integrated to obtain the estimate of the actual instantaneous phase angle θ_{PLL} used for transformation of the voltage measurements into the PLL-oriented SRF.

For modelling of the PLL, the voltage vector **v***o* in the VSM-oriented SRF must be transformed into the PLL-oriented SRF according to (11).

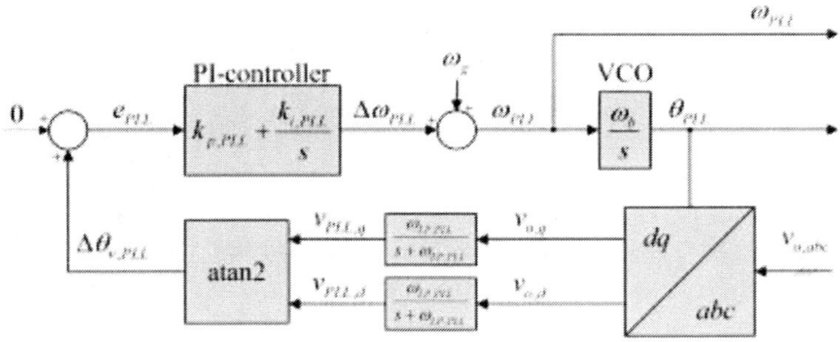

Figure 5. Phase locked loop.

The states of the applied first order low-pass filters in the PLL, defining the filtered voltage $\mathbf{v}_{PLL}$, can be expressed by (12), where the cut-off frequency of the applied low pass filters is given by $\omega_{LP,\,PLL}$:

$$\frac{d\mathbf{v}_{PLL}}{dt} = -\omega_{LP,PLL} \cdot \mathbf{v}_{PLL} + \omega_{LP,PLL} \cdot \mathbf{v}_o e^{-j(\delta\theta_{PLL} - \delta\theta_{VSM})} \tag{12}$$

The integrator state ε_{PLL} of the PI controller can then be defined by:

$$\frac{d\varepsilon_{PLL}}{dt} = \tan^{-1}\left(\frac{v_{PLL,q}}{v_{PLL,d}}\right) \tag{13}$$

In the same way as explained for the SRF modelling of the VSM swing equation, a speed deviation $\delta\omega_{PLL}$ with respect to the grid frequency is defined for the PLL according to(14). The corresponding phase angle displacement, $\delta\theta_{PLL}$, of the PLL is then defined by(15):

$$\delta\omega_{PLL} = k_{p,PLL} \cdot \tan^{-1}\left(\frac{v_{PLL,q}}{v_{PLL,d}}\right) + k_{i,PLL} \cdot \varepsilon_{PLL} \tag{14}$$

$$\frac{d\delta\theta_{PLL}}{dt} = \delta\omega_{PLL} \cdot \omega_b \tag{15}$$

In accordance with the definitions introduced for the VSM swing equation, the actual per unit frequency ωPLL detected by the PLL is given by (16). The phase angle used in the implementation of the PLL, for transformation of the measured three-phase voltage measurements into the PLL-oriented SRF, is then defined by θPLL according to (17):

$$\omega_{PLL} = \delta\omega_{PLL} + \omega g \tag{16}$$

$$\frac{d\theta_{PLL}}{dt} = \omega_{PLL} \cdot \omega_b \tag{17}$$

Virtual Impedance and Voltage Controllers

As indicated in Fig. 1, the voltage amplitude reference $\hat{v}^{r*}$ resulting from the reactive power droop controller in Fig. 3 is passed through a virtual impedance before it is used as a reference for controlling the voltage $\mathbf{v}o$ at the filter capacitors. This virtual impedance can be considered as an emulation of the quasi-stationary characteristics of the synchronous impedance in a traditional SM. The virtual impedance will influence the steady-state and dynamic operation of the VSM, and it will be shown how it can be used to shape the dynamic characteristics of the system. Since power flowing through the virtual inductance will cause a phase angle displacement between the grid voltage and the virtual inertia position of the VSM, it will also reduce the sensitivity of the VSM to small disturbances in the grid. The influence from the virtual resistance rv and inductance lv on the capacitor voltage reference vector $\mathbf{v}_o^*$ is defined on basis of the current $\mathbf{i}o$ according to[33] and [34]:

$$\mathbf{v}_o^* = \hat{v}^{r*} - (r_v + j \cdot \omega_{VSM} \cdot l_v) \cdot \mathbf{i}_o \tag{18}$$

The resulting *d*- and *q* -axis voltage components $v^{*}_{o,d}$ and $v^{*}_{o,q}$ are used directly as references for the decoupled SRF PI voltage controllers as shown in the left part of Fig. 6.

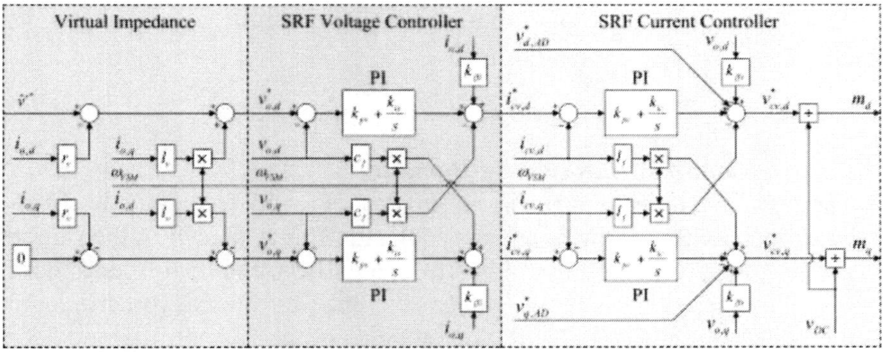

Figure 6. Virtual inductance, voltage control and current control.

The detailed structure of the SRF PI controllers for the filter capacitor voltage is shown in the middle of Fig. 6, and is producing the reference values $\mathbf{i}^{*}_{cv}$ for the converter currents[27]. These current references can be expressed by (19), where the PI controller gains are defined by *kpv* and *kiv*. A gain factor *kffi* that can be set to 1 or 0 is used to enable or disable the feed-forward of measured currents flowing into the grid. It should also be noted that the decoupling terms of the voltage controller are based on the per unit speed of the VSM inertia as defined in (6). The states ξ are defined to represent the integrators of the PI voltage controllers as given by (20):

$$\mathbf{i}^{*}_{cv} = k_{pv}(\mathbf{v}^{*}_{o} - \mathbf{v}_{o}) + k_{iv}\xi + j \cdot c_1 \cdot \omega_{VSM} \cdot \mathbf{v}_o + k_{ffi} \cdot i_o \tag{19}$$

$$\frac{d\xi}{d} = \mathbf{v}^{*}_{o} - \mathbf{v}_{o} \tag{20}$$

The current references from the voltage controllers should be limited to avoid over-currents in case of voltage drops, fault conditions or other severe transients. This also implies that the voltage controllers must be protected from windup conditions in case the current references are saturated. However, the required limitations and anti-windup techniques for the investigated VSM scheme are similar to what is needed in conventional droop-based control schemes with cascaded SRF voltage and current controllers, as for instance discussed in [28]. Since these limitations are not influencing the dynamics of the control scheme within the normal operating range, further details will not be discussed here.

Current Controllers and Active Damping

The applied inner loop current controllers are conventional SRF PI controllers with decoupling terms [27] and [35], as shown in the right side part of Fig. 6. The output voltage reference from the controller is defined by (21), where the resulting voltage reference for the converter is denoted by $\mathbf{v}^*_{CV}$. The proportional and integral gains of the PI controller are defined by *kpc* and *kic*, and a gain factor *kffv* is used to disable or enable the voltage feed-forward in the output of the current controllers. The states γ are defined to represent the integrators of the PI controllers according to (22):

$$\mathbf{v}^*_{CV} = k_{pc}(\mathbf{i}^*_{CV} - \mathbf{i}_{CV}) + k_{ic} \cdot \gamma + j \cdot l_1 \cdot \omega_{VSM} \cdot \mathbf{i}_{CV} + k_{ffv} \cdot \mathbf{v}_o - \mathbf{v}^*_{AD} \quad (21)$$

$$\frac{d\gamma}{dt} = \mathbf{i}^*_{CV} - \mathbf{i}_{CV} \quad (22)$$

In (21), the voltage reference for the converter also includes an active damping term $\mathbf{v}^*_{AD}$ designed for suppressing LC oscillations in the filter [36]. The implementation of the active damping algorithm is shown in Fig. 7, and is based on high pass filtering of the measured voltage **v***o*, obtained from the difference between **v***o* and the low pass filtered value of the same voltage. The resulting high pass filtered signal is then scaled by the gain *kAD* according to (23) and subtracted from the output of the current controllers to cancel detected oscillations in the capacitor voltages:

$$\mathbf{v}^*_{AD} = k_{AD}(\mathbf{v}_o - \varphi) \quad (23)$$

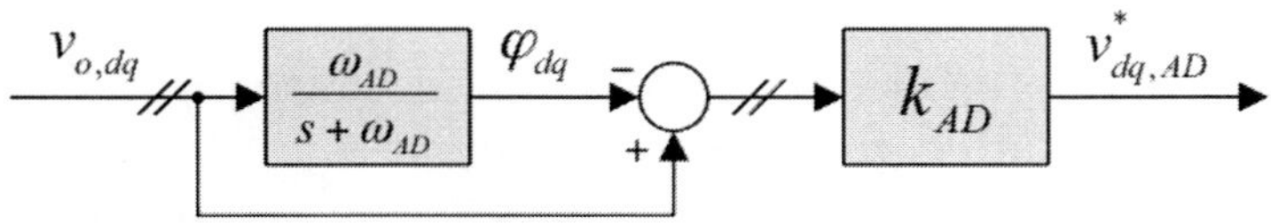

Figure 7. Implementation of active damping.

The corresponding internal states **φ** of the low pass filters used for the active damping are defined by (24), where *ωAD* is the cut-off frequency:

$$\frac{d\boldsymbol{\varphi}}{dt} = \omega_{AD} \cdot \mathbf{v}_o - \omega_{AD} \cdot \boldsymbol{\varphi} \tag{24}$$

For the practical implementation of the VSC control system, the voltage reference $\mathbf{v}^*_{cv}$ resulting from the current controller and the active damping is divided by the measured DC-link voltage to result in the modulation index **m** as shown to the right of Fig. 6. Neglecting the switching operation of the converter and any delay due to the PWM implementation, the instantaneous average value of the per unit converter output voltage is given by the product of the modulation index and the actual DC-voltage. Under this assumption, the output converter voltage will be approximately equal to the voltage reference as summarized by (25)[37]:

$$\mathbf{m} = \frac{\mathbf{v}^*_{cv}}{v_{DC}}, \quad \mathbf{v}_{cv} = \mathbf{m} \cdot v_{DC} \rightarrow \mathbf{v}_{cv} \approx \mathbf{v}^*_{cv} \tag{25}$$

Thus, the AC side operation of the converter will be effectively decoupled from any dynamics in the DC voltage, and it is not necessary to further discuss or model the DC side of the converter for achieving an accurate representation of the dynamics on the AC side. It should be noted that the actual source or storage unit connected to the DC link of the converter might still impose restrictions on the allowable power exchange during various operating conditions. However, to maintain generality and avoid detailed discussion of particular application-specific limitations it will be assumed that the power requested from the AC side is always available at the DC link of the converter.

Electrical System Equations

The electrical system included in the model according to Fig. 1 consists of a set of filter inductors connected to the converter, a shunt capacitor bank representing the capacitance of the LC filter, and a Thévenin equivalent of the grid. This simple structure is assumed to achieve a simple model that mainly includes the dynamics of the converter control system and its interaction with the equivalent grid voltage. However, a more complex AC grid topology can be easily included in the model for both simulations and analysis. Considering an instantaneous average model of the converter, the SRF state space equations of the electrical system can be established as given by (26)[27] and [35]:

$$\begin{aligned}
\frac{d\mathbf{i}_{cv}}{dt} &= \frac{\omega_b}{l_f}\mathbf{v}_{cv} - \frac{\omega_b}{l_f}\mathbf{v}_o - \left(\frac{r_{l_f}\omega_b}{l_f} + j\cdot\omega_g\omega_b\right)\mathbf{i}_{cv} \\
\frac{d\mathbf{v}_o}{dt} &= \frac{\omega_b}{c_f}\mathbf{i}_{cv} - \frac{\omega_b}{c_f}\mathbf{i}_g - j\cdot\omega_g\omega_b\cdot\mathbf{v}_o \\
\frac{d\mathbf{i}_o}{dt} &= \frac{\omega_b}{l_g}\mathbf{v}_o - \frac{\omega_b}{l_g}\mathbf{v}_g - \left(\frac{r_g\omega_b}{l_g} + j\cdot\omega_g\omega_b\right)\mathbf{i}_o
\end{aligned} \tag{26}$$

In these equations **icv** is the filter inductor current, **vcv** is the converter output voltage, **vo** is the voltage at the filter capacitors, **ig** is the current flowing into the grid equivalent and **vg**is the grid equivalent voltage. The inductance and equivalent resistance of the filter inductor is given by *lf* and *rlf*, the filter capacitor is *cf*, while the grid inductance and resistance are given by *lg* and *rg*. The per unit grid frequency is given by ωg, while the base angular grid frequency is defined by ωb. It should be noted that the state space model from (26) can represent the electrical system in any SRF, but in this case the system will always be modelled in the SRF defined by the VSM swing equation.

It should also be noted that the presented model only represents the case of grid-connected operation, while the investigated VSM scheme is inherently suitable for stand-alone operation. In this case the operational frequency will only be determined by the actual load in the system, the power-frequency droop gain and the power and frequency references for the VSM. Further details on modelling of the investigated VSM implementation in islanded operation, and corresponding analysis of the dynamic characteristics in stand-alone mode can be found in [38].

Non-linear System Model in Grid-connected Operation

All equations needed for detailed modelling of the VSM configuration in grid-connected operation have been presented in the previous sub-sections, and can be reduced to a model on state-space form with 19 distinct state variables and 6 input signals, with the state vector $\mathbf{x}$ and the input vector $\mathbf{u}$ defined by (27). The resulting non-linear state-space model of the overall system is given by (28):

$$\mathbf{x} = \begin{matrix} \big[v_{o,d} \quad v_{o,q} \quad i_{cv,d} \quad i_{cv,q} \quad \gamma_d \quad \gamma_q \quad i_{o,d} \quad i_{o,q} \quad \varphi_d \quad \varphi_q \dots \\ \dots v_{PLL,d} \quad v_{PLL,q} \quad \varepsilon_{PLL} \quad \delta\theta_{VSM} \quad \xi_d \quad \xi_q \quad q_m \quad \delta\omega_{VSM} \quad \delta\theta_{PLL} \big]^T \end{matrix}$$

$$\mathbf{u} = \left[p^* \quad q^* \quad \hat{v}_g \quad \hat{v}^* \quad \omega^* \quad \omega_g \right]^T \tag{27}$$

$$\frac{dv_{o,d}}{dt} = \omega_b \omega_g v_{o,q} + \frac{\omega_b}{c_f} i_{cv,d} - \frac{\omega_b}{c_f} i_{o,d}$$

$$\frac{dv_{o,q}}{dt} = -\omega_b \omega_g v_{o,d} + \frac{\omega_b}{c_f} i_{cv,q} - \frac{\omega_b}{c_f} i_{o,q}$$

$$\frac{di_{cv,d}}{dt} = \frac{\omega_b(k_{ffv} - 1 - k_{AD} - k_{pc}k_{pv})}{l_f} v_{o,d} - \frac{\omega_b c_f k_{pc}}{l_f} \omega_g v_{o,q} - \frac{\omega_b(k_{pc} + r_f)}{l_f} i_{cv,d} + \frac{\omega_b k_{ic}}{l_f} \gamma_d + \frac{\omega_b k_{pc}(k_{ffi} - k_{pv} r_v)}{l_f} i_{o,d} + \frac{\omega_b k_{pc} k_{pv} l_v}{l_f} \omega_g i_{o,q}$$
$$+ \frac{\omega_b k_{AD}}{l_f} \varphi_d + \frac{\omega_b k_{iv} k_{pc}}{l_f} \xi_d - \frac{\omega_b k_{pc} k_{pv} k_q}{l_f} q_m - \omega_b i_{cv,q} \delta\omega_{VSM} + \frac{\omega_b k_{pc} k_{pv} l_v}{l_f} i_{o,q} \delta\omega_{VSM} - \frac{\omega_b c_f k_{pc}}{l_f} v_{o,q} \delta\omega_{VSM} + \frac{\omega_b k_{pc} k_{pv} k_q}{l_f} q^* + \frac{\omega_b k_{pc} k_{pv}}{l_f} \hat{v}^*$$

$$\frac{di_{cv,q}}{dt} = \frac{\omega_b c_f k_{pc}}{l_f} \omega_g v_{o,d} + \frac{\omega_b(k_{ffv} - 1 - k_{AD} - k_{pc}k_{pv})}{l_f} v_{o,q} - \frac{\omega_b(k_{pc} + r_f)}{l_f} i_{cv,q} + \frac{\omega_b k_{ic}}{l_f} \gamma_q - \frac{\omega_b k_{pc} k_{pv} l_v}{l_f} \omega_g i_{o,d} + \frac{\omega_b k_{pc}(k_{ffi} - k_{pv} r_v)}{l_f} i_{o,q}$$
$$+ \frac{\omega_b k_{AD}}{l_f} \varphi_q + \frac{\omega_b k_{iv} k_{pc}}{l_f} \xi_q + \omega_b i_{cv,d} \delta\omega_{VSM} - \frac{\omega_b k_{pc} k_{pv} l_v}{l_f} i_{o,d} \delta\omega_{VSM} + \frac{\omega_b c_f k_{pc}}{l_f} v_{o,d} \delta\omega_{VSM}$$

$$\frac{d\gamma_d}{dt} = -k_{pv} v_{o,d} - c_f \omega_g v_{o,q} - i_{cv,d} + (k_{ffi} - k_{pv} r_v) i_{o,d} + k_{pv} l_v \omega_g i_{o,q} + k_{iv} \xi_d - k_{pv} k_q q_m + k_{pv} l_v i_{o,q} \delta\omega_{VSM} - c_f v_{o,q} \delta\omega_{VSM} + k_{pv} k_q q^* + k_{pv} \hat{v}^*$$

$$\frac{d\gamma_d}{dt} = c_f \omega_g v_{o,d} - k_{pv} v_{o,q} - i_{cv,q} - k_{pv} l_v \omega_g i_{o,d} + (k_{ffi} - k_{pv} r_v) i_{o,q} + k_{iv} \xi_q - k_{pv} l_v i_{o,d} \delta\omega_{VSM} + c_f v_{o,d} \delta\omega_{VSM}$$

$$\frac{di_{o,d}}{dt} = \frac{\omega_b}{l_g} v_{o,d} - \frac{\omega_b r_g}{l_g} i_{o,d} + \omega_b \omega_g i_{o,q} + \frac{\omega_b \hat{v}_g \cos(\delta\theta_{VSM})}{l_g}$$

$$
\begin{aligned}
\frac{di_{o,q}}{dt} &= \frac{\omega_b}{l_g} v_{o,q} - \omega_b \omega_g i_{o,d} - \frac{\omega_b r_g}{l_g} i_{o,q} + \frac{\omega_b \hat{v}_g \sin(\delta\theta_{VSM})}{l_g} \\
\frac{d\varphi_d}{dt} &= \omega_{AD} v_{o,d} - \omega_{AD} \varphi_d \\
\frac{d\varphi_q}{dt} &= \omega_{AD} v_{o,q} - \omega_{AD} \varphi_q \\
\frac{dv_{PLL,d}}{dt} &= \omega_{LP,PLL} v_{o,d} \cos(\delta\theta_{PLL} - \delta\theta_{VSM}) + \omega_{LP,PLL} v_{o,q} \sin(\delta\theta_{PLL} - \delta\theta_{VSM}) - \omega_{LP,PLL} v_{PLL,d} \\
\frac{dv_{PLL,q}}{dt} &= -\omega_{LP,PLL} v_{o,d} \sin(\delta\theta_{PLL} - \delta\theta_{VSM}) + \omega_{LP,PLL} v_{o,q} \cos(\delta\theta_{PLL} - \delta\theta_{VSM}) - \omega_{LP,PLL} v_{PLL,q} \\
\frac{d\varepsilon_{PLL}}{dt} &= \tan^{-1}\left(\frac{v_{PLL,q}}{v_{PLL,d}}\right) \\
\frac{d\delta\theta_{VSM}}{dt} &= \omega_b \delta\omega_{VSM} \\
\frac{d\delta\xi_d}{dt} &= -v_{o,d} - r_v i_{o,d} + l_v \omega_g i_{o,q} - k_q q_m + l_v i_{o,q} \delta\omega_{VSM} + k_q q^* + \hat{v}^* \\
\frac{d\delta\xi_q}{dt} &= -v_{o,q} - l_v \omega_g i_{o,d} - r_v i_{o,q} - l_v i_{o,d} \delta\omega_{VSM} \\
\frac{dq_m}{dt} &= -\omega_f i_{o,q} v_{o,d} - \omega_f i_{o,d} v_{o,q} - \omega_f q_m
\end{aligned}
$$

$$
\begin{aligned}
\frac{d\delta\omega_{VSM}}{dt} &= -\frac{1}{T_a} i_{o,d} v_{o,d} - \frac{1}{T_a} i_{o,q} v_{o,q} + \frac{k_d k_{p,PLL}}{T_a} \tan^{-1}\left(\frac{v_{PLL,q}}{v_{PLL,d}}\right) + \frac{k_d k_{i,PLL}}{T_a} \varepsilon_{PLL} - \frac{k_d + k_\omega}{T_a} \delta\omega_{VSM} + \frac{1}{T_a} p^* + \frac{k_\omega}{T_a} \omega^* - \frac{k_\omega}{T_a} \omega_g \\
\frac{d\delta\theta_{PLL}}{dt} &= \omega_b k_{p,PLL} \tan^{-1}\left(\frac{v_{PLL,q}}{v_{PLL,d}}\right) + \omega_b k_{i,PLL} \varepsilon_{PLL}
\end{aligned}
\tag{28}
$$

The steady state operating point of the system under any combinations of input signals can be found by solving this nonlinear system model with derivative terms set to zero.

Small Signal Model of the Reference VSM

Since the state-space model from (28) is nonlinear, classical stability assessment techniques based on eigenvalues are not directly applicable. Thus, in this section, a corresponding linearized small-signal state-space model is derived in the form given by:

$$
\Delta\dot{\mathbf{x}} = \mathbf{A} \cdot \Delta\mathbf{x} + \mathbf{B} \cdot \Delta\mathbf{u}. \tag{29}
$$

where the prefix Δ denotes small-signal deviations around the steady-state operating point [30]. The values of the state variables at this linearization point are denoted by subscript '0' when they appear in the matrices. For convenience of notation, the dynamic matrix A is expressed through four sub-matrices according to:

$$\begin{bmatrix} \Delta\dot{\mathbf{x}}_1 \\ \Delta\dot{\mathbf{x}}_2 \end{bmatrix} = \begin{bmatrix} \mathbf{A}_{11} & \mathbf{A}_{12} \\ \mathbf{A}_{21} & \mathbf{A}_{22} \end{bmatrix} \cdot \begin{bmatrix} \Delta\mathbf{x}_1 \\ \Delta\mathbf{x}_2 \end{bmatrix} + \mathbf{B}\cdot\Delta\mathbf{u} \tag{30}$$

where $\mathbf{A}_{11}$, $\mathbf{A}_{12}$, $\mathbf{A}_{13}$ and $\mathbf{A}_{22}$ are given by (31), (32), (33) and (34), while the **B** matrix is given by (35):

$$\mathbf{A}_{11} = \begin{bmatrix} 0 & \omega_b\omega_{g,0} & \frac{\omega_b}{c_f} & 0 & 0 & 0 & -\frac{\omega_b}{c_f} & 0 & 0 & 0 \\ -\omega_b\omega_{g,0} & 0 & 0 & \frac{\omega_b}{c_f} & 0 & 0 & 0 & -\frac{\omega_b}{c_f} & 0 & 0 \\ \frac{\omega_b(k_{ffv}-1-k_{AD}-k_{pc}k_{pv})}{l_f} & -\frac{\omega_b c_f k_{pc}\omega_{g,0}}{l_f} & -\frac{\omega_b(k_{pc}+r_f)}{l_f} & 0 & \frac{\omega_b k_{ic}}{l_f} & 0 & \frac{\omega_b k_{pc}(k_{ffi}-k_{pv}r_v)}{l_f} & \frac{\omega_b k_{pc}k_{pv}l_v\omega_{g,0}}{l_f} & \frac{\omega_b k_{AD}}{l_f} & 0 \\ \frac{\omega_b c_f k_{pc}\omega_{g,0}}{l_f} & \frac{\omega_b(k_{ffv}-1-k_{AD}-k_{pc}k_{pv})}{l_f} & 0 & -\frac{\omega_b(k_{pc}+r_f)}{l_f} & 0 & \frac{\omega_b k_{ic}}{l_f} & -\frac{\omega_b\omega_g k_{pc}k_{pv}l_v}{l_f} & \frac{\omega_b k_{pc}(k_{ffi}-k_{pv}r_v)}{l_f} & 0 & \frac{\omega_b k_{AD}}{l_f} \\ -k_{pv} & -c_f\omega_{g,0} & -1 & 0 & 0 & 0 & k_{ffi}-k_{pv}r_v & k_{pv}l_v\omega_{g,0} & 0 & 0 \\ c_f\omega_{g,0} & -k_{pv} & 0 & -1 & 0 & 0 & -k_{pv}l_v\omega_{g,0} & k_{ffi}-k_{pv}r_v & 0 & 0 \\ \frac{\omega_b}{l_g} & 0 & 0 & 0 & 0 & 0 & -\frac{\omega_b r_g}{l_g} & \omega_b\omega_{g,0} & 0 & 0 \\ 0 & \frac{\omega_b}{l_g} & 0 & 0 & 0 & 0 & -\omega_b\omega_{g,0} & -\frac{\omega_b r_g}{l_g} & 0 & 0 \\ \omega_{AD} & 0 & 0 & 0 & 0 & 0 & 0 & 0 & -\omega_{AD} & 0 \\ 0 & \omega_{AD} & 0 & 0 & 0 & 0 & 0 & 0 & 0 & -\omega_{AD} \end{bmatrix} \tag{31}$$

$$\mathbf{A}_{12} = \begin{bmatrix} 0 & 0 & 0 & 0 & 0 & 0 & 0 & 0 & 0 \\ 0 & 0 & 0 & 0 & 0 & 0 & 0 & 0 & 0 \\ 0 & 0 & 0 & 0 & \frac{\omega_b k_{iv}k_{pc}}{l_f} & 0 & -\frac{\omega_b k_{pc}k_{pv}k_q}{l_f} & \frac{\omega_b(-l_f i_{cv,q,0}+k_{pc}k_{pv}l_v i_{o,q,0}-c_f k_{pc}v_{o,q,0})}{l_f} & 0 \\ 0 & 0 & 0 & 0 & 0 & \frac{\omega_b k_{iv}k_{pc}}{l_f} & 0 & \frac{\omega_b(l_f i_{cv,d,0}-k_{pc}k_{pv}l_v i_{o,d,0}+c_f k_{pc}v_{o,d,0})}{l_f} & 0 \\ 0 & 0 & 0 & 0 & k_{iv} & 0 & -k_{pv}k_q & k_{pv}l_v i_{o,q,0}-c_f v_{o,q,0} & 0 \\ 0 & 0 & 0 & 0 & 0 & k_{iv} & 0 & -k_{pv}l_v i_{o,d,0}+c_f v_{o,d,0} & 0 \\ 0 & 0 & 0 & \frac{\omega_b\hat{v}_{g,0}\sin(\delta\theta_{VSM,0})}{l_g} & 0 & 0 & 0 & 0 & 0 \\ 0 & 0 & 0 & \frac{\omega_b\hat{v}_{g,0}\cos(\delta\theta_{VSM,0})}{l_g} & 0 & 0 & 0 & 0 & 0 \\ 0 & 0 & 0 & 0 & 0 & 0 & 0 & 0 & 0 \\ 0 & 0 & 0 & 0 & 0 & 0 & 0 & 0 & 0 \end{bmatrix} \tag{32}$$

$$
\mathbf{A}_{21} = \begin{bmatrix}
\omega_{LP,PLL}\cos(\delta\theta_{PLL,0}-\delta\theta_{VSM,0}) & \omega_{LP,PLL}\sin(\delta\theta_{PLL,0}-\delta\theta_{VSM,0}) & 0 & 0 & 0 & 0 & 0 & 0 & 0 & 0 \\
-\omega_{LP,PLL}\sin(\delta\theta_{PLL,0}-\delta\theta_{VSM,0}) & \omega_{LP,PLL}\cos(\delta\theta_{PLL,0}-\delta\theta_{VSM,0}) & 0 & 0 & 0 & 0 & 0 & 0 & 0 & 0 \\
0 & 0 & 0 & 0 & 0 & 0 & 0 & 0 & 0 & 0 \\
0 & 0 & 0 & 0 & 0 & 0 & 0 & 0 & 0 & 0 \\
-1 & 0 & 0 & 0 & 0 & 0 & -r_v & \omega_g l_v & 0 & 0 \\
0 & -1 & 0 & 0 & 0 & 0 & -\omega_g l_v & -r_v & 0 & 0 \\
-\omega_f i_{o,q,0} & \omega_f i_{o,d,0} & 0 & 0 & 0 & 0 & \omega_f v_{o,q,0} & -\omega_f v_{o,d,0} & 0 & 0 \\
-\frac{i_{o,d,0}}{T_a} & -\frac{i_{o,q,0}}{T_a} & 0 & 0 & 0 & 0 & -\frac{v_{o,d,0}}{T_a} & -\frac{v_{o,q,0}}{T_a} & 0 & 0 \\
0 & 0 & 0 & 0 & 0 & 0 & 0 & 0 & 0 & 0
\end{bmatrix} \quad (33)
$$

$$
\mathbf{A}_{22} = \begin{bmatrix}
-\omega_{LP,PLL} & 0 & 0 & -\omega_{LP,PLL}\begin{pmatrix} v_{o,q,0}\cos(\delta\theta_{PLL,0}-\delta\theta_{VSM,0}) \\ -v_{o,d,0}\sin(\delta\theta_{PLL,0}-\delta\theta_{VSM,0}) \end{pmatrix} & 0 & 0 & 0 & 0 & \omega_{LP,PLL}\begin{pmatrix} v_{o,q,0}\cos(\delta\theta_{PLL,0}-\delta\theta_{VSM,0}) \\ -v_{o,d,0}\sin(\delta\theta_{PLL,0}-\delta\theta_{VSM,0}) \end{pmatrix} \\
0 & -\omega_{LP,PLL} & 0 & \omega_{LP,PLL}\begin{pmatrix} v_{o,d,0}\cos(\delta\theta_{PLL,0}-\delta\theta_{VSM,0}) \\ +v_{o,q,0}\sin(\delta\theta_{PLL,0}-\delta\theta_{VSM,0}) \end{pmatrix} & 0 & 0 & 0 & 0 & -\omega_{LP,PLL}\begin{pmatrix} v_{o,d,0}\cos(\delta\theta_{PLL,0}-\delta\theta_{VSM,0}) \\ +v_{o,q,0}\sin(\delta\theta_{PLL,0}-\delta\theta_{VSM,0}) \end{pmatrix} \\
0 & \frac{1}{v_{PLL,d,0}} & 0 & 0 & 0 & 0 & 0 & 0 & 0 \\
0 & 0 & 0 & 0 & 0 & 0 & 0 & \omega_b & 0 \\
0 & 0 & 0 & 0 & 0 & 0 & -k_q & l_v i_{o,q,0} & 0 \\
0 & 0 & 0 & 0 & 0 & 0 & 0 & -l_v i_{o,d,0} & 0 \\
0 & 0 & 0 & 0 & 0 & 0 & -\omega_f & 0 & 0 \\
0 & \frac{k_d k_{p,PLL}}{T_a v_{PLL,d,0}} & \frac{k_d k_{i,PLL}}{T_a} & 0 & 0 & 0 & 0 & -\frac{k_d + k_\omega}{T_a} & 0 \\
0 & \frac{\omega_b k_{p,PLL}}{v_{PLL,d,0}} & \omega_b k_{i,PLL} & 0 & 0 & 0 & 0 & 0 & 0
\end{bmatrix}
$$

(34)

$$
\begin{bmatrix}
0 & 0 & 0 & 0 & 0 & \omega_b v_{o,q,0} \\
0 & 0 & 0 & 0 & 0 & -\omega_b v_{o,d,0} \\
0 & \frac{\omega_b k_{pc} k_{pv} k_q}{l_f} & 0 & \frac{\omega_b k_{pc} k_{pv}}{l_f} & 0 & \frac{\omega_b k_{pc}(k_{pv} l_v i_{o,q,0} - c_f v_{o,q,0})}{l_f} \\
0 & 0 & 0 & 0 & 0 & \frac{\omega_b k_{pc}(-k_{pv} l_v i_{o,d,0} + c_f v_{o,d,0})}{l_f} \\
0 & k_{pv} k_q & 0 & k_{pv} & 0 & k_{pv} l_v i_{o,q,0} - c_f v_{o,q,0} \\
0 & 0 & 0 & 0 & 0 & -k_{pv} l_v i_{o,d,0} + c_f v_{o,d,0} \\
0 & 0 & -\frac{\omega_b \cos(\delta\theta_{VSM,0})}{l_g} & 0 & 0 & \omega_b i_{o,q,0}
\end{bmatrix}
$$

$$
\mathbf{B} = \begin{bmatrix}
0 & 0 & \frac{\omega_b \sin(\delta\theta_{VSM,0})}{l_g} & 0 & 0 & \omega_b i_{o,d,0} \\
0 & 0 & 0 & 0 & 0 & 0 \\
0 & 0 & 0 & 0 & 0 & 0 \\
0 & 0 & 0 & 0 & 0 & 0 \\
0 & 0 & 0 & 0 & 0 & 0 \\
0 & 0 & 0 & 0 & 0 & 0 \\
0 & 0 & 0 & 0 & 0 & 0 \\
0 & k_q & 0 & 1 & 0 & l_v i_{o,q,0} \\
0 & 0 & 0 & 0 & 0 & -l_v i_{o,d,0} \\
0 & 0 & 0 & 0 & 0 & 0 \\
\frac{1}{T_a} & 0 & 0 & 0 & \frac{k_\omega}{T_a} & -\frac{k_\omega}{T_a} \\
0 & 0 & 0 & 0 & 0 & 0
\end{bmatrix} \tag{35}
$$

DYNAMIC OPERATION OF THE REFERENCE VSM IMPLEMENTATION

The functional behaviour of the reference VSM implementation is illustrated in this section by means of numerical simulations of a few relevant cases. A dynamic model of the investigated system and its corresponding small-signal state-space model have been implemented in Matlab Simulink with the set of parameters listed in Table 1. The parameters of the PI current and voltage controller in this table are obtained by the Modulus Optimum (MO) and Symmetrical Optimum (SO) criteria, respectively, according to the approach explained in [14] and [39], assuming a switching frequency of 5 kHz. Similarly, the PI controller parameters of the PLL are calculated by the SO criterion assuming a low pass filter with crossover frequency of 500 rad/s. The inertia time constant of the VSM is selected to emulate a SM with an intermediate inertia representative for a distributed generation unit, while the other parameters are selected by experience-based trial-and-error.

Table 1. Parameters of investigated VSM configuration.

Parameter	Value	Parameter	Value
Rated voltage *VS,LL,RMS*	690 V	Filter inductance *lf*	0.08 pu
Rated power *Sb*	2.75 MVA	Filter resistance *rlf*	0.003 pu
Rated angular frequency *ωb*	2π*50 Hz	Filter capacitance *cf*	0.074 pu
VSM Inertia constant *Ta*	2 s	Grid inductance *lg*	0.20 pu
VSM Damping coefficient *kd*	400	Grid resistance *rg*	0.01 pu
Current controller gain: *kpc*, *kic*	1.27, 14.3	Grid voltage r_{lf}	1.0 pu
Voltage controler gain: *kpv*, *kiv*	0.59, 736	Active damping filter *ωAD*	50 rad/s
Power reference *p**	0.5 pu	Active damping gain *kAD*	0.5 pu
Frequency droop gain *kω*	20 pu	Virtual Inductance *lv*	0.2 pu
Reactive power reference *q**	0.0 pu	Virtual Resistance *rv*	0.0 pu
Reactive power droop gain *kq*	0.2 pu	PLL filter $\omega_{LP,PLL}$	500 rad/s
Reactive power filter *ωf*	1000 rad/s	PLL proportional gain *kp,PLL*	0.084
Voltage reference $\hat{v}^*$	1.02 pu	PLL integral gain *ki,PLL*	4.69

Dynamic Response to Change in Loading

In the first simulated case, the dynamic response of the system is examined for a step in the power reference input to the VSM from 0.5 pu to 0.7 pu when the grid frequency and the reference frequency are both equal to 1.0 pu. This case is equivalent to a sudden increase in the input power or torque on the shaft of a SM connected to an infinite bus.

The power reference and the resulting electrical power from the VSM are plotted in Fig. 8, where it is shown that the VSM with the selected parameters exhibits a smooth transient response and reaches steady state conditions in approximately 1 s without any overshoot. The step change in the reference triggers also a dynamic response in the rotating speed of the virtual inertia, as shown in Fig. 9. Indeed, the excess "mechanical" power input is accumulated in the virtual inertia of the VSM, resulting in an increasing speed during the first part of the transient. This leads to an increase in the phase angle between the VSM-oriented reference frame and the grid voltage vector as shown in the upper part of Fig. 10, until the electrical power output from the machine balances its "mechanical" input.

When the electrical power reaches the input power and the steady-state power balance of the system is restored, the rotational speed of the virtual inertia returns to the synchronous speed of the grid source, just as for a traditional SM.

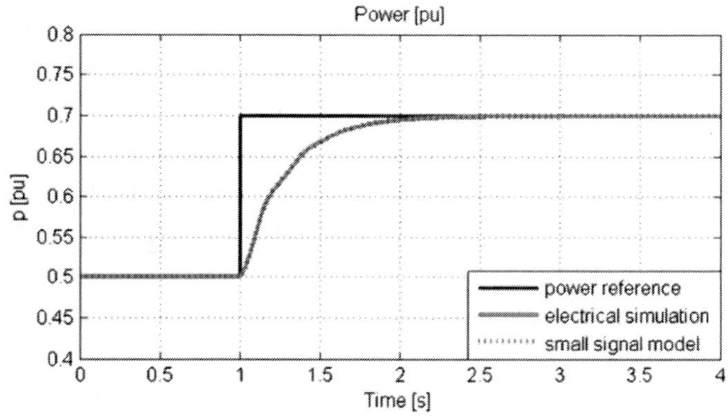

Figure 8. Response of nonlinear model and linearized small-signal model to a step in the power reference input.

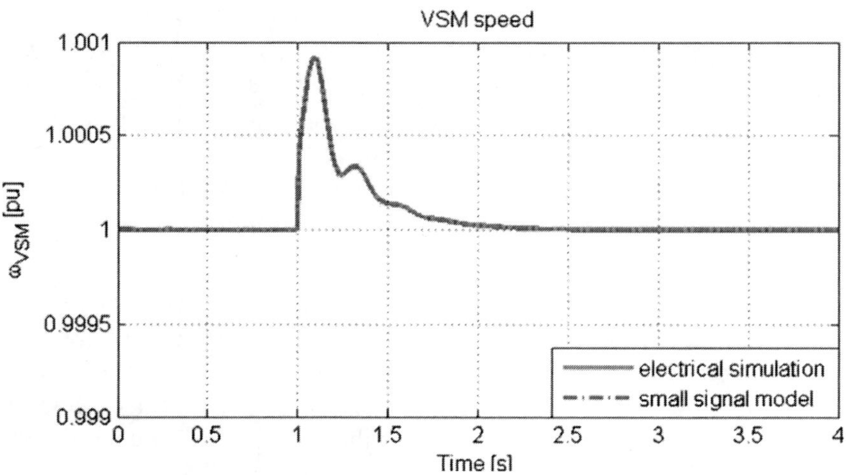

Figure 9. Response of the VSM speed to a step in the power input.

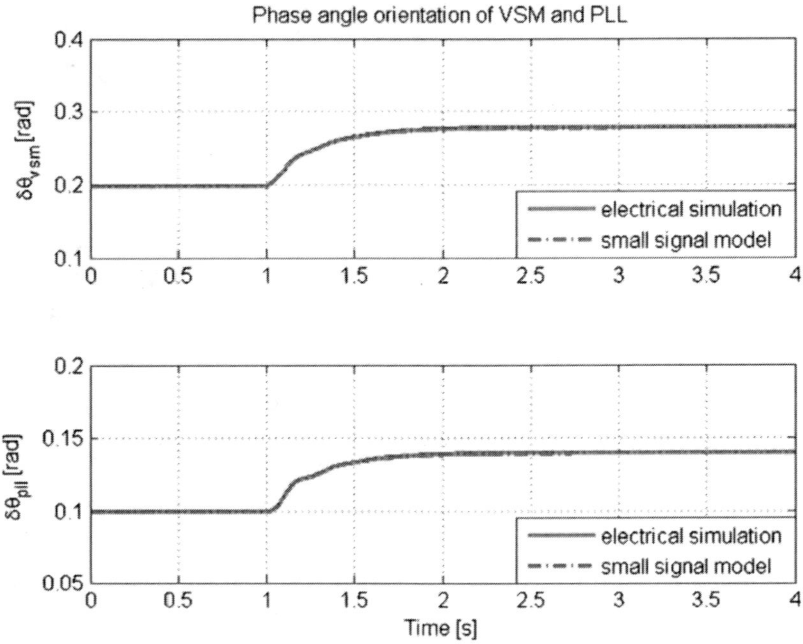

Figure 10. Phase angle displacements of the VSM and PLL reference frames with respect to the grid voltage in response to the step in power reference.

The transient behaviour shown in the presented curves exhibits the same general characteristics as for a conventional SM, but with a more damped response. Indeed, the VSM replicates the behaviour of a classical SM, but its parameters do not have to comply with any physical design constraint. Thus, the parameters of the VSM can be selected with more freedom, without considering any efficiency aspects. In particular, power losses due to the damping effects of the VSM appear only in the control system and not in any physical circuit.

The reference frame orientation of the PLL with respect to the grid voltage vector is shown in the lower part of Fig. 10, illustrating how the phase angle of the voltage at the filter capacitors is changing with the power flow due to the grid impedance. However, it is noticeable how the phase angle of the VSM changes more than for the PLL, due to the virtual inductance included in the VSM. Thus, the VSM phase angle can be considered as equivalent to the phase angle of the internal voltage behind the synchronous reactance of a traditional SM, while the PLL is tracking the phase angle of the voltage at the terminals of the VSM.

The reactive power flow from the VSM is shown in Fig. 11, to illustrate the very small change in the reactive power flow due to the droop controller's response to the change of operating conditions when the active power flow is changed. The plot also indicates small oscillation at a relatively high frequency. This response is mainly due to the response of the LC filter and the measurement filter used in the droop controller.

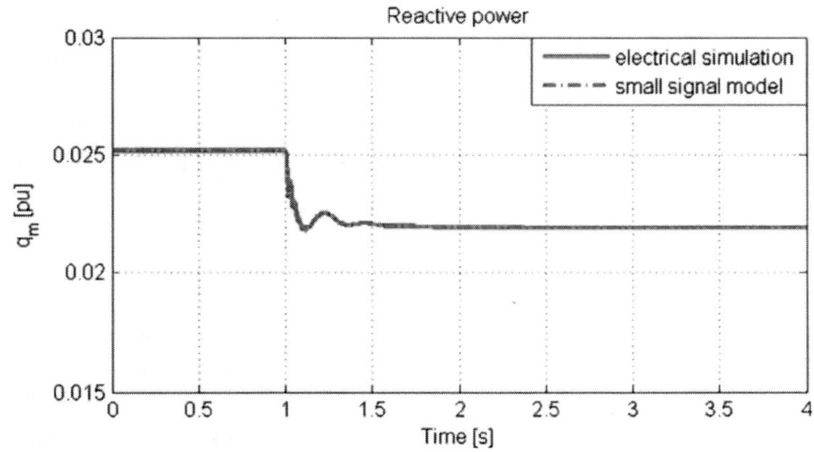

Figure 11. Response of reactive power flow to the step in active power reference.

The presented figures include results from simulation of the non-linear model of the VSM as well as results from simulating the same events with the linearized small-signal state-space model in the same plots. It can be observed that all the curves indicate an excellent match between the two models, for both the fast and slow transients in the system, as long as the operating point stays close enough to the linearization point. This verifies the validity of the linearized model around the steady-state operating point, and indicates that it can be used for investigating system stability and the influence of parametric variations by traditional techniques for linear system analysis.

Response to Change in the Grid Frequency

The results from an additional simulation case are shown in Fig. 12 and Fig. 13, where the grid frequency of the system is ramped down from 1.0 pu to 0.995 pu as shown in the upper part of Fig. 12. Due to this change of grid frequency, the frequency droop gain specified by $k\omega$ is activated and the VSM increases its power output in steady state. The VSM speed follows the grid frequency, releasing energy from its inertia corresponding to the change of speed, and settles at the new operating frequency of the grid with increased power output to contribute to the frequency control of the power system. This can also be seen from the power flow in Fig. 12 and the corresponding converter currents in Fig. 13.

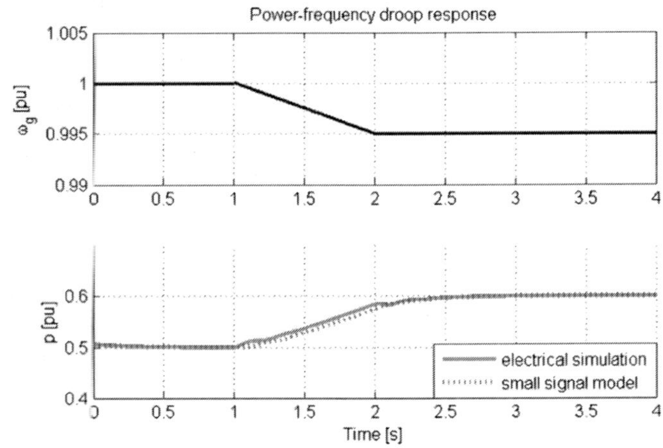

Figure12. Response of the VSM power flow to a change in the grid frequency.

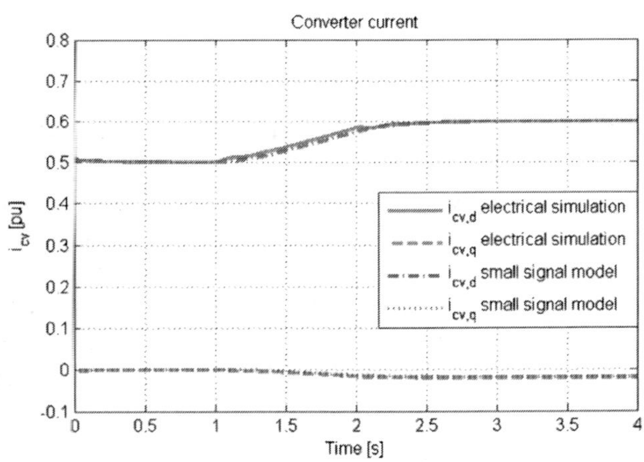

Figure 13. Converter currents as a response to the change of operating conditions when the grid frequency changes.

However, for the linearized small signal model, it should be remembered that the speed deviation from the linearization point in steady state must be zero for the phase angle specified by (5) to settle to a constant value. This implies that the linearized model of the VSM connected to an infinite bus cannot correctly represent the dynamic transition into a new operating point with a different grid frequency than at the linearization point. Thus, a small deviation between the non-linear and linearized models can be observed while the frequency is changing and for a short transient after the frequency has settled to its new steady-state value, as shown by the curves in Fig. 12 and Fig. 13. This is because the grid frequency is not a state variable in the developed small-signal model. Still, the contribution from the change of grid frequency through the B-matrix from (35) is able to represent the corresponding change of power flow also for the linearized model. Thus, the results presented in Fig. 12 and Fig. 13 are demonstrating that the VSM is operating as intended, while at the same time revealing some of the limitations of the developed small-signal model.

System Eigenvalue Analysis

Since the developed linearized small-signal model has been shown to accurately represent the investigated system for small deviations around the linearization point, the eigenvalues of the **A** matrix can be calculated to systematically identify all the modes of the system. The resulting system eigenvalues for the operating point corresponding to the conditions given in Table 1 are listed in Table 2.

Table 2. System eigenvalues

$\lambda_1 = -500$	$\lambda_{12} = -224$
$\lambda_{2,3} = -1460 \pm j\,4498$	$\lambda_{13,14} = -6.8 \pm j\,26.4$
$\lambda_{4,5} = -1272 \pm j\,4329$	$\lambda_{15} = -50.8$
$\lambda_{6,7} = -2262 \pm j\,225$	$\lambda_{16} = -50.6$
$\lambda_8 = -1002$	$\lambda_{17} = -37.0$
$\lambda_9 = -470$	$\lambda_{18} = -11.2$
$\lambda_{10,11} = -19.5 \pm j\,245$	$\lambda_{19} = -11.2$

For assessing the system stability, the slow and poorly damped poles are of main interest. It can be noticed from the eigenvalues in Table 2, that the system has several real poles close to the origin and two pairs of complex conjugate poles with small real parts and low oscillation frequency. There are also two pairs of high frequency poles, and one pair of fast poles close to the real axis.

As already seen, the system is stable with the parameters and operating conditions specified in Table 1. Since the small signal model has been shown to accurately represent the system around the linearization point, the system stability for the full operation range of the VSM has been investigated by plotting the system poles when the power reference p^* is swept from −1.0 pu to 1.0 pu. This corresponds to all the steady state operating conditions of the system that can be found by solving the nonlinear model from (28) as a function of the power reference while all other parameter values are as specified in Table 1. The result is shown in Fig. 14, where it can be seen that none of the system poles are moving much when the steady state operating conditions are changed. This indicates that the VSM will exhibit similar dynamic response as already discussed for its entire allowed operating range.

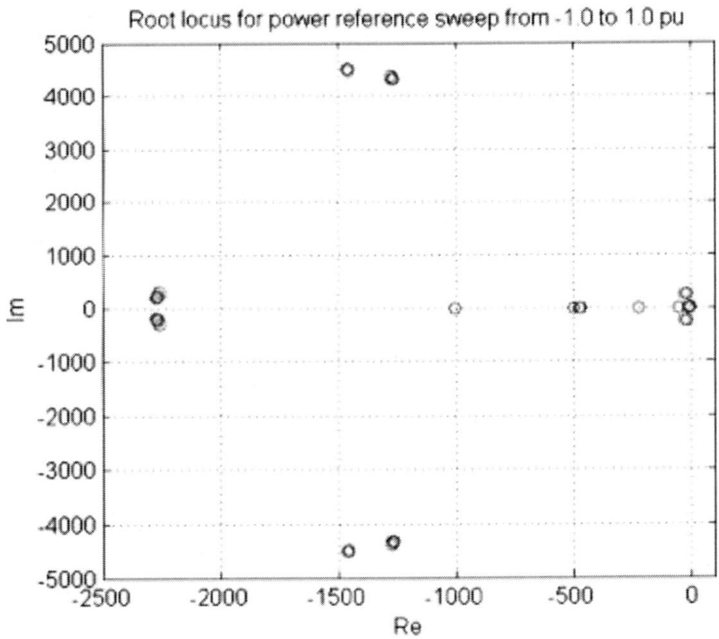

Figure 14. Root locus of all system poles when the power reference p^* is swept through the full operating range of the VSM.

Although the system is stable for the full range of operating conditions with the parameters and operating conditions from Table 1, it should be noted that the stability can still be sensitive to variations in the system parameters. For instance, an investigation of a similar configuration reported in [14] indicated that stable operation of the system will be challenging if the switching frequency and thus the bandwidth of the inner loop current controllers is low. In such a case, or due to other changes in the parameters of the inner loop controllers, the poles with the highest imaginary parts, which are associated with the LC filter of the system, can easily move into the right half-plane and cause instability.

There are also other potential parameters that can cause instability in the system under certain conditions. As an example, the trajectory of the poles closest to the imaginary axis are shown in Fig. 15 when the reactive power droop gain *kq* is swept from 0 to 1.0. The change of the pole location is indicated with a colour gradient, starting from blue when *kq*is 0 and changing towards red as *kq* increases. It can be seen from the figure that one pair of complex conjugate poles are moving towards the imaginary axis, and are causing instability when *kq* becomes too high. Similar investigations of stability limitations can easily be carried out for any of the system parameters, and can be used to support the tuning of the controllers as long as one single or a few clearly identified parameters are responsible for the stability problems in a particular system configuration. Such studies can be of high importance, considering the utilization of the VSM in a SmartGrid scenario with large expected variations in grid configurations, operating conditions and system parameters.

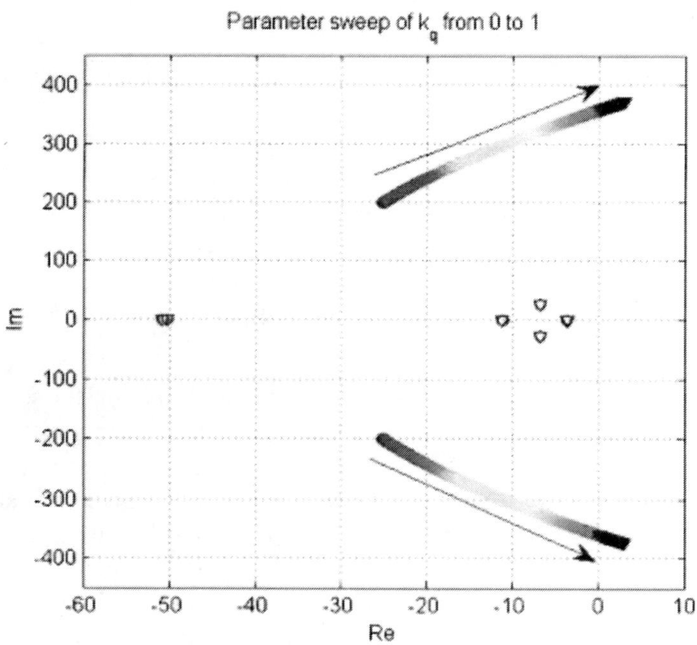

Figure 15. Root locus of the poles closest to the imaginary axis when the reactive power droop gain k_q is swept from 0 to 1.

Parametric Sensitivity of Critical Poles

Investigation of the system stability by ad hoc variations of the controller parameters is challenging for a high order system as the investigated VSM, especially if several parameters are influencing the same pole. To more easily identify the parameters that should be modified to improve the stability and dynamic performance of the system, the parameter sensitivity of pole locations can instead be studied.

The parameter sensitivity of the system poles is defined as the derivative of the eigenvalues with respect to the system parameters. For a dynamic system of order *N* and with a set of *K* tunable parameters, the sensitivities define a sensitivity matrix of *N* by *K* complex elements. The relative sensitivity $\alpha_{n,k}$ of the parameter ρk with respect to the eigenvalue *n* can be expressed by (36), where Ψ_n^T and Φ_n are the left and right eigenvectors associated to the eigenvalue λ_n [14] and [30]:

$$\alpha_{n,k} = \frac{\partial \lambda_n}{\partial \rho_k} = \frac{\Phi_n^T \frac{\partial \mathbf{A}}{\partial \rho_k} \Psi_n}{\Phi_n^T \Psi_n} \tag{36}$$

The real part of these sensitivities is directly associated to the derivatives of the pole location along the real axis with respect to each parameters. Thus, a positive value means that the investigated pole will be moved towards the right by an increase of the corresponding parameter. Similarly, the imaginary part of the sensitivity is associated with the derivative of the pole location along the imaginary axis. However, since the real parts of the pole locations determine the stability and corresponding time constant of the associated system mode, only the real part of the sensitivity matrix is investigated here.

An example of parametric sensitivity analysis for the pole causing instability in Fig. 15 is shown in Fig. 16. From this figure it can be seen that the investigated pole is mainly sensitive to the resistance in the grid and the virtual resistance of the VSM. However, the grid resistance cannot be influenced by the VSM and the virtual resistance should preferably be kept at a low value to achieve a good decoupling of the active and reactive power control of the VSM. Also the actual filter inductance of the converter and the virtual inductance of the VSM have some impact on this pole, but as the grid inductance cannot be influenced by the VSM only the virtual inductance can be modified to improve the stability and dynamic response of the system. The parameter sensitivity also proves that the reactive power droop gain *kq* has an influence on the pole, and that reducing the droop gain would help to move the pole towards the left. It can also be noticed that an increase of the voltage controller gain *kpv* can bring the pole towards the left, while all other adjustable parameters have negligible influence on the pole location. Thus, the main parameters that can be used to influence the location of this pole are the droop gain, the voltage controller gain and the virtual inductance.

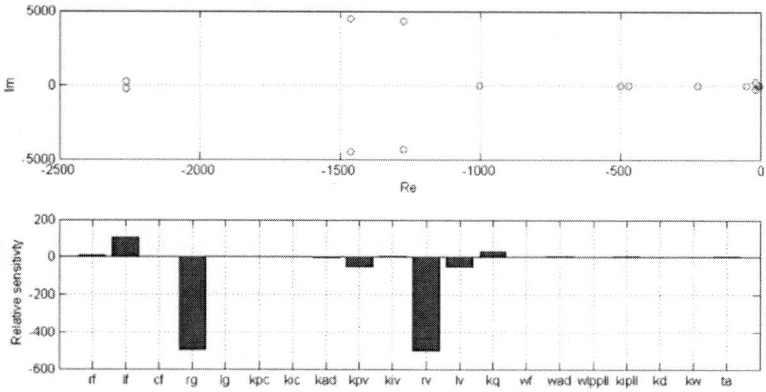

Figure 16. Parametric sensitivity of the pole resulting in instability with high values of k_q.

Another example of a parameter sensitivity analysis is shown in Fig. 17, for the pole with the highest imaginary part. This is one of the poles that can cause instability due to resonance in the LC filter, even if it is very well damped in this case due to the implemented active damping algorithm. It can, however, be seen that it is sensitive to the physical parameters of the LC filter and the grid, and it is also clear that the location of this pole can be influenced by the proportional gains of the voltage and current controllers, as well as the gain of the active damping algorithm *kAD*.

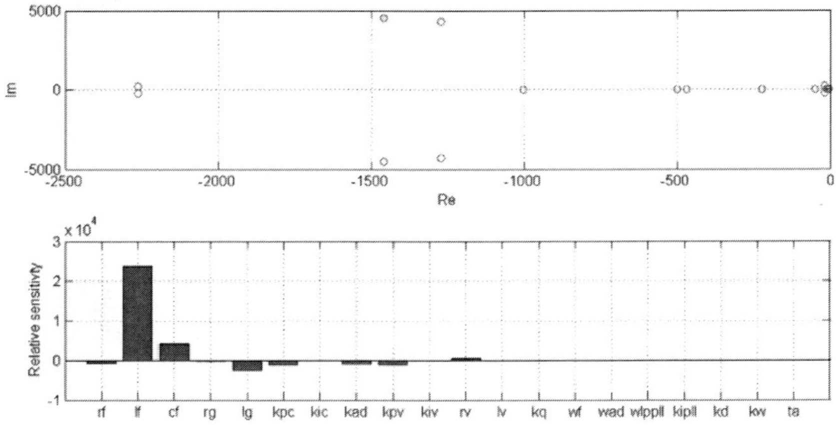

Figure 17. Parametric sensitivity of the pole with the highest imaginary part.

In case of stability problems in the system, such parameter sensitivity analysis can be easily performed for any of the implicated poles. However, it should be remembered that the change of a parameter to improve the stability conditions or dynamic response associated with one pole, might move another pole in the wrong direction. If that is the case, manual tuning of the controller parameters might be challenging even with support from parameter sensitivity analysis. However, as proposed in [14], it is possible to design automated iterative procedures that can help to bring the pole locations of the system into stable conditions with improved dynamic response. Such approaches might be necessary if a high order system like the investigated VSM is interfaced with other dynamic models and the interaction between the states in the various parts of the overall system model is causing instabilities or poorly damped dynamic responses. The presented VSM model and the tools applied to analyze the properties of the investigated configuration provide a starting point for studying the operation of VSM performance in a larger scale system as will be the case in a SmartGrid scenario.

CONCLUSION

The concept of a Virtual Synchronous Machine (VSM) as an approach for controlling power electronics converters to replicate the behavioural characteristics of Synchronous Machines (SMs) has been introduced during the last decade. This paper has highlighted the inherent advantages of the VSM as a possible alternative for releasing the potential advantages of distributed autonomous control actions of power electronics converters in the SmartGrid context. An implementation of the VSM concept based on an emulated swing equation providing references for cascaded voltage and current controllers has been presented in detail. In particular, a nonlinear mathematical model of the investigated VSM scheme, and its linearized small-signal equivalent, has been derived. These models have been simulated numerically in order to verify and illustrate the behaviour of the VSM and a few of its inherent features. Finally, the system eigenvalues and their parametric sensitivities have been analyzed for the linearized system in order to provide further insight into the dynamic properties and stability characteristics of this VSM implementation. Thus, in addition to the detailed description of a VSM implementation suitable for power system applications in the SmartGrid context, the paper has also presented and verified the models needed for accurately representing the impact of this VSM implementation on the power system in larger scale small-signal power system stability studies.

ACKNOWLEDGEMENT

The work of SINTEF Energy Research in this paper was supported by the project “Releasing the Potential of Virtual Synchronous Machines – ReViSM,” through the Blue Sky instrument of SINTEF Energy Research as a Strategic Institute Programme (SIP) funded by the National Basic Funding Scheme of Norway.

REFERENCES

1. CIRED Working Group on SmartGrids, SmartGrids on the Distribution Level – Hype or Vision? CIRED's Point of View, Final Report, 2013, Available at: http://www.cired.at/pdf/CIRED WG SmartGrids FinalReport.pdf
2. European Center for Power Electronics, ECPE, European Power Electronics and Drives Association, EPE, Position Paper on Energy Efficiency – The Role of Power Electronics, 2007, Available at: http://www.ecpe.org/securedl/0/1391290694/35ad20b5395221115c886f006df4d0f595ed7174/fileadmin/user upload/Public Relations/ECPE Publications/ECPE Position Paper Energy Efficiency.pdf

3. J. Morren, J. Pierik, S.W.H. de Haan, Inertial response of variable speed wind turbines, Electr. Power Syst. Res. 76 (July (11)) (2006) 980–987.
4. H.-P. Beck, R. Hesse, Virtual synchronous machine, in: Proc. of the 9th Int. Conf. on Electrical Power Quality and Utilisation, Barcelona, Spain, 9–11 October, 2007, 6 pp.
5. J. Zhu, C.D. Booth, G.P. Adam, A.J. Roscoe, C.G. Bright, Inertia emulation control strategy for VSC-HVDC transmission systems, IEEE Trans. Power Syst. 28 (May (2)) (2013) 1277–1287.
6. Z. Linn, Y. Miura, T. Ise, Power system stabilization control by HVDC with SMES using virtual synchronous generator, IEEJ J. Ind. Appl. 1 (2) (2012) 102–110.
7. T. Shintai, Y. Miura, T. Ise, Reactive power control for load sharing with virtual synchronous generator control, in: Proc. of the IEEE 7th Int. Power Electronics and Motion Control Conference – ECCE Asia, Harbin, China, 2–5 June 2012, 2012, pp. 846–853.
8. J. Driesen, K. Visscher, Virtual synchronous generators, in: Proc. of the IEEE Power and Energy Society 2008 General Meeting: Conversion and Delivery of Energy in the 21st Century, Pittsburgh, PA, USA, 20–24 July, 2008, 3 pp.
9. Q.-C. Zhong, G. Weiss, Synchronverters: inverters that mimic synchronous generators, IEEE Trans. Ind. Electron. 58 (April (4)) (2011) 1259–1267.
10. S. D'Arco, J.A. Suul, Virtual Synchronous Machines – classification of implementations and analysis of equivalence to droop controllers for microgrids, in: Proc. of IEEE PowerTech Grenoble 2013, Grenoble, France, 16–20 June 2013, 2013, 7 pp.
11. K. Visscher, S.W.H. De Haan, Virtual Synchronous Machines (VSG's) for frequency stabilization in future grids with a significant share of decentralized generation, in: Proc. of the CIRED Seminar 2008: SmartGrids for Distribution, Frankfurt, Germany, 23–24 June 2008, 2008, 4 pp.
12. K. Sakimoto, Y. Miura, T. Ise, Stabilization of a power system with a distributed generator by a virtual synchronous generator function, in: Proc. of the 8th Int. Conference on Power Electronics – ECCE Asia, Jeju, Korea, 30 May–3 June 2011, 2011, 8 pp.
13. H. Bevrani, T. Ise, Y. Miura, Virtual synchronous generators: a survey and new perspectives, Int. J. Electr. Power Energy Syst. 54 (2014) 244–254.
14. S. D'Arco, J.A. Suul, O.B. Fosso, Control system tuning and stability analysis of Virtual Synchronous Machines, in: Proc. of the 2013 IEEE Energy Conversion Congress and Exposition, ECCE 2013, Denver, CO, USA, 15–19 September 2013, 2013, pp. 2664–2671.
15. S. D'Arco, J.A. Suul, O.B. Fosso, Small-signal modelling and parametric sensitivity of a Virtual Synchronous Machine, in: Proc. of the 18th Power Systems Computational Conference, PSCC, Wrocław, Poland, 18–22 August 2014, 2014, 9 pp.

16. F. Blaabjerg, R. Teodorescu, M. Liserre, A.V. Timbus, Overview of control and grid synchronization for distributed power generation systems, IEEE Trans. Ind. Electron. 53 (October (5)) (2006) 1398–1409.
17. J. Rocabert, A. Luna, F. Blaabjerg, P. Rodríguez, Control of power converters in AC microgrids, IEEE Trans. Power Electron. 27 (November (11)) (2012) 4734–4749.
18. D. Pudjianto, C. Ramsay, G. Strbac, Virtual power plant and system integration of distributed energy resources, IET Renew. Power Gener. 1 (March (1)) (2007) 10–16.
19. E.A. Setiawan, Concept and Controllability of Virtual Power Plant (Dr. Ing thesis), Kassel University, Kassel, Germany, 2007.
20. J. Hansen et al., Providing Flexibility with a Virtual Power Plant, Final Demo Report, Deliverable No: 10.3, Twenties Project. Available at: http://www.twenties-project.eu/system/files/D10.3 Providing%20flexibility%20with%20a%20virtual%20power%20plant.pdf
21. T.L. Vandoorn, J.D.M. De Kooning, B. Meersman, L. Vandevelde, Review of primary control strategies for islanded microgrids with power-electronic interfaces, Renew. Sustain. Energy Rev. 19 (2013) 613–628.
22. T.L. Vandoorn, B. Zwaenpoel, J.D.M. De Kooning, B. Meersman, L. Vandevelde, Smart microgrids and virtual power plants in a hierarchical control structure, in: Proc. of the 2nd IEEE PES Int. Conference and Exhibition on Innovative Smart Grid Technologies, ISGT Europe 2011, Manchester, UK, 5–7 December 2011, 2011, 7 pp.
23. J. Machowski, J.W. Bialek, J.R. Bumby, Power System Dynamics and Stability, Wiley, Chichester, UK, 1997 (Chapters 2 and 5).
24. R. Hesse, D. Turschner, H.-P. Beck, Micro grid stabilization using the Virtual Synchronous Machine (VISMA), in: Proc. of the Int. Conference on Renewable Energies and Power Quality, ICREPQ'09, Valencia, Spain, 15–17 April 2009, 2009, 6 pp.
25. Y. Chen, R. Hesse, D. Turschner, H.-P. Beck, Dynamic properties of the Virtual Synchronous Machine (VSIMA), in: Proc. of the Int. Conference on Renewable Energies and Power Quality, ICREPQ'11, Las Palmas, Spain, 13–15 April 2011, 2011, 5 pp.
26. Y. Chen, R. Hesse, D. Turschner, H.-P. Beck, Investigation of the Virtual Synchronous Machine in the Island Mode, in: Proc. of the 2012 3rd IEEE Innovative Smart Grid Technologies Conference – Europe, ISGT Europe, Berlin, Germany, 15–17 October 2012, 2012, 6 pp.
27. N. Pogaku, M. Prodanovic, ´ T.C. Green, Modeling, analysis and testing of autonomous operation of an inverter-based microgrid, IEEE Trans. Power Electron. 22 (March (2)) (2007) 613–625.

28. S. D'Arco, G. Guidi, J.A. Suul, Embedded limitations and protections for droop-based control schemes with cascaded loops in the synchronous reference frame, in: Proc. of the 2014 Int. Power Electronics Conference, IPEC-Hiroshima 2014 – ECCE Asia, Hiroshima, Japan, 18–21 May 2014, 2014, pp. 1544–1551.
29. S. D'Arco, J.A. Suul, Equivalence of Virtual Synchronous Machines and frequency-droops for converter-based MicroGrids, IEEE Trans. Smart Grid 5 (January (1)) (2014) 394–395.
30. P. Kundur, Power System Stability and Control, McGraw-Hill, New York, 1994.
31. V. Kaura, V. Blasko, Operation of a phase locked loop system under distorted utility conditions, IEEE Trans. Ind. Appl. 33 (January/February (1)) (1997) 58–63.
32. H. Kolstad, Control of an Adjustable Speed Hydro utilizing Field Programmable Devices (Ph.D. thesis), Norwegian University of Science and Technology, 2002.
33. A. Haddadi, G. Joos, Load Sharing of Autonomous Distribution-level Microgrids, in: Proc. 2011 IEEE PES General Meeting: The Electrification of Transportation and the Grid of the Future, Detroit, MI, USA, 24–28 July 2011, 2011, 9 pp.
34. J. He, Y.W. Li, Analysis, design, and implementation of virtual impedance for power electronics interfaced distributed generation, IEEE Trans. Ind. Appl. 47 (November/December (6)) (2011) 2525–2538.
35. V. Blasko, V. Kaura, A new mathematical model and control of a three-phase AC–DC voltage source converter, IEEE Trans. Power Electron. 12 (January (1))(1997) 116–123.
36. M. Malinowski, M.P. Kazmierkowski, S. Bernet, New simple active damping of resonance in three-phase PWM converter with LCL filter, in: Proc. of the 2005 IEEE Int. Conference on Industrial Technology, ICIT 2005, Hong Kong, 14–17 December 2005, 2005, pp. 861–865.
37. N. Kroutikova, C.A. Hernandez-Aramburo, T.C. Green, State-space model of grid-connected inverters under current control mode, IET Electr. Power Appl. 1 (May (3)) (2007) 329–338.
38. S. D'Arco, J.A. Suul, O.B. Fosso, Small-signal modelling and parametric sensitivity of a Virtual Synchronous Machine in islanded operation, IntJ Elec Power (2014), accepted for publication.
39. C. Bajracharya, M. Molinas, J.A. Suul, T.M. Undeland, Understanding of tuning techniques of converter controllers for VSC-HVDC, in: Proc. of Nordic Workshop on Power and Industrial Electronics, NORPIE 2008, Espoo, Finland, 9–11 June, 2008, 2008, 8 pp.

CITATION

Salvatore D'Arco, Jon Are Suul, Olav B. Fosso, A Virtual Synchronous Machine implementation for distributed control of power converters in SmartGrids, Electric Power Systems Research, Volume 122, May 2015, Pages 180-197, ISSN 0378-7796, http://dx.doi.org/10.1016/j.epsr.2015.01.001.

CHAPTER 3

Performance of PSS to Damp Power Oscillations Caused by an HVDC Link in AC-DC Power System

Mokhtar Benasla[1], *Tayeb Allaoui*[1], *Mostefa Brahami*[2]

[1]Department of Electrical Engineering, L2GEGI Laboratory, IbnKhaldoun University, Tiaret, Algeria,

[2]Department of Electrical Engineering, ICEPS Laboratory, DjillaliLiabes University, SidiBel Abbes, Algeria

ABSTRACT

HVDC (High Voltage Direct Current) systems are increasingly being applied to improve power system operation and controllability. However, inappropriate setting of HVDC controller may have a detriment effect on the system performance. Generally, PSS (Power System Stabilizer) is known as a simple concept, easy to perform, and computationally effective to enhance damping of power system oscillations through excitation control of synchronous generator. This paper examines the effectiveness of the PSS to enhance the dynamic performance of AC-DC power systems and to compensate the negative damping of HVDC system. The dynamic performance is evaluated by examining the system response to various disturbances. In order to ensure the reliability of the simulation test results as well as the performance of the PSS, detailed HVDC modeling is adopted using Sim Power Systems toolbox in the MATLAB, and some important conclusions are drawn.

INTRODUCTION

The development of power systems follows the requirements to transmit power from generation to the consumers. With an increased demand for energy and the construction of new generation plants, the size and complexity of power systems have grown. With an increasing size of the interconnected systems, the technical and economical advantages diminish. This is related to problems regarding load flow, power oscillations and voltage quality. If power is to be transmitted through the interconnected system over longer distances, transmission needs to be supported [1]. HVDC (High Voltage Direct Current) provides the necessary features to avoid technical problems in the power systems, and it increases the transmission capacity and system stability very efficiently. HVDC can also be applied as a hybrid AC-DC solution in synchronous AC systems either as Back-to-Back for grid power flow control (elimination of congestion and loop flows) or as long-distance point-to-point transmission [2]. However, it has caused some problems in the applications [3,4]. For example, in 1977, the Square Butter HVDC system in USA suffered from the subsynchronous oscillation (SSO). The following studies showed that this dynamic instable phenomenon occurred at the side of the HVDC rectifier, and was mostly attributed to the negative damping torque caused by the inappropriate control of a HVDC link [5]. Since then, many analog simulators and computer based simulation programs have been developed to study the SSO caused by HVDC links [5].

So, it is necessary to study how to decouple the interactions and solve the coordinate problems of the controllers. Both the excitation control of generators and the HVDC are important to improve the stability of power systems and restrain the low frequency oscillation.

Conventional power systems are known to be subject to low frequency oscillations and due to the complexity and increasing dimensions associated with modern power systems. Several modes of oscillations have appeared in the dynamic response of the power system. If these oscillations persist, they can present limitation on power system performance.

Power system stabilizers (PSSs) are used as an effective way to enhance the damping of these oscillations [6].

The use of PSS in power system has been both economical and successful in improving the power system stability, and is expected to be installed on many generators connected to the system [7].

With the growing use of the HVDC technology in China power grid, unfortunately only a few of the generators in the China System are provided with power system stabilizers (PSS). The combination of this condition and the radial structure is likely the cause of poorly damped oscillations in central China [8]. In this paper a detailed study has been carried out in Sim Power Systems toolbox in the MATLAB environment to examine the effectiveness of the PSS to enhance the dynamic performance of AC-DC power systems and to compensate the negative damping of HVDC system.

MODEL DESCRIPTION

Test Power System

Figure 1 shows a model configuration of electric power systems with DC transmission. The model consists of single synchronous generator-infinite bus system and one DC link which is connected onto the bus of synchronous generator. The simple configuration of Figure 1 is based on a practical power system in Japan [9]. The mechanical power supplied by turbine is considered invariant for the duration of transient simulation runs.

HVDC Link

The HVDC considered in this paper (Figure 2) is of conventional type based on the CIGRE benchmark system [10,11]. The rectifier and the inverter are 12-pulse converters using two 6-pulse thyristor bridges connected in series. The transformer tap changers are not simulated and fixed taps are assumed.

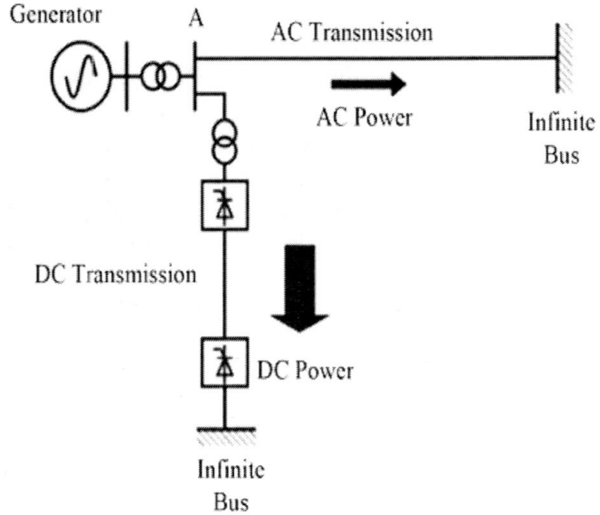

Figure 1. Electric power system with DC transmission.

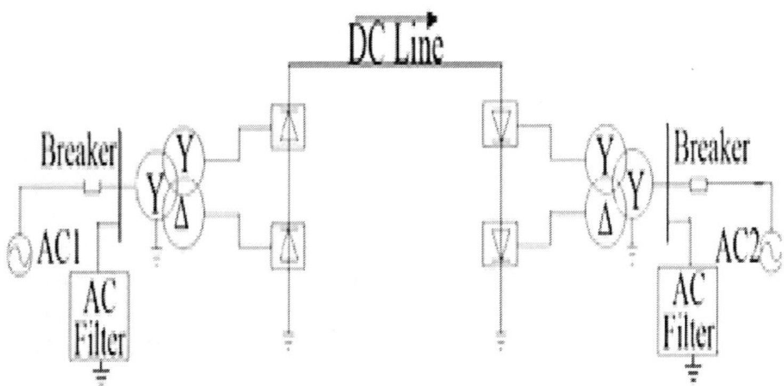

Figure 2. HVDC system.

Reactive power required by the converters is provided by a set of capacitor banks plus 11th, 13th and high pass filters on each side. A series reactor is also included between the two HVDC stations to make the DC current smooth.

By controlling the firing angle for the rectifier α_{Rec}, the power through the HVDC is controlled. The current controller is shown in Figure 3. The limited current I_{ord} reference is generated using the Voltage Dependent Current Limit (VDCL) unit. These units provide current reference values during steady and transient state conditions respectively. In order to maintain the operation of the AC system, VDCL limits the current in the

DC line, if the DC voltage decreases, e.g. due to an AC system disturbance [10]. When normal operation has returned and the DC voltage recovered, current returns to its steady-state level Iref (1 pu).

AUTOMATIC VOLTAGE REGULATOR (AVR)

The principal function of the excitation system is to furnish dc power (direct current and voltage) to the generator field, creating the magnetic field. The excitation system also provides control and protective equipment that regulates the generator electrical output.

One desirable characteristic of an excitation system is its ability to produce high levels of excitation voltage (ceiling) very rapidly, following a change in terminal voltage.

The automatic voltage regulator (AVR) provides for control of the terminal voltage of the generator by changing the generator field voltage to make the system more transient stable during disturbances.

There are a variety of designs for the AVR, including various means of ensuring stable response to transient changes in terminal voltage.

Figure 4 shows The IEEE AVR Type I [12]. Where Vt is the generator terminal voltage or a remote-bus regulated voltage and Se is the ceiling function. The parameters of this AVR are shown in Appendix.

The contribution of the AVR on the transient stability is shown by the following simulations:

At time $t = 1$ s a three-phase to ground fault of 130 ms duration occurs at the AC transmission line near bus A as shown in Figure 1.

Figure 5 shows the transient responses for this fault condition. Comparison of results between both cases with AVR and without AVR for the similar nature and duration of faults indicate:

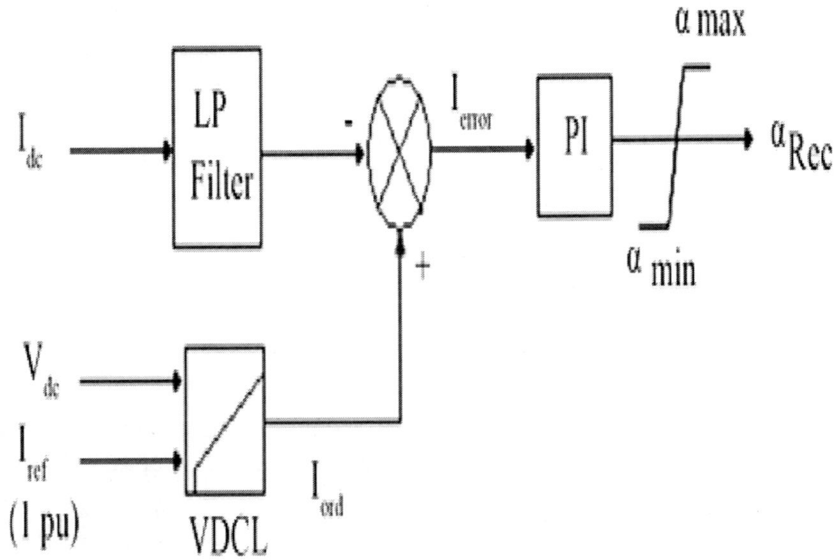

Figure 3. Current controller.

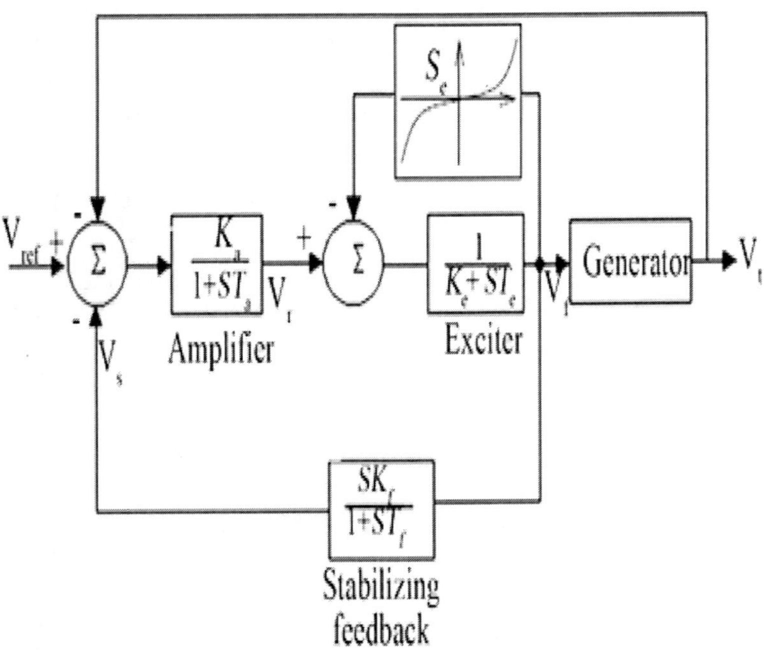

Figure 4. Automatic voltage regulator IEEE type 1.

It can be observed that, the system becomes unstable when the AVR is not in operation (without AVR). This is due to the poor damping of the oscillations. However, When the AVR is applied; the system becomes more oscillatory but remains stable.

The mighty action of the AVR to improve the transient stability has an important negative contribution unfortunately on the small-signal stability of the system (Figure 5(c)). However the current of excitation cannot immediately change because of the time constants of the regulator.

Figure 6 shows the adverse effect of the AVR on system oscillations. This type of oscillation is called local mode oscillation and its behavior is mainly limited to the local area in the vicinity of the power plant, and it seldom influences the rest of the system. It has been known that the local oscillation is likely to occur when power is transmitted over long-distance transmission lines from a power plant at a remote location [7].

This category of oscillations can exist simultaneously with another mode of oscillations as the control mode oscillations, this type of oscillations are associated with the controls of generating units and other equipment. Poorly tuned controls of excitation systems, prime movers, static var compensators, and HVDC converters are the usual causes of instability of control modes [13].

The performance of power systems decrease with the size and complexity of the networks. Should power be transmitted through the interconnected system over longer distances, transmission needs to be supported. The choice between transmission alternatives is made on the basis of cost and controllability. The original justification for HVDC systems was its lower cost for long electrical distances, (i.e. the favorable case for the local-mode oscillations).

It is well documented that inappropriate setting of HVDC controller could cause dynamic instability of an AC-DC system [5].

Figure 7 shows the negative effect of the HVDC on the system considered in Figure 1. The main reason of the instability caused by the HVDC control system is due to the existence of the negative damping.

These oscillations may sustain and grow to cause system separation if no adequate damping is available.

However, the capability of an HVDC link to rapidly modulate the power flow, in response to control signals such as frequency deviation and rate of change of AC power, has been utilized for some time to improve the dynamic stability of AC-DC systems [14].

The ability to enhance power system stability based on HVDC power modulation depends on several factors including the location of the HVDC link converters, the feedback signal used as input to the control, as well as the internal characteristics of the AC system.

However, The HVDC with classical DC power control even with supplementary damping signal does not contribute to system synchronizing torque and may increase the risk of instability. A good example is the black-outs of the AC-DC system of the western USA in summer '96 [15].

HVDC systems were also found to interact adversely with torsional vibrations (the subsynchronous torsional interaction problem), especially when augmented with supplemental modulation controls to damp power swings [13]. Furthermore, the excessive DC modulated power are usually not desired due to the increased reactive power requirements in the inverter side AC system and its effect on voltage level, as well as the possibility of adverse influence on DC recovery and operation performance during the initial stages of system post-fault recovery [4].

POWER SYSTEM STABILIZER (PSS)

There are several ways to damp the low-frequency oscillations. Among them, the use of Power System Stabilizer (PSS) is the most effective and robust [6].

A power system stabilizer (PSS) is frequently, but not always, included in an excitation system. It is designed to modulate the AVR input in such a manner as to contribute damping to oscillations. The input to the PSS may be generator rotor speed ($\Delta\omega$), electrical power (ΔP), or other signals [13].

The output signal of the PSS is injected into the summing junction of the exciter block (Figure 8) in the generator to damp out low frequency oscillations of the power system.

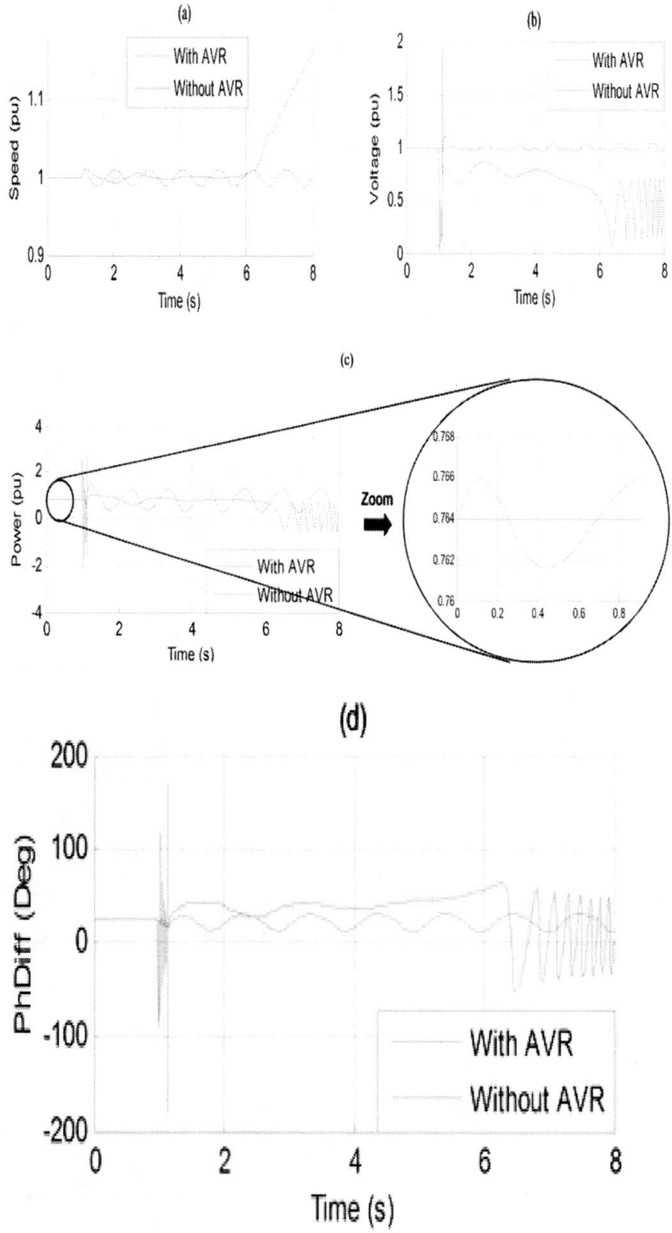

Figure 5. (a) Generator speed. (b) Generator terminal voltage. (c) Generator active power output. (d) Transmission angle (PhDiff) between two ends.

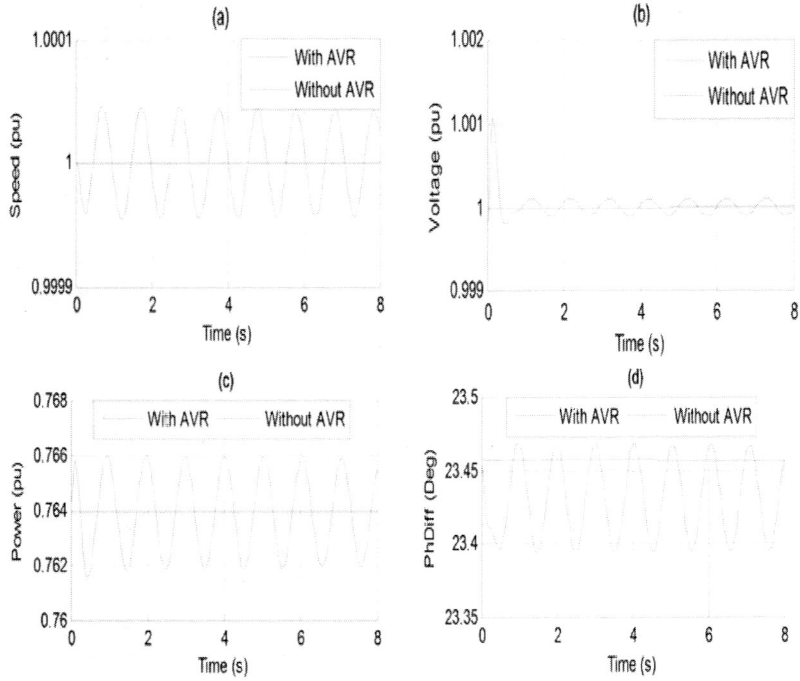

Figure 6. (a) Generator speed. (b) Generator terminal voltage. (c) Generator active power output. (d) Transmission angle (PhDiff) between two ends.

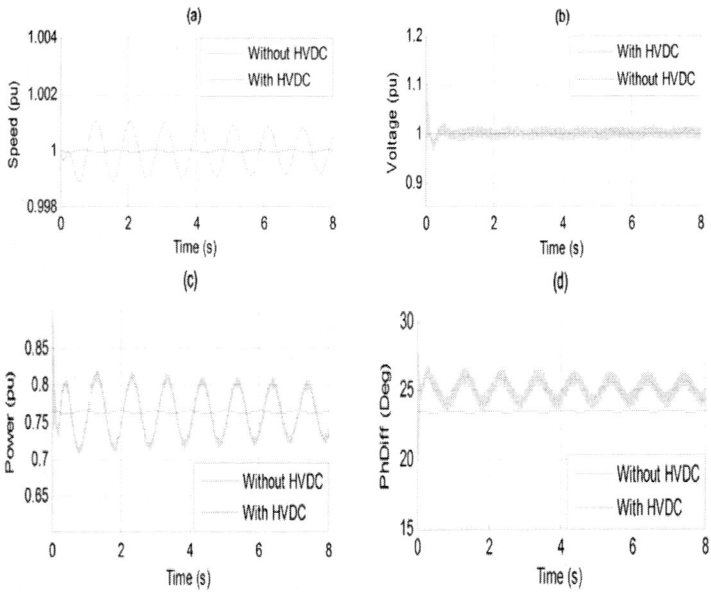

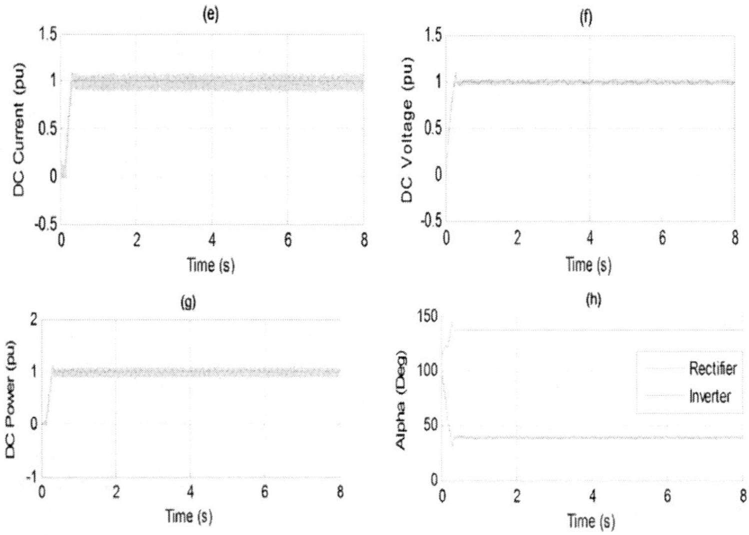

Figure 7. (a) Generator speed. (b) Generator terminal voltage. (c) Generator active power output. (d) Transmission angle (PhDiff) between two ends. (e) DC Current. (f) DC Voltage. (g) DC power transfer (Pdc). (h) Rectifier and inverter firing angle order.

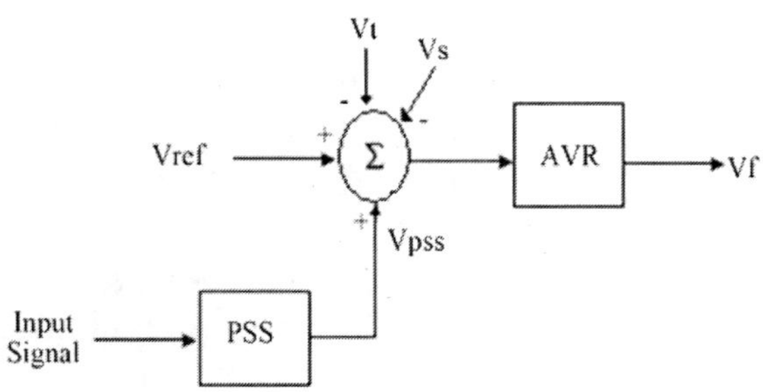

Figure 8. AVR with PSS.

It is the most economical way to mitigate this kind of low frequency oscillations. Furthermore, PSS design by method of phase compensation is based on knowledge of the physical aspect of power systems, and is more robust and relatively easier than other methods. The input signal of the PSS used in this paper is the electrical power signal. Its block diagram is shown in Figure 9. The parameters of this PSS are shown in Appendix.

In a multi-machine power system, it is important to determine the best location for the application of power system stabilizers [16]. Several techniques have been proposed in the literature to determine the best location for applying power system stabilizers. Participation factor method is one of the methods in detecting the contribution of various generators in each mode and for suitable location for applying the PSSs. Some authors have proposed sensitivity of PSS effect (SPE) as a measure for determining the suitable location of PSSs [16], so that the dominant generator with the greatest influence on both the power system stability, and the low-frequency oscillation becomes the candidate for PSS installation.

In a AC-DC power system the low-frequency oscillation occurred at the side of the HVDC rectifier[5]. So the generator at the rectifier side is the best location for the application of power system stabilizers.

An appropriate parameter design is very important in order for a PSS to operate effectively. In general, these parameters are set with the single machine infinite bus model. So before deployment in the AC-DC power system, it is worth checking the performance of the PSS in the system without HVDC.

The results of Figure 10 show the effectiveness of the PSS device when added to conventional voltage regulators AVR.

It is clear from these figures that PSS improves the stability performance of the example power system without HVDC, and the low frequency oscillations are significantly damped in comparison to the results without PSS.

The effect of PSS to improve the performance transient stability is shown by simulation a three phase to ground fault (at 1 s) near bus A (Figure 1, without HVDC) of duration 130 ms and 180 s in Figures 11 and 12, respectively.

It can be clearly seen that the PSS enhances the stability performance of the power system. Figure 12(a) gives the variation of generator speed. As can be seen in the figure, the system without PSS falls out of synchronism. However, when the PSS is applied, the system remains stable. Therefore, power system stabilizers (PSSs) are employed to enhance the dynamic and transient performance of power systems.

Once the performance of PSS is evaluated without HVDC, the relevant question is: it is possible to get an effective damping of oscillation modes with the same PSS (with the same parameter setting) if the system is equipped with an HVDC link (Figure 1). Figure 13 gives a better answer to this question.

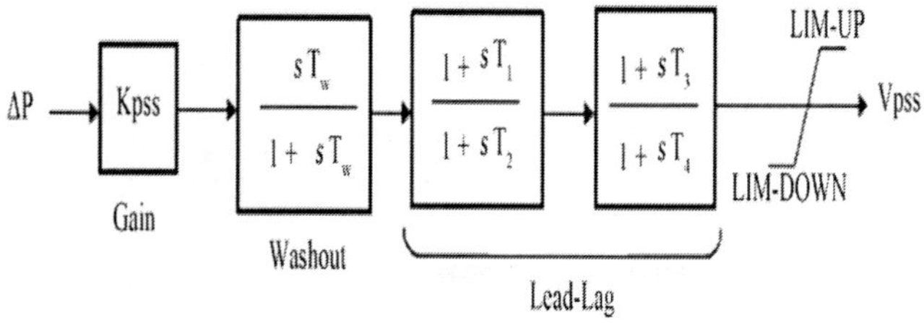

Figure 9. Power system stabilizer.

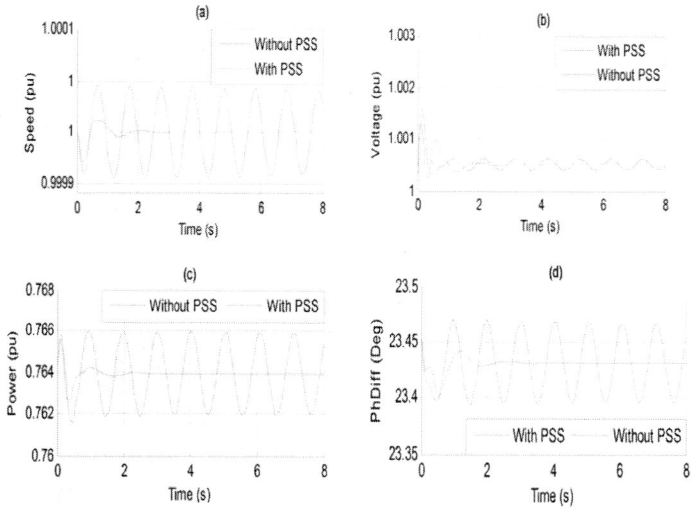

Figure 10. (a) Generator speed. (b) Generator terminal voltage. (c) Generator active power output. (d) Transmission angle (PhDiff) between two ends.

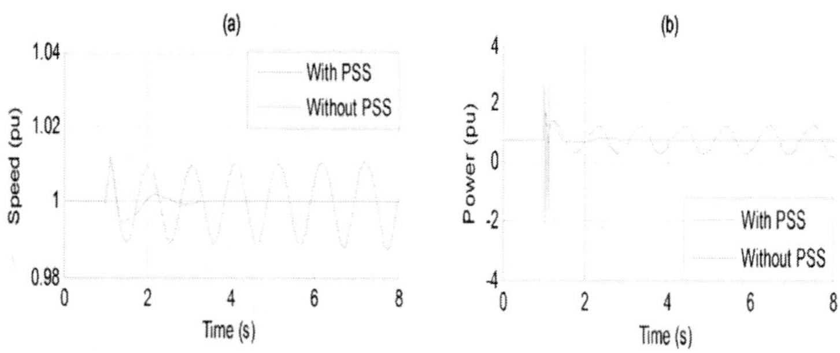

Figure 11. (a) Generator speed. (b) Generator active power output.

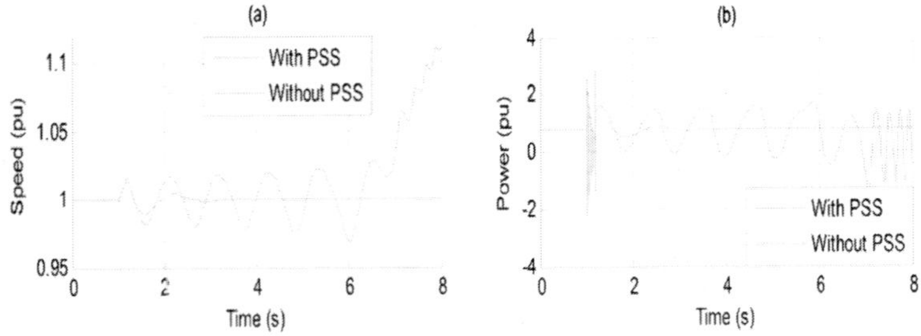

Figure 12. (a) Generator speed. (b) Generator active power output.

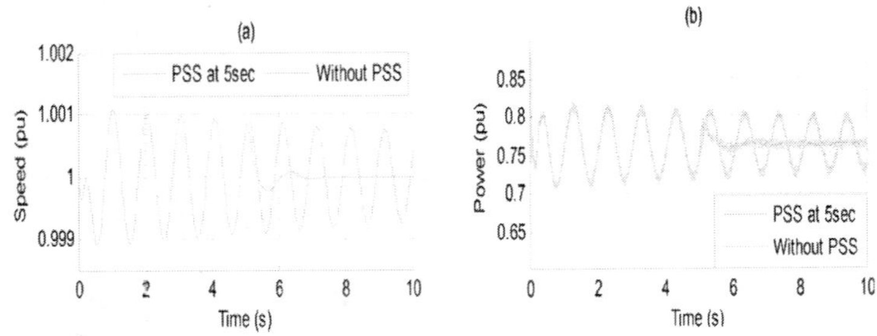

Figure 13. (a) Generator speed. (b) Generator active power output.

It can be seen from this figure that the installation of the PSS (at 5 sec) provides faster response and can enhance the system dynamic performance well and lines can transmit the power stably.

A three-phase to ground fault occurs (at 3s) near bus A (Figure 1) of 130ms duration in order to test the system damping with a short-circuit fault. The simulation results are shown in Figure 14. It is clear from these figures that PSS improves the stability performance of the example power system, and the oscillations are significantly damped in comparison to the results without PSS.

So, the HVDC can be used now to improve the transient stability by modulation of the transmitted active power with an acceptable level of reliability.

CONCLUSIONS

The steady state and transient simulations show that the HVDC system can cause dynamic instability of an AC-DC power system due to the negative interaction between AC and DC systems and the problem of poor damping. The effectiveness of the application of PSS to compensate the negative damping of HVDC system has been examined by detail simulation. The results show that the PSS improves the stability performance of the example AC-DC power system, and the oscillations are significantly damped.

The research will bring important and significant reference for further operation and stability control of HVDC and AC system. Further research still remained on the topics such as the coordinated control strategy between PSS and HVDC damping modulation for the damping of all the critical modes of the system, best location of the PSS and also the tuning of PSS parameters in multi-machine AC-DC power systems and so on.

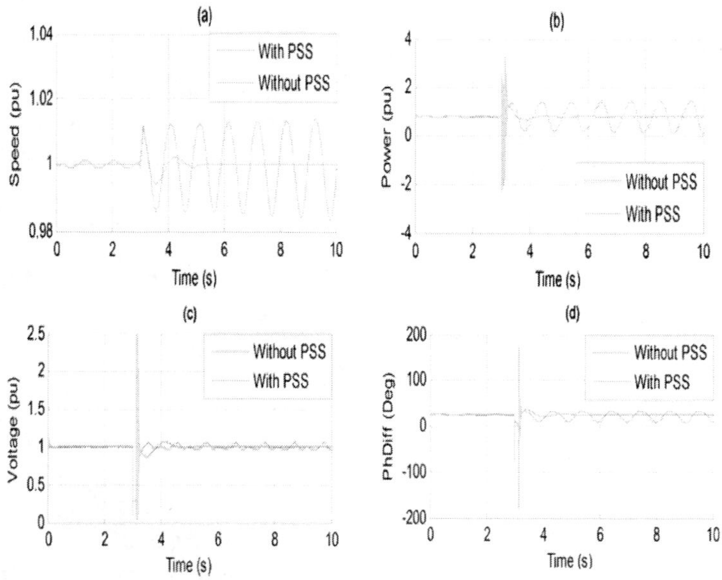

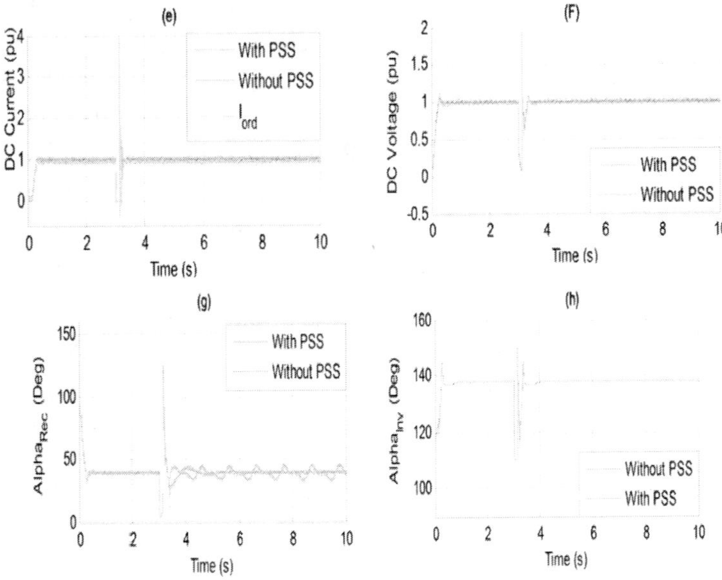

Figure 14. (a) Generator speed. (b) Generator active power output. (c) Generator terminal voltage. (d) Transmission angle (PhDiff) between two ends. (e) DC Current. (f) DC Voltage. (g) Rectifier firing angle order and (h) Inverter firing angle order.

Therefore, the control of HVDC has the potential for future application to power systems.

REFERENCES

1. G. Beck, D. Povh, D. Retzmann and E. Teltsch, "Global Blackouts—Lessons Learned," Power Transmission Division, Siemens AG, 2011.
2. W. Breuer, D. Retzmann and K. Uecker, "Highly Efficient Solutions for Smart and Bulk Power Transmission of Green Energy," Power Transmission Division, Siemens AG, 2011.
3. D.-J. Kim, H.-K. Nam and Y.-H. Moon, "A Practical Approach to HVDC System Control for Damping Subsynchronous Oscillation Using the Novel Eigenvalue Analysis Program," IEEE Transactions on Power Systems, Vol. 22, No. 4, 2007, pp. 1926-1934.http://dx.doi.org/10.1109/TPWRS.2007.907974
4. X.-M. Mao, Y. Zhang, L. Guan and X. C. Wu, "Researches on Coordinated Control Strategy for Inter-Area Oscillations in AC/DC Hybrid Grid with Multi-Infeed HVDC," IEEE/PES Transmission and Distribution Conference & Exhibition: Asia and Pacific, Dalian, 2005, pp. 1-6.

5. T. Yu, K. Chan and Z. Ren, “Damping Analysis of Primary and Auxiliary Control of HVDC Systems,” IEEE/ Power Engineering Society General Meeting, Montreal, 2006.
6. G. Liu, Z. Xu, Y. Huang and W. Pan, “Analysis of Inter-Area Oscillations in the South China Interconnected Power System,” Electric Power Systems Research, Vol. 70, No. 1, 2004, pp. 38-45.
7. A. R. Fereidouni, B. Vahidi, T. Hoseini and M. Tahmasbi, “Improvement of Low Frequency Oscillation Damping by Allocation and Design of Power System Stabilizers in the Multi-Machine Power System,” International Journal of Electrical Power & Energy Systems, Vol. 52, 2013, pp. 207-220.
8. K. Högberg, M. Ericsson, A. Kumar, K. Lindén and W. Weibing, “Stability Improvement for Central China System.” library.abb.com/GLOBAL/.../ICPS01Stability.pdf.
9. Y. Susuki, T. Hikihara and H. Chiang, “Discontinuous Dynamics of Electric Power System with DC Transmission: A Study on DAE System,” IEEE Transactions on Circuits and Systems, Vol. 55, No. 2, 2008, pp. 697-707.
10. V. K. Sood, “HVDC and FACTS Controllers Applications of Static Converters in Power Systems,” Kluwer Academic Publishers, 2004.
11. M. Benasla, T. Allaoui, H, Chaib and N. Cherif, “BackUp Protection to Enhance Operation of HVDC System,” The International Conference on Electronics & Oil: From Theory to Applications, Algeria, 2013.
12. F. Milano, “Power System Modelling and Scripting,” Springer-Verlag, London, 2010.http://dx.doi.org/10.1007/978-3-642-13669-6
13. R. G. Farmer, “Power System Dynamics and Stability,” The Electric Power Engineering Handbook Series, CRC Press LLC, 2001.
14. Transmission and Distribution Committee of the IEEE Power Engineering Society, “IEEE Guide for Planning DC Links Terminating at AC Locations Having Low Short-Circuit Capacities,” IEEE Std 1204-1997.
15. E. Hammad, “Stability and Control of HVDC and AC Transmissions in Parallel,” IEEE Transactions on Power Delivery, Vol. 14, No. 4, 1999, pp. 1545-1554.http://dx.doi.org/10.1109/61.796252
16. Karimpoura, R. Asgharianb and O. P. Malik, “Determination of PSS Location Based on Singular Value Decomposition,” International Journal of Electrical Power & Energy Systems, Vol. 27, No. 8, 2005, pp. 535-541.

CITATION

M. Benasla, T. Allaoui and M. Brahami, "Performance of PSS to Damp Power Oscillations Caused by an HVDC Link in AC-DC Power System," Energy and Power Engineering, Vol. 6 No. 1, 2014, pp. 1-9. doi: 10.4236/epe.2014.61001.

CHAPTER 4

Power Quality Improvement Using Switch Mode Regulator

Raju Ahmed and Mohammad Jahangir Alam

Electrical and Electronic Engineering Department, Dhaka University of Engineering and Technology (DUET), Gazipur, Bangladesh

INTRODUCTION

Power quality describes the quality of voltage and current. It is an important consideration in industries and commercial applications. Power quality problems commonly faced are transients, sags, swells, surges, outages, harmonics and impulses [1]. Among these voltage sags and extended under voltages have large negative impact on industrial productivity, and could be the most important type of power quality variation for many industrial and commercial customers [1-5].

Voltage sags is mainly due to the fault occurring in the transmission and distribution system, loads like welding and operation of building construction equipment, switching of the loaded feeders or equipments. Both momentary and continuous voltage sags are undesirable in complex process controls and household appliances as they use precision electronic and computerized control.

Major problems associated with the unregulated long term voltage sags include equipment failure, overheating and complete shutdown. Tap changing transformers with silicon-controlled rectifiers (SCR) are usually used as a solution of continuous voltage sags [6]. They require large transformer with many SCRs to control the voltage at the load which lacks the facility of adjusting to momentary changes. Some solutions have been suggested in the past to encounter the problems of voltage sag [7-11]. But these proposals have not been realized practically to replace conventional tap changing transformers.

Now a day's various power semiconductor devices are used to raise power quality levels to meet the requirements [12]. Several AC voltage regulators have been studied as a solution of voltage sags [13-18]. In [13] the input current was not sinusoidal, in [14-16] the efficiency of the regulator was not analyzed and in [17-18] the input power factor was very low and the efficiency is also found poor. Compact and fully electronic voltage regulators are still unavailable practically.

Dynamic Voltage Restorer (DVR) is sometimes used to regulate the load side voltage [19-21]. The DVR requires energy storage device to compensate the voltage sags. Flywheels, batteries, superconducting magnetic energy storage (SMES) and super capacitors are generally used as energy storage devices. The rated power operation of DVR depends on the size and capacity of energy storage device which limits its use in high power applications. Whereas, switching regulator needs no energy storage devices, therefore, can be used both in low power and high power applications.

The objective of this chapter is to describe the operation and design procedure of a switch mode AC voltage regulator. Firstly, some reviews of the regulators are presented then the procedure of design and analysis of a switch mode regulator is described step by step. Simulation software ORCAD version 9.1 [22] is used to analyze the regulator. The proposed regulator consists mainly two parts, power circuit and control circuit. The power circuit consist two bi-directional switches which serve as the freewheeling path for each other. A signal generating control circuit is to be associated with the power circuit for getting pulses of the switches. In the control circuit, a commercially available pulse width modulator IC chip SG1524B is used, thus circuit is compact and more viable.

REVIEW OF VOLTAGE REGULATORS

Switching-Mode Power Supply (SMPS)

A switching-mode power supply (SMPS) is switched at very high frequency. Conversion of both step down and step up of voltage is possible using SMPS. Uses of SMPSs are now universal in space power applications, computers, TV and industrial units. SMPSs are used in DC-DC, AC-AC, AC-DC, DC-AC conversion for their light weight, high efficiency and isolated multiple outputs with voltage regulation. Main parts of a Switching-mode power supply are:

(a) Power circuit,
(b) Control circuit.

Figure 1 shows the block diagram of a SMPS. The power circuit is mainly the input, output side with the switching device. The switching device is continuously switched at high frequency by the gate signal from the control circuit to transfer power from input to the output. The control circuit of a SMPS basically generates high frequency gating pulses for the switching devices to control the output voltage. Switching is performed in multiple pulse width modulation (PWM) fashion according to feedback error signal from the load. High frequency switching reduces filter requirements at the input and output sides of the converter. Simplest PWM control uses multiple pulse modulations generated by comparing a DC with a high frequency carrier triangular wave.

The PWM control circuit is commonly available as integrated form. The designer can select the switching frequency by choosing the value of RC to set oscillator frequency. As a rule of thumb to maximize the efficiency, the oscillation period should be about 100 times longer than the switching time of the switching device such as Transistor, Metal oxide semiconductor field-effect transistor (MOSFET), Insulated gate bipolar transistor (IGBT). For example, if a switch has a switching time of 0.5 us, the oscillator period would be 50 us, which gives the maximum oscillation frequency of 20 KHz. This limitation is due to the switching loss in the switching devices. The switching loss of switching devices increases with the switching frequency. In addition, the core loss of inductor limits the high frequency operation.

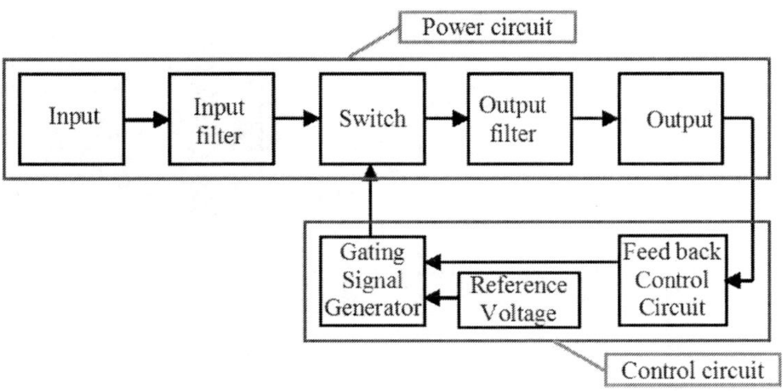

Figure 1. Block diagram of Switching-mode power supply (SMPS).

DC-DC Converter

Figure 2 illustrates the circuit of a classical linear power converter. Here power is controlled by a series linear element; either a resister or a transistor is used in the linear mode. The total load current passes through the series linear element. In this circuit greater the difference between the input and the output voltage, more is the power loss in the controlling device (linear element). Linear power conversion is dissipative and hence is inefficient.

The circuit of Fig. 3 illustrates basic principle of a DC-DC switching-mode power converter. The controlling device is a switch. By controlling the duty cycle, (the ratio of the time in on positions to the total time of on and off position of a switch) the power flow to the load can be controlled in a very efficient way. Ideally this method is 100% efficient. In practice, the efficiency is reduced as the switch is non-ideal and losses occur in power circuits. Hence, one of the prime objectives in switch mode power conversion is to realize conversion with the least number of components having better efficiency and reliability. The DC output voltage to the load can be controlled by controlling the duty cycle of the rectangular wave supplied to the base or gate of the switching device. When the switch is on, it has only a small saturation voltage drop across it. In the off condition the current through the switch is zero.

The output of the switch mode power conversion circuit (Fig. 3) is not pure DC. This type of output is applicable in some cases such as oven heating without proper filtration. If constant DC is required, then output of converter has to be smoothed out by the addition of low-pass filter.

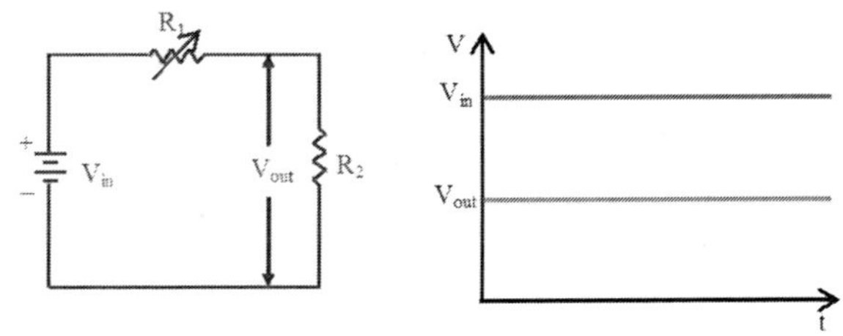

Figure 2. Linear (dissipative) DC-DC power conversion circuit.

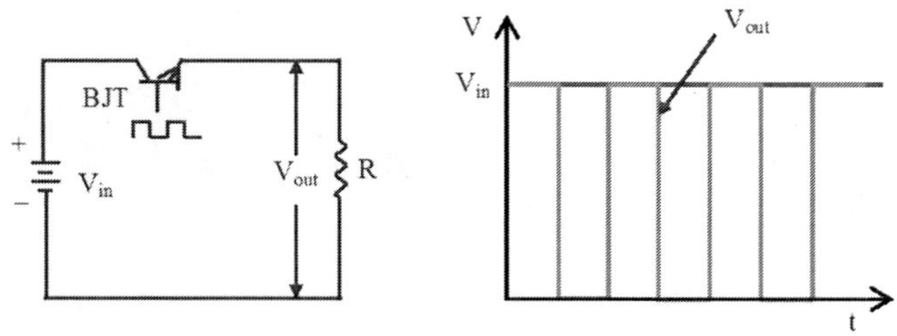

Figure 3. Switching-mode (non dissipative) DC-DC power conversion circuit.

Types of DC-DC Converter

There are four basic topologies of switching DC-DC regulators:

a) Buck regulator
b) Boost regulator
c) Buck-Boost regulator and
d) Cûk regulator.

The Circuit diagram of four basic DC-DC switching regulators is shown in Fig. 4. The expression of output voltage for the four types of DC-DC regulators are as follows:

For Buck regulator, $V_{out} = kV_{in}$, For Boost regulator, $V_{out} = \frac{V_{in}}{1-k}$

For Buck- Boost regulator and Cûk regulator, $V_{out} = \frac{-kV_{in}}{1-k}$

Where k is the duty cycle, the value of k is less than 1. For Buck regulator output voltage is always lower than input voltage, for Boost regulator output voltage is always higher than input voltage. For Buck-Boost regulator and Cûk regulator output voltage is higher than input voltage when the value of k is higher than 0.5, and output voltage is lower than input voltage when the value of k is lower than 0.5. When k is equal to 0.5 output voltage is same as input voltage. In Buck-Boost and Cûk regulator, the polarity of output voltage is opposite to that of the input voltage, therefore theses regulators are also called inverting regulators.

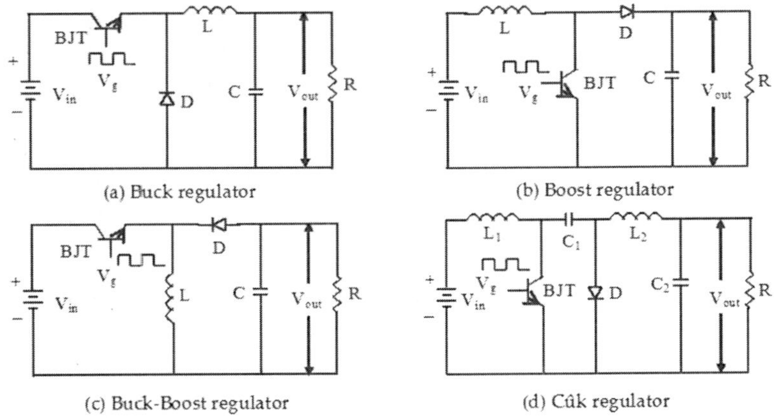

Figure 4. Circuit diagram of DC-DC regulator, (a) Buck regulator, (b) Boost regulator, (c) Buck-Boost regulator, (d) Cûk regulator.

AC-AC Converter

The AC voltage regulator is an appliance by which the AC output voltage can be set to a desired value and can be maintained constant all the time irrespective of the variations of input voltage and load. This subject is vast and the field of application extends from very large power systems to small electronic apparatus. Naturally, the types of regulators are also numerous. The design of the regulators depends mainly on the power requirements and degree of stability.

The AC voltage can be regulated by the following ways.

a) Solid-state tap changer and steeples control by variac
b) Solid-tap changer using anti-parallel SCRs
c) Voltage regulation using servo system
d) Phase controlled AC regulator
e) Ferro-resonant AC regulator
f) Switch mode AC regulator

Solid-State Tap Changer and Stepless Control By Variac

The voltage regulations by tap-changing switches are used in many industrial applications where the maintenance of output voltage at a constant value is not very stringent, such as ordinary battery chargers, electroplating rectifiers etc. For smaller installation, off-load tap changing switches are used and for large installation on-load tap changing switches are used. The switches are generally incorporated at the secondary of the transformer. For a low voltage high current load, the switches are provided on the primary side of the transformer due to economical reason. For line voltage correction, taps are provided on the primary of the transformer. For three-phase transformers three pole tap changing switches are used.

In off-load tap changer, the output is momentarily cut-off from the supply. It is therefore used for low capacity equipment and where the momentary cut-off of the supply is not objectionable for the load. The major limitation of the off-load tap changing switches is the occurrences of arcs at the contact points during the change-over operation. This shortens the life of off-load rotary switches, particularly of high current ratings. In Fig. 5 (a), three four-position switches of an off load tap changer are shown, such that the minimum of X volts per step are available at the output.

The voltage is corrected by tap-charging switches in steps. Where stepless control is required, variable autotransformers or variacs are used. The normal variac consists of a toroidal coil wound on a laminated iron ring. The insulation of the wire is removed from one of the end faces and the wire is grounded to ensure a smooth path for the carbon brush. Carbon brush is used to limit the circulating current, which flows between the short-circuited turns.

A Buck-Boost transformer is sometimes used for AC voltage regulation when the output voltage is approximately the same as the mean input voltage as shown in Fig. 5(b). In this case if the output voltage is less than or greater than the desired value, it can be increased or decreased to the desired value by adding a suitable forward or reverse voltage with the input through the Buck-Boost transformer.

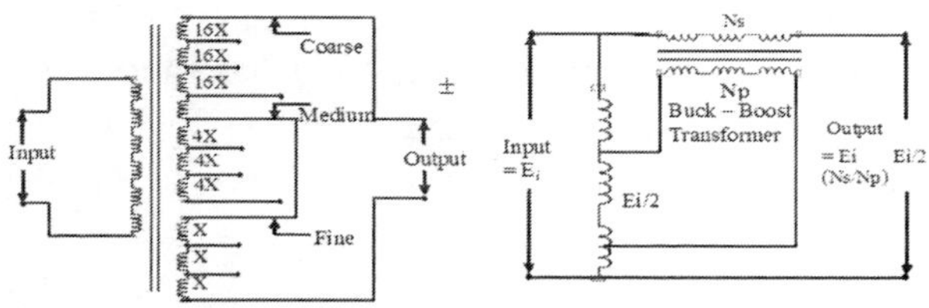

Figure 5. Circuit diagram of AC voltage controller using (a) Off load tap changer and (b) Buck-Boost transformer and variac.

Solid Tap Changer Using Anti-Parallel SCRS

Anti-parallel SCRs combinations can replace the voltage sensitive relay in the tap-changing regulator. Figure 6 shows a tap changer with three taps which can be connected to the load through three anti-parallel switches. When the SCR1-SCR2 switch is fired, tap 1 is connected to the load. Similarly taps 2 and 3 can be connected to the load through the SCR3-SCR4 and CSR5-SCR6 switches respectively. Thus, any number of taps can be connected to the load with similar SCR switches. When one group of SCRs operates for the whole cycle and other groups are off, the voltage corresponding to the tap of that group appears at the load. Changeover from one loop to the other is done simply by shifting the firing pulses from one group of SCRs to the other.

With resistive load, the load current becomes zero and the SCRs stop conduction as soon as the voltage reverses its polarity. Therefore, when one group is fired, the other groups are commutated automatically. With reactive loads, the situation is complicated by the fact that the zero current angle depends on the load power factor. This means that the SCR conducts a finite value of current at the time of reversal of line voltage. This results in either preventing a tap change due to reverse bias on the SCR to be triggered or causing a short circuit between the taps through two SCRs.

Voltage Regulation Using Servo System

Voltage regulators using servo systems are quite common. Both single and three-phase types are available. The rating of this type of regulator is quite high and is more economical for high power rating. This regulator normally consist a variac driven by a servomotor, a sensing unit and a voltage and power amplifier to drive the motor in a reversible way. Various types of driving motor may be used for regulating the unit, such as direct current, induction and synchronous motors. However, in all cases, the motor must come to rest rapidly to avoid overrun and hunting. The amount of overrun may be reduced by dynamic braking in the case of a DC motor or by disconnecting the motor from the variac by a clutch as soon as the signal from the measuring unit ceases. The main disadvantage of this type of regulator is the low life of the contact points of the relays.

Phase Controlled Ac Voltage Regulator

Voltage regulators using SCRs are quit common. The load voltage is regulated by controlling the firing instants of the SCRs. There are various circuits for single phase and three phase regulators using SCRs. Though the output voltage can be precisely controlled by this method, the harmonic introduced in the load voltage are quit large and this circuit is used for applications where the output voltage waveform need not be strictly sinusoidal. The circuit arrangement for a single phase SCR regulator is shown in Fig. 6 and Fig. 7.

Ferro-Resonant Ac Voltage Regulator

The concept of the stabilization of AC voltage using a saturated transformer is rather old. The basic circuit arrangement consists of a linear reactor or transformer T1 and a nonlinear saturated reactor or transformer T2 connected in series as shown in Fig. 8. Since the two elements T1 and T2 are in series, the current through them is the same. Transformer T2 is operated under saturated. The voltage division between the two is according to their impedances. Due to nonlinear characteristics of T2 the percentage change of voltage across it is much smaller compared to the percentage change of input voltage. If a suitable voltage proportional to the current is subtracted from the voltage across T2 a practically constant output voltage can be obtained. The circuit arrangement shown is Fig. 8(a) has some drawbacks such as, no load input current is high, and good output voltage stability cab be achieved only at a particular load. Hence some modifications are necessary to improve its performance. The major modification is to place a capacitor across the saturated transformerT2 that is shown in Fig. 8(b).

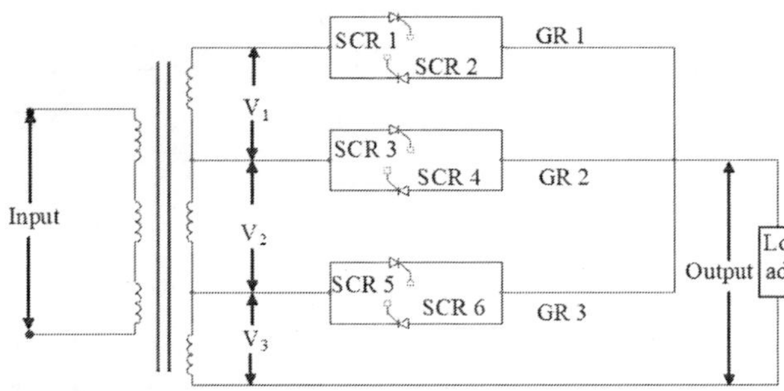

Figure 6. Circuit diagram of solid tap changer using anti-parallel SCRs.

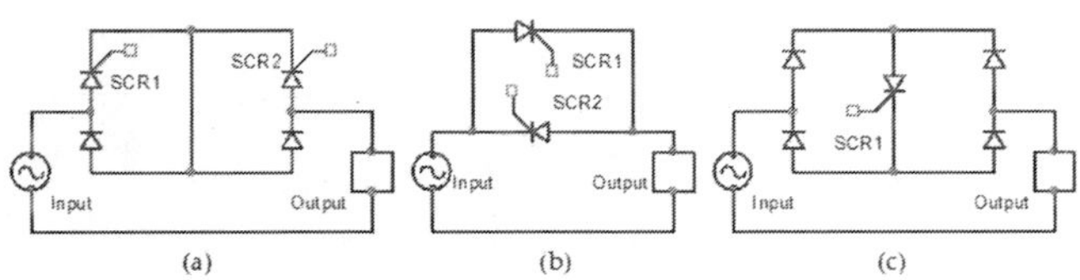

Figure 7. Phase controlled AC voltage regulator, (a) Using back to back SCR and diode and (b) using inverse parallel SCR (c) Using diode-bridge and single SCR.

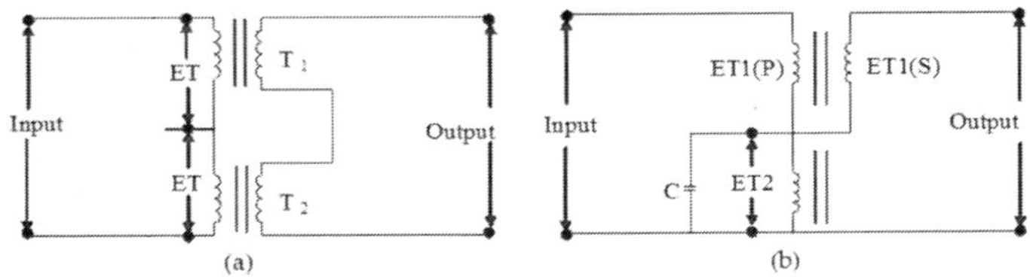

Figure 8. Fero-resonant AC voltage regulator.

The value of the capacitor is such that it resonant with the saturated inductance of T2 at some point. The characteristics of the circuit is such that a small change in voltage across T2 causes the circuit to go out of resonance consequently a large change in input current and power factor. For a change in the input voltage, the change in voltage across the resonant circuit is small but the change in voltage at T1 is large, and by suitable proportioning of the voltage, a good degree of stabilization is achieved for the variation of input voltage as well as load current. The simple Ferro-resonant regulator has the following disadvantages:

a) The output voltage changes with frequency.
b) Since the core operates in saturation and output is derived from the tank circuit, the core volume is large, the core losses are high and external magnetic field is also high.

Switch Mode AC Voltage Regulator

In switch mode AC voltage regulator, the switching devices are continuously switching on and off with high frequency in order to transfer energy from input to output. The high operating frequency results in the smaller size of the switch mode power supplies since the size of power transformer inductors and filter capacitors is inversely proportional to the frequency. The SMPS are more complicated and more expensive, their switching current can cause noise problems, and simple designs can have a poor power factor.

Four common types of switch mode converters are used in DC-DC conversion. They are Buck, Boost, Buck-Boost and C^UK converters. Researches are trying to modify these DC regulators to regulate AC voltages. Buck- Boost and Cûk converter configuration has been investigated for voltage regulation [17-18]. But in every case it is found that the input power factor is very low and the efficiency is poor.

DESIGN AND ANALYSIS OF SWITCHING-MODE AC VOLTAGE REGULATOR

Operation Principle of Switching-Mode AC Voltage Regulator

Operation of Power Circuit

Voltage sag is an important power quality problem, which may affect domestic, industrial and commercial customers. Voltage sags may either decrease or increase in the magnitude of system voltage due to faults or change in loads. Momentary and sustained over voltage and under voltage may cause the equipment to trip out, which is highly undesirable in certain application. In order to maintain the load voltage constant in case of any fluctuation of input voltage or variation of load some regulating device is necessary.

In this chapter the principle of operation of high frequency switching AC voltage regulator, design of its filter circuit and snubber circuit are described. Performance of the regulator is also analyzed using simulation software ORCAD version 9.1. Switch-mode power supplies (SMPS) incorporate power handling electronic components which are continuously switching on and off with high frequency in order to provide the transfer of electric energy from input to output. The design of AC voltage regulator depends on power requirement, degree of stability and efficiency. Solid state AC regulator using phase control technique are not new and are widely used in many application such as heating and lighting control etc. These regulators are not suited for critical loads because the output waveforms are truncated sine waves, which contain large percentage of distortion. The input power factor is low. These drawbacks are largely overcome and the voltage can be efficiently controlled by means of a solid-state AC regulator using PWM technique.

The power circuit of the proposed AC voltage regulator is shown in Fig. 9. The circuit operation can be explained with the help of Fig. 10. During positive half cycle of the input voltage, at mode 1, when switch-1 is on and switch-2 is off, the current passes through diode D1, switch-1, diode D4 and through the inductor and the energy is stored in the inductor. At mode 2, when switch-1 is off and switch-2 is on, the energy stored in the inductor is transferred through diode D8, switch-2 and diode D5. At mode 1, power is transferred from source and at mode 2, power is not transferred from the source, so by controlling the on and off duration of switch-1 output power can be controlled.

During negative half cycle of the input voltage, at mode 1, when switch-1 is on and switch-2 is off the current passes through the inductor, diode D3, switch-1, and diode D2 and the energy is stored in the inductor.

At mode 2, when switch-1 is off and switch-2 is on the energy stored in the inductor is transferred through diode D6, switch-2 and diode D7.

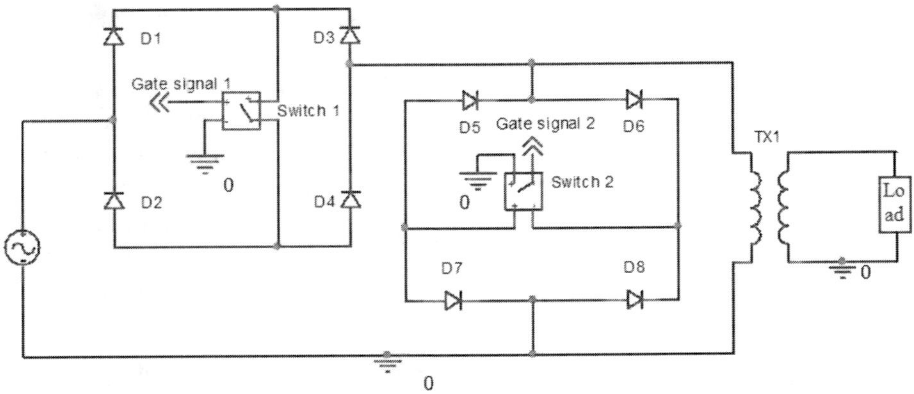

Figure 9. Power circuit of the proposed AC voltage regulator.

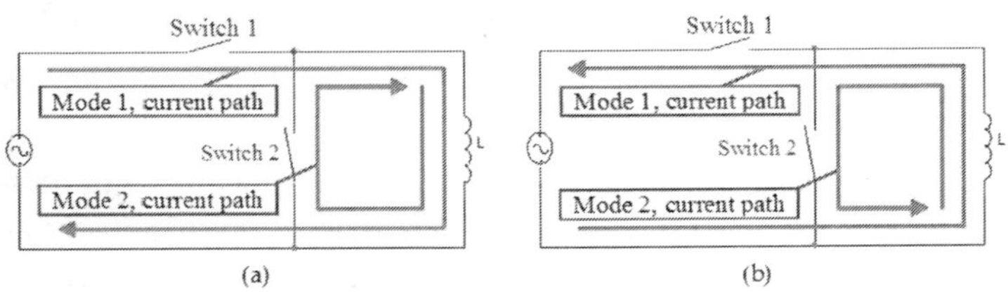

Figure 10. Operation of the power circuit of AC voltage regulator (a) Operation during positive half cycle (b) Operation during negative half cycle.

Operation of Control Circuit

The gate signal generating circuit for a manually controlled AC voltage regulator is shown in Fig. 11. The control circuit incorporates an Operational Amplifier (OPAMP) whose positive input is a variable DC voltage V1, and negative input is a fixed saw-tooth signal V2. In this circuit the OPAMP acts as a comparator, output of the OPAMP depends on the difference of the two inputs. The negative input (saw-tooth wave) is kept constant and positive input (DC voltage) is varied. So output pulse width depends on DC input voltage of OPAMP i.e. when DC input is higher the output of comparator will be wider and when DC input is lower the output of comparator will be narrower.

The outputs of OPAMP are used to turn on/off the switches of the power circuit of the regulator to regulate the output voltage. The output of OPAMP is directly passed through limiter-1 which is the gate signal for switch-1 and after inverting the output of the comparator is passed through the limiter-2 which is the gate signal for switch-2. The function of the limiter is to limit the output of comparator from 0 to 5 V. When switch-1 of the power circuit is on then switch-2 should be off. So the gate signal generating circuit is arranged in such a way that when gate signal of switch-1 is on then gate signal for switch-2 is off and vice versa.

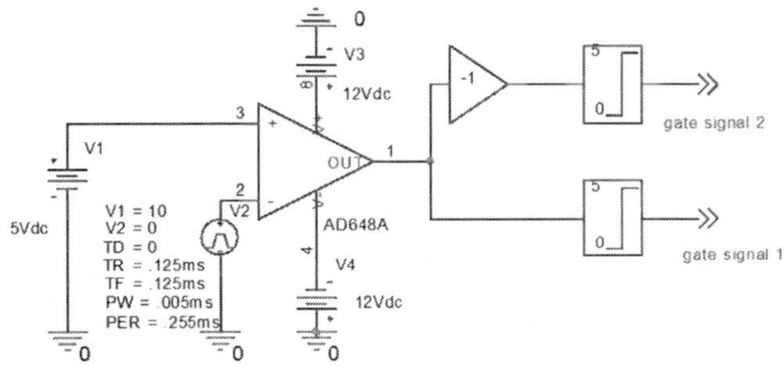

Figure 11. Gate signal generation circuit of manually controlled AC voltage regulator.

Manually Controlled AC Voltage Regulator

When the stability is not very stringent, manually controlled AC voltage regulator is generally preferred from economic considerations.

The basic circuit of a manually controlled AC voltage regulator is shown in Fig. 12. When any change in output voltage occurs due to change in input voltage or change in load, the voltage can be regulated to the desirable value by changing the DC voltage of the gate signal generating circuit manually. The power circuit of the proposed regulator shown in Fig. 12 is implemented using ideal switches; later part of this chapter the ideal switches is replaced by practical switches. The regulator proposed in this chapter is employed to regulate the output voltage to 300V (peak) for variations of input voltage from 200V (peak) to 400V (peak), also for variation of load from 100 ohm to 200 ohm. However, the output voltage can be set to any desirable value according to requirement. The values of all voltages and currents indicated in this chapter are in peak values.

The input current and output voltage waveforms of the manually controlled AC voltage regulator as shown in Fig. 12, is shown in Fig. 13, when the input voltage is 300V and output voltage is also 300V. The spectrum of the input current and output voltage as shown in Fig. 13 is shown in Fig. From the waveforms it is seen that the waveforms are not smooth, sinusoidal and from the spectrum it is seen that due to high frequency switching the significant number and amount of harmonics occur. The switching frequency is selected to 4 KHz. Harmonics occurs at switching frequency and its odd multiple frequencies. So, filters are required at input and output side to filter out these harmonics to get desired sinusoidal waveforms.

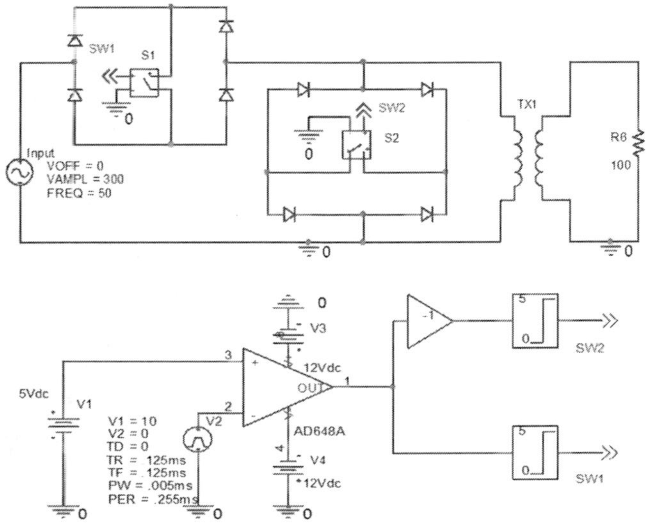

Figure 12. Fundamental circuit of manually controlled AC voltage regulator (ideal switch implementation).

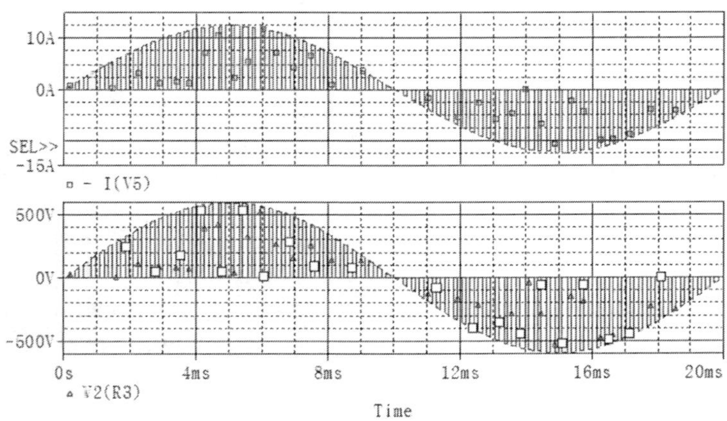

Figure 13. Input current and output voltage waveforms of the regulator shown in Fig. 12. -I(V5): Input current, V(R3:2): Output voltage.

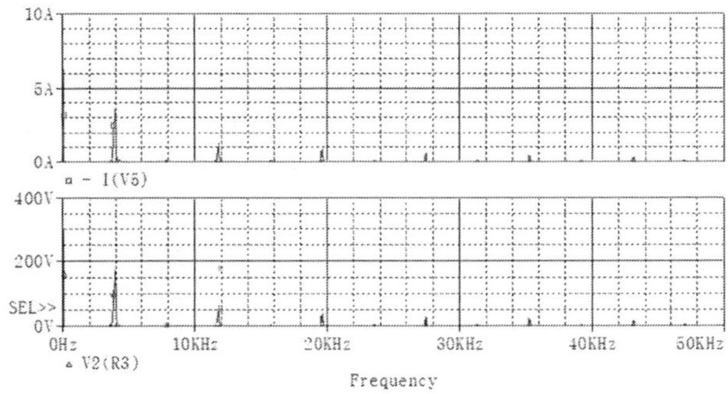

Figure 14. Spectrum of input current and output voltage waveforms. - I(V5):Input current V(R3:2): Output voltage.

Filter Design

Output Filter Design

For getting smooth output voltage, a low pass LC filter of proper L and C value is needed at the output of this regulator. The output filter circuit and the corresponding AC sweep are shown in Fig. 15. From this circuit we can write,

$$V_0 = \frac{V_{in} \times J/\omega C}{J(\omega L - 1/\omega C)} \quad \text{or} \quad \frac{V_0}{V_{in}} = \frac{1}{\omega C(\omega L - 1/\omega C)}$$

The input to the filter is high frequency modulated 50 Hz AC input. The switching signal that modulates the 50 Hz signal is taken to be 4 KHz in this case. So, we will have to make a filter that would pass signal up to 1 KHz (say) and attenuate all other frequencies. This would result a nearly sinusoidal output voltage. In the LC filter section we choose a capacitor of 5μF and determine the value of inductor for a cutoff from AC sweep analysis through ORCAD simulation. We found the value of the inductor to be 30 mH.

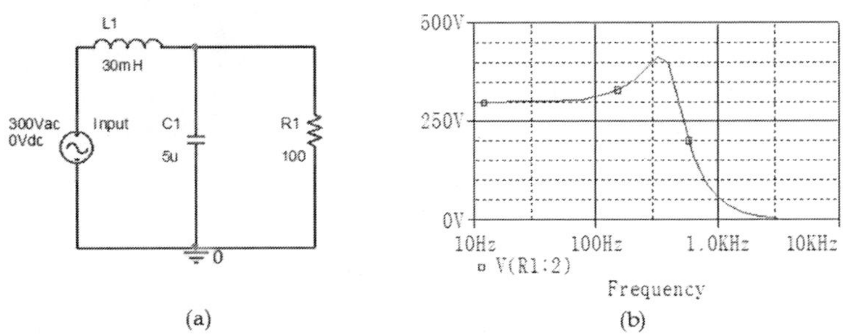

Figure 15. Output voltage filter and AC sweep analysis (a) Output voltage filter (b) AC sweep analysis.

Input Filter Design

A low pass LC filter of proper L and C value is needed at the input of the regulator to filter out some of the harmonics from the supply system. The input filter circuit and the corresponding AC sweep are shown in Fig. 16. Input current contains harmonics at switching frequency 4 KHz and its odd multiple. In order to remove harmonics above 1 KHz, we choose a capacitor of 5μF and determine the value of inductor for a cutoff from AC sweep analysis through OrCAD simulation. We found the value of the inductor to be 30 mH.

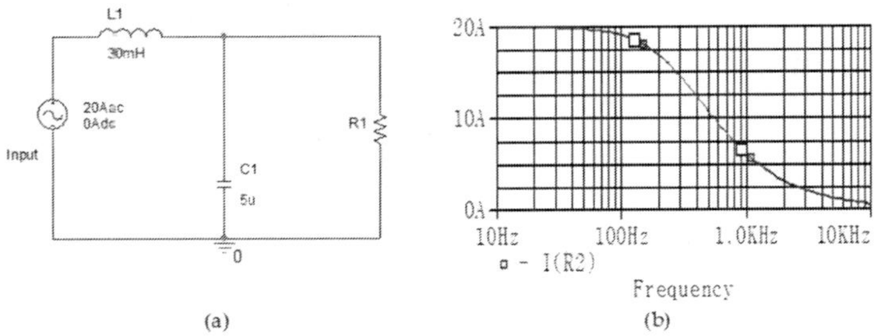

Figure 16. Input current filter and corresponding AC sweep analysis (a) Input current filter (b) AC sweep analysis.

Free Wheeling Path and Surge Voltage across Switching Devices

The power circuit of the proposed regulator with input and output filter is shown in Fig. 17. In an inductor, current does not change instantaneously. When the switches of power circuit switched on and off the current into the inductor of input and output filter are changed abruptly. Abrupt change of current causes a high di/dt resulting high voltage which is equal to Ldi/dt. These voltages appear across the switches as surge. Usually providing freewheeling path in restricts such occurrence.

Surge Voltage across Switches

In the proposed circuit, two switches serve as the freewheeling path for each other. However, for very short period when one switch is turned off and other is turned on, an interval elapses due to delay in the switching time. As a result, freewheeling during this interval is disrupted in the proposed circuit. If the current in any inductive circuit is abruptly disrupted, a high Ldi/dt across the switch appears due to the absence of freewheeling path. High spiky surge voltage appears across the switches during these short intervals as shown in Fig. 18.

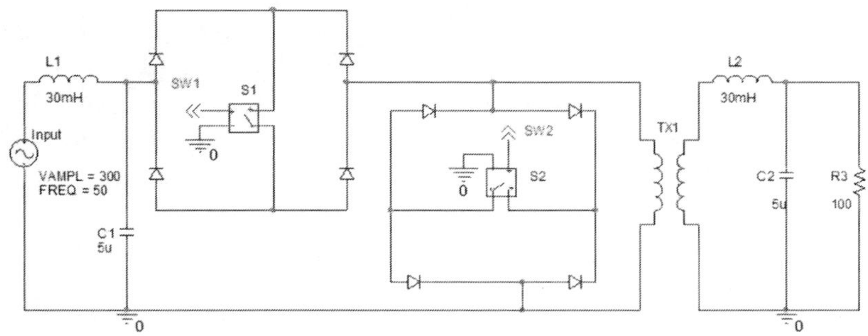

Figure 17. The power circuit of the proposed AC voltage regulator with input and output filters.

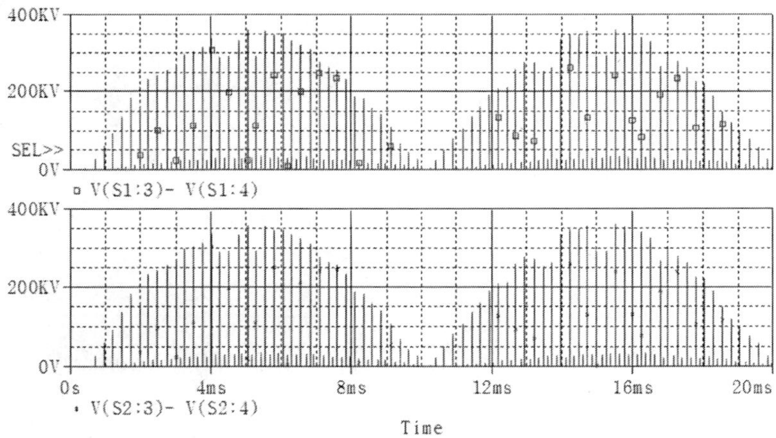

Figure 18. Voltage across switches with filters and without snubbers. V(S1:3)-V(S1:4): Voltage across switch-1, V(S2:3)-V(S2:4): Voltage across switch-2.

These spiky voltages across the switches may be excessively high, about thousand of kilovolt and which may destroy the switches during the operation of the circuit. Remedial measures should be taken to prevent this phenomenon to make the circuit commercially viable. In the proposed circuit RC snubbers are used for suppressing surge voltage across the switches. The power circuit of the proposed regulator with input output filter and snubbers is shown in Fig. 19.

Snubber enhances the performance of the switching circuits and results in higher reliability, higher efficiency, higher switching frequency, smaller size and lower EMI. The basic intent of a snubber is to absorb energy from the reactive elements in the circuit. The benefits of this may include circuit damping, controlling the rate of change of voltage or current or clamping voltage overshoot. The waveforms of voltages across switches with input output filters and snubbers are shown in Fig. 20.

Use of snubbers reduces the spiky voltage across the switches to a tolerate limit for practical application of the AC voltage regulator.

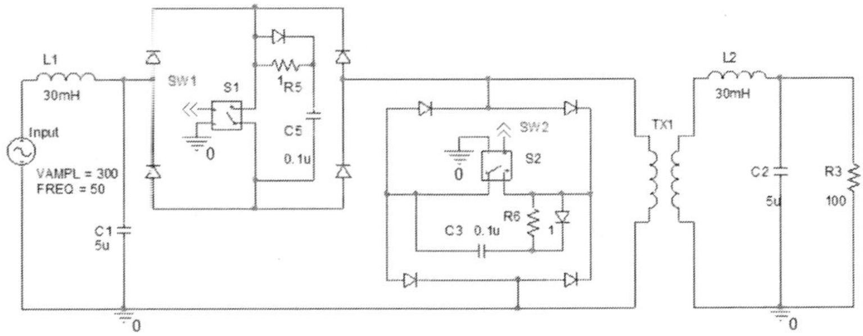

Figure 19. The power circuit of the proposed AC voltage regulator with input output filters and snubbers.

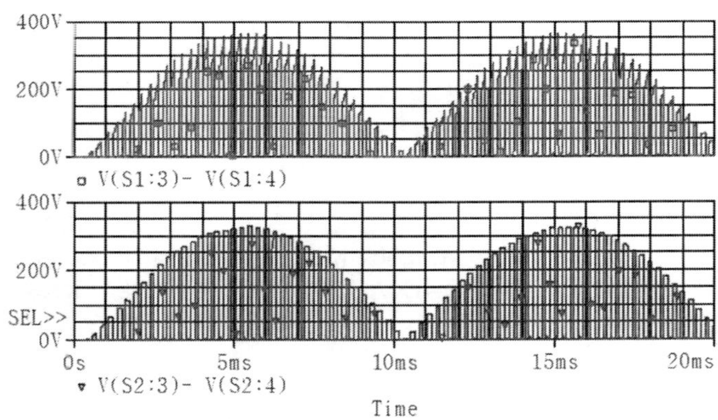

Figure 20. Voltage across switches with filters and snubbers. V(S1:3)-V(S1:4): Voltage across switch-1, V(S2:3)-V(S2:4): Voltage across switch-2.

Proposed AC Voltage Regulator with Practical Switches

In the previous section, we have studied the regulator using ideal S-break switches which have been operated by the pulses from the limiter. But for practical application, real switches are essential which are to be controlled by the pulses having ground isolation. The proposed AC voltage regulator circuit with practical switches is shown in Fig. 21. The ideal S-break switch is replaced by IGBT.

In the proposed regulator chip SG1524B is used to control the gate signal. Signal from the chip is fed to the Limiter and finally to the optocoupler. The output of the optocoupler is used to control the on off time of the IGBTs. The function of the Limiter is to limit the output voltage of the gate signal generating IC from 0 to 6 volts. Optocoupler is used to generate signaling voltage with ground isolation.

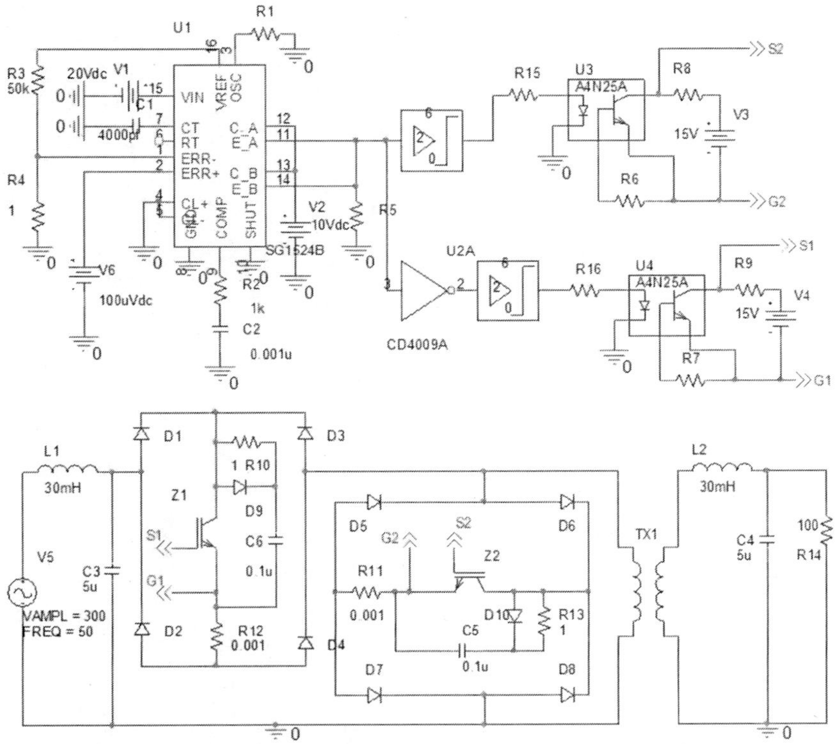

Figure 21. Manually controlled AC voltage regulator circuit with practical switches.

Chip Sg1524b for Generation of Gate Signal

The Block diagram of the internal circuitry of the chip SG1524B is shown in Fig. 22. By controlling the error signals of the error amplifier the duty cycle of the gate signal to the regulator can be controlled. Thus it is a very suitable device for using in the regulator circuits.

Results of Proposed AC Voltage Regulator (Practical Switch Implementation

The waveforms of the input and output voltages of the proposed regulator are shown in Fig. 23 and Fig. 24. Fig. 23 shows the input and output voltages waveform when the input voltage is 200V and output voltage is 300V. Fig. 24 shows the input and output voltages waveform when the input voltage is 400V and output voltage is 300V. Fig. 25 and Fig. 26 show the input and output current waveforms corresponding to Fig. 23 and Fig. 24.

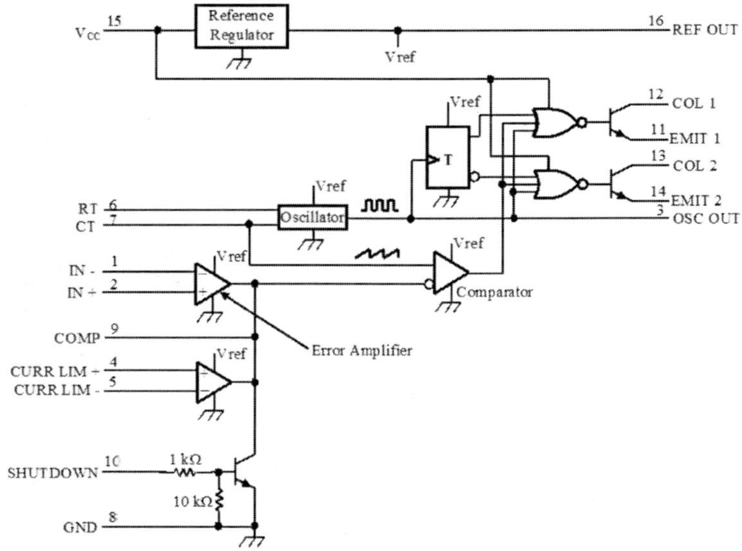

Figure 22. Block diagram of IC chip SG1524B.

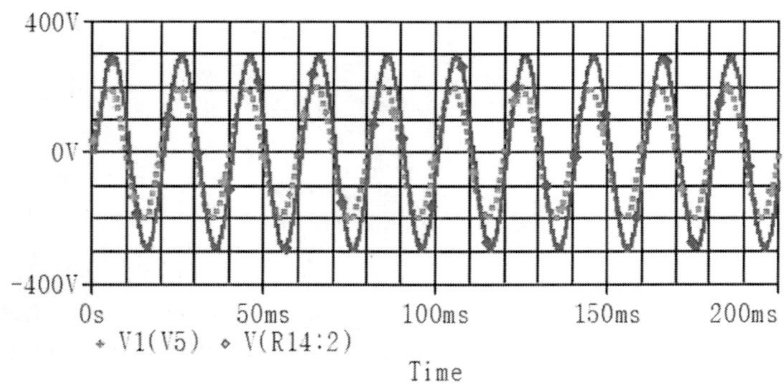

Figure 23. Input and output voltage waveforms, Input 200V output 300V. V1(V5): Input voltage – dotted line, V(R14:2): Output voltage – solid line.

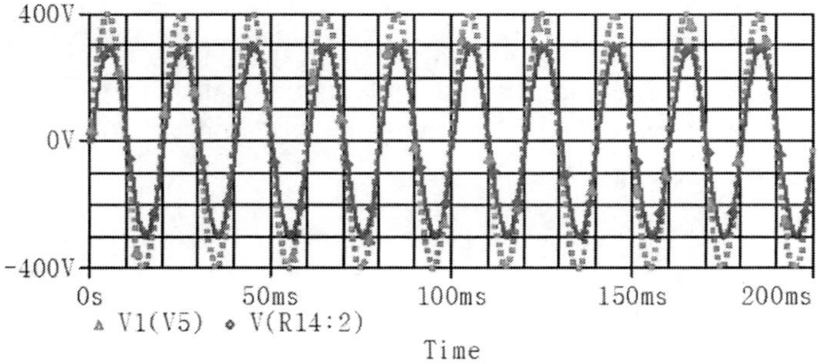

Figure 24. Input and output voltage waveforms, Input 400V output 300V. V1(V5): Input voltage – dotted line, V(R14:2): Output voltage – solid line.

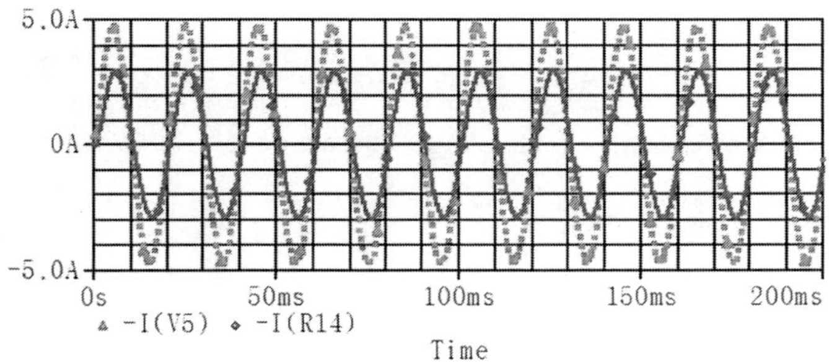

Figure 25. Input and output current waveforms for input 200V output 300V. -I(V5): Input current - dotted line, -I(R14): Output current – solid line.

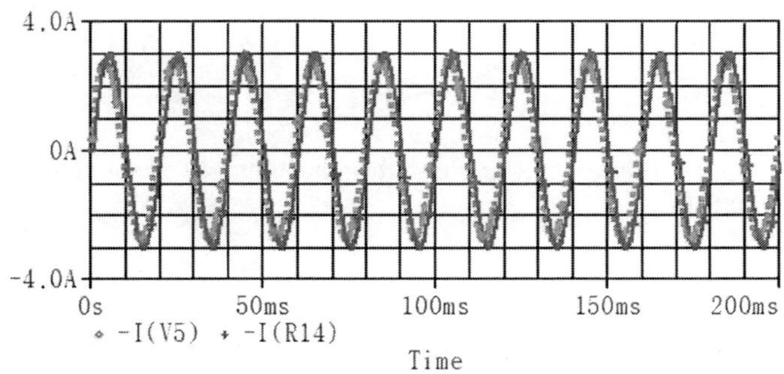

Figure 26. Input and output current waveforms for input 400V output 300V. -I(V5): Input current – dotted line, -I(R14): Output current – solid line.

From the waveforms shown in Fig. 23 to Fig. 26, it is seen that the waveforms of output voltage and input current is perfectly sinusoidal. The variation of output voltage of the proposed regulator with the duty cycle is shown in Fig. 27. The value of input voltage is kept constant to 300V. From Fig. 23 it is seen that the variation of output voltage with duty cycle is almost linear. The variation of duty cycle with the variation of input voltage from 200V to 400V to maintain the output voltage constant to 300V is shown in Fig. 28.

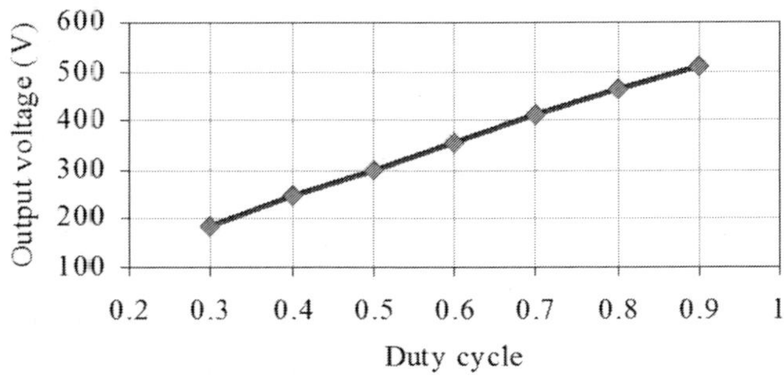

Figure 27. Variation of output voltage with duty cycle. Input voltage is 300V.

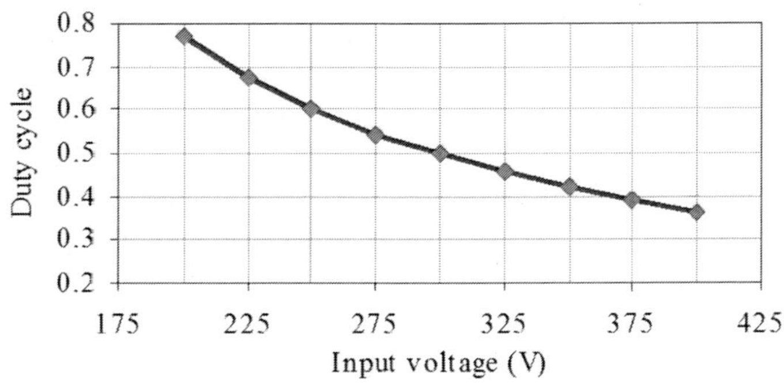

Figure 28. Variation of duty cycle with input voltage to maintain output voltage constant to 300V.

AUTOMATIC CONTROLLED AC VOLTAGE REGULATOR

In manually controlled AC voltage regulator control, the output voltage is sensed with a voltmeter connected at the output; the decision and correcting operation is made by a human judgment. The manual control may not be feasible always due to various factors. In automatic voltage regulators, all functions are performed by instruction, and give much better performance, so far as stability, speed of correction, consistency, fatigue, etc. are concerned.

There are two types of automatic control voltage regulator, discontinuous control and continuous control. The automatic control system consists of a sensing or measuring unit and a power control or regulating unit. The sensing unit compares the output voltage or the controlled variable with a steady reference and gives an output proportional to their difference called the error signal. The error voltage is amplified, integrated or differentiated or modified whenever necessary. The processed error voltage is fed to the main control unit to have required corrective action.

In the discontinuous type of control, the measuring unit is such as to produce no signal as long as the voltage is within certain limits. When the voltage goes outside this limit, a signal is produced by the measuring unit until the voltage is again brought within this limit. In this type of measuring or sensing unit, the correcting voltage is independent of percentage of error. When the voltage is brought back to this limit, the signal from the measuring unit is zero and the regulating unit remains at its new position until another signal is received from the measuring unit.

In continuous control, the measuring unit produces a signal with amplitude proportional to the difference between the fixed reference and the controlled voltage. The output of the measuring unit is zero when the controlled voltage or a fraction of it is equal to the reference voltage. The regulating or the controlling unit, which is associated with the continuous measuring unit, gives a correcting voltage proportional to the output of the measuring unit. The principle of operation of a continuous control AC voltage regulator is described in this section.

Control and Gate Signal Generating Circuit for Controlled AC Voltage Regulator

Figure 29 shows the circuit of the proposed automatic controlled AC voltage regulator including the control and gate signal generating circuit. A fraction of the output voltage after capacitor voltage dividing and rectifying is passed through an OPAMP buffer. Buffer is used to remove the loading effect. Output voltage of the buffer is same as its input voltage. The output voltage of the buffer is further reduces using resistive voltage divider and taken as the negative input of the error amplifier of the PWM voltage regulating IC SG1524B.

The positive input of the error amplifier is taken from the reference voltage of the chip, after voltage dividing using 50K and 1 ohm resistance. The positive input of the error amplifier is fixed and the negative input is error signal which will vary according to the output voltage. Since the error signal is applied to the negative input of the error amplifier, the duty cycle will be increased if the error signal is decreased and vice versa.

When the output voltage increases above the set value which is 300V either due to change in input voltage or load, the error signal will be increased, therefore the duty cycle will decrease. As a result less power will be transferred from the input to output, and output voltage start to decrease until it reaches to the set value.

When the output voltage decreases below the set value either due to change in input voltage or load then error signal will be decreased which will increase the duty cycle. As a result, more power will be transferred from the input to output, and output voltage start to increase until it reaches to the set value.

When the output voltage is same as the set value than the negative and positive input of the error amplifier will be same as a result the duty cycle will remain same and output voltage will remain unchanged. In this way the proposed regulator will maintain output voltage constant, irrespective of the variation of input voltage and load.

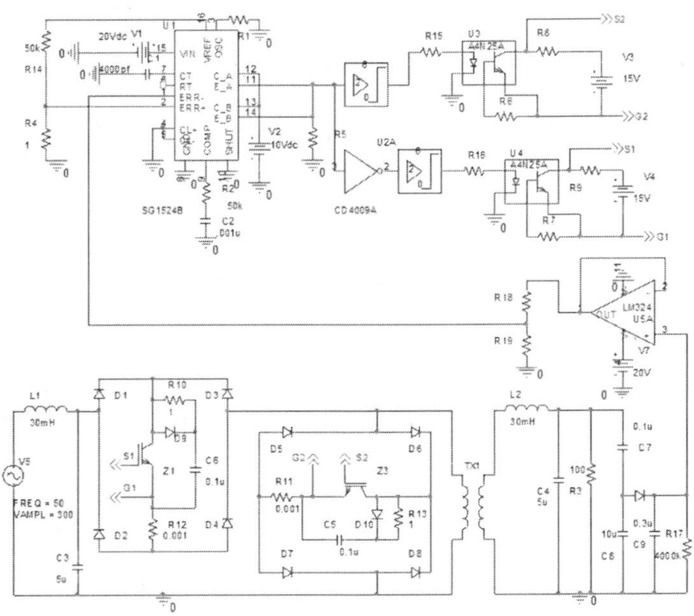

Figure 29. Automatic controlled AC voltage regulator circuit with practical switches.

Results of Automatic Controlled AC Voltage Regulator

Figure 30 shows the input and output voltage waveforms of the proposed automatic controlled AC voltage regulator when the input voltage is 250V and output voltage is 300V. Figure 31 shows the input and output voltage waveforms of the proposed regulator when the input voltage is 350V and output voltage is 300V. Figure 32 and Fig. 33 shows the waveforms of the input current and output currents corresponding to the waveforms of Fig. 30 and Fig. 31 for a load of 100V

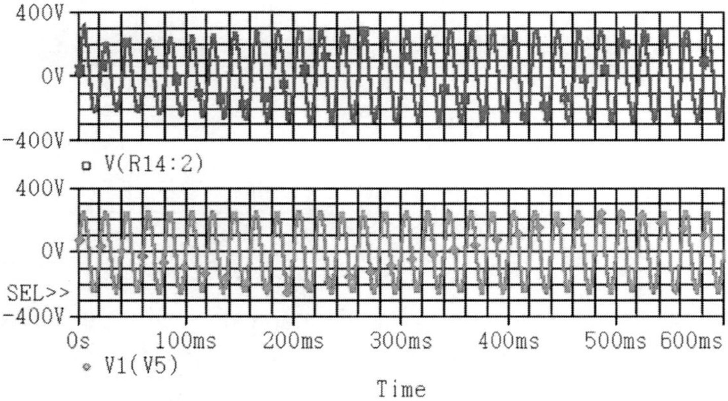

Figure 30. Input and output voltage waveforms for input 250V and output 300V. V1(V5): Input voltage- bottom figure, V(R14:2): Output voltage – top figure.

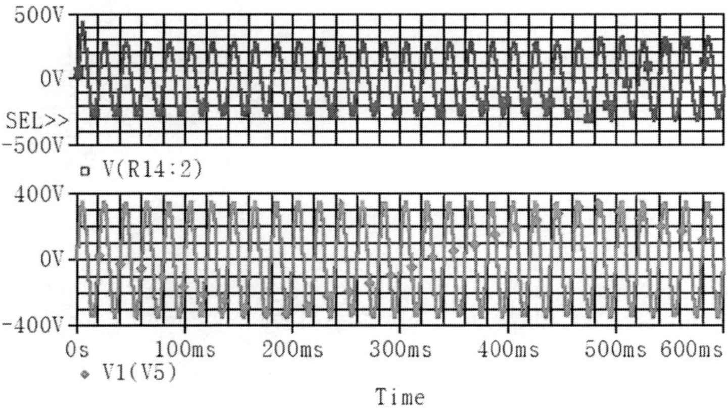

Figure 31. Input and output voltage waveforms for input 350V and output 300V. V1(V5): Input voltage-bottom figure, V(R14:2): Output voltage – top figure.

Table 1 summarizes the result of the proposed regulator to regulate output voltage to 300V for variation of input voltage from 200V to 350V and load from 100 ohm to 200 ohm. In this table input current, output current, input power factor, and efficiency of the regulator are also provided. The proposed regulator can regulate the output voltage effectively, for a wide variation of input voltage and load with efficiency of more than 90% and input power factor more than 0.9.

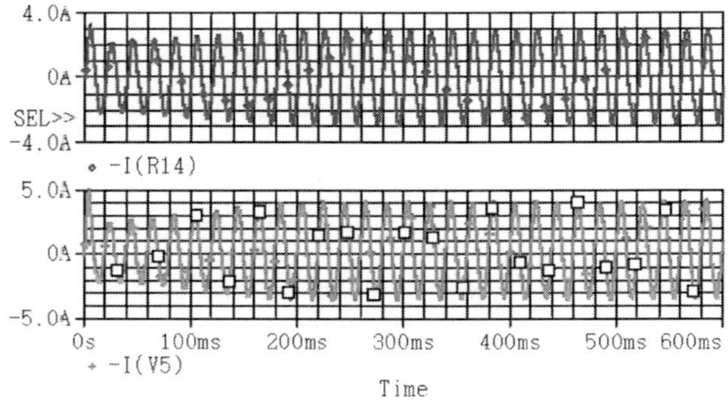

Figure 32. Input and output current waveforms for input 250V output 300V. -I(V5): Input current – bottom figure, -I(R14): Output current – top figure.

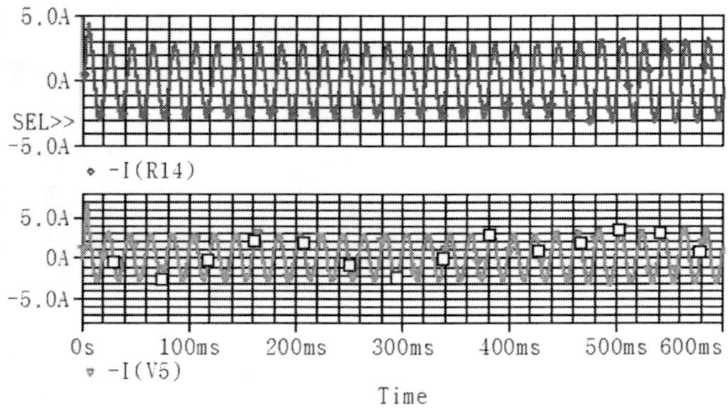

Figure 33. Input and output current waveforms for input 350V output 300V. -I(V5): Input current – bottom figure, -I(R14): Output current – top figure.

Table 1. Results of proposed automatic controlled AC voltage regulator for maintaining output 300 V

V_{in}(V)	I in (A)	Input pf	P_{in}(W)	V_{out}(V)	Load (Ω)	I_{out}(A)	P_{out}(W)	Efficiency (%)
200	4.81	1	481	295	100	2.95	435.13	90.46
225	4.3	1	483.09	298	100	2.98	444.02	91.91
250	3.92	1	489.7	300	100	3	450	91.89
275	3.6	1	495	300	100	3	450	90.91
300	3.3	1	493.79	300	100	3	450	91.13
325	3.1	0.99	498.85	302	100	3.02	456.02	91.41
350	2.95	0.98	508.41	305	100	3.05	465.13	91.49
250	2.07	0.96	248.73	300	200	1.5	225	90.46
275	1.9	0.95	248.46	300	200	1.5	225	90.56
300	1.75	0.95	248.2	300	200	1.5	225	90.65
325	1.68	0.93	253.12	302	200	1.51	228.01	90.08
350	1.6	0.91	253.77	305	200	1.53	232.56	91.64

CONCLUSION

An essential feature of efficient electronic power processing is the use of semiconductors devices in switch mode to control the transfer of energy from source to load through the use of pulse width modulation techniques. Inductive and capacitive energy storage elements are used to smooth the flow of energy while keeping losses at a lower level. As the frequency of the switching increases, the size of the capacitive and inductive elements decreases in a direct proportion. Because of the superior performance, the SMPS are replacing conventional linear power supplies.

In this chapter the design and analysis of an AC voltage regulator operated in switch mode is described in details. AC voltage regulator is used to maintain output voltage constant either for an input voltage variation or load variation to improve the power quality. If the output voltage remains constant, equipment life time increases and outages and maintenance are reduced.

At first the regulator is analyzed using ideal switches, then the ideal switches is replaced by practical switches which required isolated gate signal. The procedure of smoothing the input current and output voltage, and suppressing the surge voltage across the switches is described. A manually controlled AC voltage regulator is analyzed then the concept of operation of an automatic controlled AC voltage regulator is described. Finally an automatic controlled AC voltage regulator is designed and its performance is analyzed.

The proposed regulator can maintain the output voltage constant to 300V, when input voltage is vary from 200V to 350V also for variation of load. To maintain constant output voltage PWM control is used. By varying the duty cycle of the control circuit have achieved the goal of maintaining the constant output voltage across load. For generation of gate signal of the switches an IC chip SG1524B is used which is compact and commercially available at a very low cost. The input current of the proposed regulator is sinusoidal and the input power factor is above 0.9. From simulation results it is seen that the efficiency of the proposed regulator is more than 90%.

REFERENCE

1. P. B Steciuk, and J. R Redmon, Voltage sag analysis peaks customer serviceIEEE Comput. Appl. Power, 94851Oct. 1996
2. M. F Mcgranaghan, D. R Mueller, and M. J Samotyj, Voltage sags
3. M. H. J Bollen, The influence of motor reacceleration on voltage sagsIEEE Transactions on Industrial Application, 31667674July/Aug. 1995
4. M. H. J Bollen, Characterization of voltage sags experienced by three phase adjustable-speed drivesIEEE Transactions on Power Delivery, 1216661671Oct. 1997
5. H. G Sarmiento, and E Estrada, A voltage sag study in an industry with adjustable speed drivesIEEE Industry Applications Magazine21619Jan./Feb. 1996
6. N Kutkut, R Schneider, T Grand, and D Divan, AC voltage regulation technologies," Power Quality Assurance, 9297July/Aug. 1997
7. D Divan, P Sulherland, and T Grant, Dynamic sag corrector: A new concept in power conditioning," Power Quality Assurance, 4248Sept./Oct. 1998
8. AElnadt, and M. A Magdy, Salama, "Unified approach for mitigating voltage sag and voltage flicker using the DSTATCOM," IEEE Transactions on Power Delivery, 302April 2005
9. S. M Hietpas, and R Pecan, Simulation of a three-phase boost converter to compensate for voltage sags, "inProceeding. IEEE 1998Rural Electric Power Conference, B4B1B4-7, Apr. 1998.
10. S. M Hietpas, and Mark Naden, "Automatic voltage regulator using an AC voltage-voltage converter," IEEE Transactions on Industrial Application, 363338Jan.\Feb. 2000
11. F. Z Peng, Lihue Chen, and Fan Zhang, "Simple topologies of PWM AC-AC convertersIEEE Power Electronics Letters11March 2003

12. G Venkataramanan, B. K Johnson, and A Sundaram, An AC-AC power converter for custom power applicationsIEEE Transactions on Power Delivery1116661671July 1996
13. V Nazquez, A Velazquez, C Hernandez, E Rodríguez, and R Orosco, A Fast, AC Voltage Regulator,", CIEP 2008. 11th IEEE International Power Electronics Congress, 162166Aug. 2008
14. J Nan, T Hou-jun, L Wei, and Y Peng-sheng, Analysis and control of Buck-Boost Chopper type AC voltage regulator,", IPEC'09. IEEE 6th International Power Electronics and Motion Control Conference, 10191023May 2009
15. N. A Ahmed, M Miyatake, H. W Lee, and M Nakaoka, A Novel Circuit Topology of Three-Phase Direct AC-AC PWM Voltage Regulator" Industry Applications Conference, 2006st IAS Annual Meeting. Conference Record of the 2006 IEEE, 20762081Oct. 2006.
16. V Nazquez, A Velazquez, and C Hernandez, AC Voltage Regulator Based on the AC-AC Buck-Boost Converter," ISIE 2007IEEE International Symposium on Industrial Electronics, 533537Jun. 2007.
17. P. K Banerjee, Power line voltage regulation by PWM AC Buck-Boost voltage controller," A M.Sc. thesis, Department of EEE, BUET, July, 2002
18. A Hossain, AC voltage regulation by Cûk switch mode power supply," A M.Sc. thesis, Department of EEE, BUET, July, 2003
19. B. H Li, S. S Choi, D. M Vilathgamuwa, Design considerations on the line-side filter used in the dynamic voltage restorerIEE Proceedings- Generation, Transmission, and Distribution, 172001
20. Wang JingXuAiqin, ShenYueyue., A Survey on Control Strategies of Dynamic Voltage Restorer, 13th International Conference on Harmonics and Quality of Power (ICHQP), 15Sept. 28 2008-Oct. 1, 2008
21. J. G Nielsen, F Blaabjerg, A Detailed Comparison of System Topologies for Dynamic Voltage RestorersIEEE Transactions on Industry Applications415127212802005
22. OrCAD SoftwareRelease 9198519991999OrCAD, Inc., U.S.A.
23. Slobodan CûkBasics of Switched Mode Power Conversion Topologies, Magnetics, and Control," Modern Power Electronics: Evaluation, Technology, and applications, Edited by B.K. Bose, IEEE Press, 2652961992
24. M. H Rashid, Power Electronics- Circuits, Devices, and Applications, "Prenctice Hall India, Second Edition, 2000317387
25. P. C Sen, Power ElectronicsTata McGraw-Hill Publishing Company Ltd., India, 1987588614
26. R Thompson, A Thyristor Alternating Voltage RegulatorIEEE Trans. on Ind. and Gen. Application, vol. IGA 4 (1968) 2, 1621661968
27. E. J Cham, and W. R Roberts, Current Regulators for Large Rectifier Power Supplies Used on Electrochemical Processing LinesIEEE Trans. on Ind. and Gen. Application IGA-4 (1968609618

28. J. M Jr, Mealing, "A Coherent Approach to the Design of Switching Mode DC Regulators," IEEE Conf. Rec. IGA, 177185Oct. 1967
29. Unitrode, "Switching Regulated Power Supply Design Seminar Manual," Unitrode Corporation, U. S. A, 1986.
30. M. H Rashid, A Thyristor Chopper With Minimum Limits on Voltage Control of DC DrivesInternational Jurnal of Electronics, 53171811982
31. K. P Severns, and G. E Bloom, Modern DC-to-DC Switch Mode Power Converter Circuits," Van NostrandReinholdd Company, Inc., New York, U. S. A, 1983
32. S Cuk, Survey of Switched Mode Power Supplies," IEEE International Conference on Power Electronics and Variable Speed Drives, London, 83941985
33. M Ehsani, R. L Kustom, and R. E Fuja, Microprocessor Control of A Current Source DC-DC Converter," IEEE Transactions on Industrial Applications, LA1956906981983
34. R. D Middlebrook, A Continuous Model for the Tapped-Inductor Boost Converter," IEEE Power Electronics Specialists Conference Record, Culver City, CA, U. S. A, 63791975
35. Slobodan Cûk RD. Middlebrook, "A General Unified Approach to Modeling Switching DC-to-DC Converters in Discontinuous Conduction Mode," IEEE Power Electronics Specialists Conference Record, Palo Alto, CA, U. S. A, 36571977
36. R. D Middlebrook, and Slobodan Cûk, "Modeling and Analysis Methods for DC-to-DC Switching Converters," IEEE International Semiconductor Power Converter Conference Record, Lake Buena Vista, FL, U. S. A, 901111977
37. Slobodan Cûk and RD. Middlebrook, "Coupled Inductor and other Extensions of a New Optimum Topology Switching DC-to-DC Converter," IEEE Industry Applications Society Annual Meeting, Record, Los Angeles, CA, U. S. A, 111011261977
38. R. D Middlebrook, and Slobodan Cûk, "Isolation and Multiple Output Extensions of a New Optimum Topology Switching DC-to-DC Converter," IEEE Power Electronics Specialists Conference Record, Syracuse, NY, U. S. A, 2562641978
39. C Chen, and M Deepakraj, Divan, "Simple Topologies for Single Phase AC Line ConditioningIEEE Transactions on Industry Applications302406412March/April 1994
40. F Mark, McGranaghan, David R. Mueller and Marek J. Samotyj, "Voltage Sags in Industrial Systems," IEEE Transactions on Industry Applications, 292397403March/April 1993
41. S Cuk, and R. D Middlebrook, Advances in Switched Mode Power Conversion," IEEE Transactions on Industrial Electronics, Vol. IE 30. 110291983

42. N Mohan, M Tore, Undeland and William P.Robbins, "Power Electronics-Converters, Applications, and Design, " John Wiley and Sons Inc., Second ed., 1611951995
43. H Veffer, High Current, Low Inductance GTO and IGBT Snubber Capacitors," Siemens Components, June, 81851990
44. B. D Bedford, and R. G Hoft, Principles of Inverter CircuitsWiley: 1964
45. E. R Hnatek, Design of Solid-State Power SuppliesVan Nostrand Reinhold; 1971
46. A. I Pressman, Switching and Linear Power Supply," Power Converter Design, vol. I and vol. II, Hayden; 1977

CITATION

Raju Ahmed and Mohammad Jahangir Alam (2013). Power Quality Improvement Using Switch Mode Regulator, An Update on Power Quality, Dr. Dylan Lu (Ed.), ISBN: 978-953-51-1079-8, InTech, DOI: 10.5772/53234

CHAPTER 5

Wind Turbine Power: The Betz Limit and Beyond

Mahmoud Huleihil[1, 2] *and Gedalya Mazor*[1]

[1]Department of Mechanical Engineering, Sami Shamoon College of Engineering, Be'erSheva, Israel,

[2]Arab Academic Institute for Education, BeitBerl College, Kfar Sava, Israel

INTRODUCTION

With a severe energy crisis facing the modern world, the production and utilization of energy has become a vital issue, and the conservation of energy has acquired prime importance. Energy production and consumption are directly related to everyday life in much of human society, and issues of energy research are extremely important and highly sensitive. Being aware of the global warming problem, humans tend to rely more on renewable energy (RE) resources.

Benefits of Wind Energy

In [1], scientists and researchers have tried to accelerate solutions for wind energy generation design parameters. Researchers claim that a short time, society, industry, and politics will welcome the use of wind energy as a clean, practical, economical, and environmentally friendly alternative. In an effort to approach a more sustainable world, after the 1973 oil crisis RE sources began to appear on the agenda, and wind energy attracted significant interest. Because of extensive studies on this topic, wind energy has recently been applied in various industries, where it has begun to compete with other energy resources [1].

Among the various renewable energy types as highlighted by [2], wind provides an intermittent but environmentally friendly energy source that does not pollute atmosphere. Wind power calculations are initiated from the kinetic energy definition, and wind power is found to be proportional to half the air density multiplied by the cube of the wind velocity. When seeking to determine the potential usage of wind energy, wind power formulation is derived first by use of kinetic energy definition and then by basic physical definitions of power as the ratio of work over time, work as the force multiplied by the distance, and force as the change of momentum. [2].

Aerodynamics Aspects of Wind Turbines

Reviews about many of the most important aerodynamic research topics in the field of wind energy are shown in the report of a different study [3] Wind turbine aerodynamics concerns the modeling and prediction of aerodynamic forces, such as performance predictions of wind farms, as well as the design of specific parts of wind turbines, such as rotor-blade geometry. The basics of blade-element momentum theory were presented along with guidelines for the construction of airfoil data. Various theories for aerodynamically optimum rotors were discussed, and recent results on classical models were presented. State-of-the-art advanced numerical simulation tools for wind turbine rotors and wakes were reviewed, including rotor predictions as well as models for simulating wind turbine wakes and flows in wind farms [3].

Wind Power Density

Concerning power density and its relation to wind speed, the report given in [4] presented the features of wind power distributions that were analytically obtained from wind distribution functions. Simple equations establishing a relationship between mean power density and wind speed have been obtained for a given location and wind turbine. Different concepts relating to wind power distribution functions were shown—among them the power transported by the wind and the theoretical maximum convertible power from wind, according to the Betz' law. Maximum convertible power from the wind was explained within more realistic limits, including an approximate limit to the maximum power from a wind turbine, was obtained. In addition, different equations were obtained establishing relationships between mean power density and mean wind speed. These equations are simple and useful when discarding locations for wind turbine installation [4].

Wind Power Applications

The range of wind power usage is scarce. One of the most important usages is electricity. Hubbard and Shepherd [5] considered wind turbine generators, ranging in size from a few kilowatts to several megawatts, for producing electricity both singly and in wind power stations that encompass hundreds of machines. According to the researchers' claims, there are many installations in uninhabited areas far from established residences, and therefore there are no apparent environmental impacts in terms of noise. The researchers do point out, however, situations in which radiated noise can be heard by residents of adjacent neighborhoods, particularly those who live in neighborhoods with low ambient noise levels [5].

Wind power is used worldwide, not only in developed countries. Specific studies [6, 7] presented a detailed study of a Manchegan windmill while considering the technological conditions of the original Manchegan windmills. In addition, a wind evaluation of the region was carried out, the power and momentum of the windmills were calculated, and the results obtained were discussed, along with a comparison with the type of Southern Spanish windmill. These windmills were important for wheat milling and had been an important factor in the socio-economic development of rural Spain for centuries [6, 7].

Another example is considered in [8]. This study, conducted with reference to land in Syria, evaluated both wind energy potential and the electricity that could be generated by the wind. An appropriate computer program was especially prepared and designed to perform the required calculations, using the available meteorological data provided by the Syrian Atlas. The program is capable of processing the wind data for any specific area that is in accordance with the needed requirements in fields of researches and applications. Calculations in the study show that a significant energy potential is available for direct exploitation. The study also shows that approximately twice the current electricity consumption in Syria can be generated by wind resources [8].

The potential usage of wind power at Kudat and Labuan for small-scale energy demand was given in [9]. According to their statement, the acquisition of detailed knowledge about wind characteristics at a site is a crucial step in planning and estimating performance for a wind energy project. From this study, the researchers concluded that sites at Kudat and Labuan that they had considered during the study years were unsuitable for large-scale wind energy generation. However, they did confirm that small-scale wind energy could be generated at a turbine height of 100 meters [9]. In light of their findings, James and others [10] reported that the potential impact of the UK's latest policy instrument, the 2010 micro-generation tariffs, is considered applicable to both micro-wind and photovoltaics.As the researchers observed, building-mounted micro-wind turbines and photovoltaics have the potential to provide widely applicable carbon-free electricity generation at the building level. Because photovoltaic systems are well understood it is easy to predict performance using software tools or widely accepted yield estimates. Micro-wind research, however, is far more complex, and in comparison, it is poorly understood [10].

Abdeen [11] addresses another example of wind power usage. As the researcher observed, the imminent exhaustion of fossil energy resources and the increasing demand for energy were the motives for Sudanese authorities to put into practice an energy policy based on rational use of energy. The authorities also based their conclusions on exploitation of new and renewable energy sources. It was pointed out that after 1980, as the supply of conventional energy has not been able to follow the tremendous increase in production demand in rural areas of Sudan; a renewed interest for the application of wind energy has been shown in many places. Therefore, the Sudanese government began to pay more attention to wind energy utilization in rural areas. Because the wind energy resource in many rural areas is sufficient for attractive application of wind pumps, although as fuel it is insufficient, the wind pumps will be spread on a rather large scale in the near future. Wind is a form of renewable energy that is always in a non-steady state due to the wide temporal and spatial variations of wind velocity. Results suggested that wind power would be more profitably used for local and small-scale applications, especially for remote rural areas. The study finds that Sudan has abundant wind energy [11]. Another recent study [12] considered the wind power in Iran. According to the study's claims, climate change, global warming, and the recent worldwide economic crisis have emphasized the need for low carbon emissions while also ensuring economic feasibility. In their paper, the researchers investigated the status and wind power potential of the city of Shahrbabak in Kerman province in Iran. The technical and economic feasibility of wind turbine installation was presented, and the potential of wind power generation was statistically analyzed [12].

Types of Wind Turbines

There are different types of wind turbines: bare wind turbines, augmented wind turbines, horizontal axis wind turbines, and vertical axis wind turbines, just to mention a few.

Bare Wind Turbines

According to research findings as given by [13], the derivation of the efficiency of an ideal wind turbine is attributed to the three prominent scientists associated with the three principal aerodynamic research schools in Europe during the first decades of the previous century: Lanchester, Betz, and Joukowsky. According to this study, detailed reading of their classical papers had shown that Lanchester did not accept that the velocity through the disc is the average of the velocities far upstream and far downstream, by which his solution is not determined. Betz and Joukowsky used vortex theory to support Froude's result and derived the ideal efficiency of a wind turbine at the same time. This efficiency has been known as the Joukowsky limit in Russia and as the Betz limit everywhere else. As the researchers suggested, because of the contribution of both scientists, this result should be called the Betz-Joukowsky limit everywhere [13]. The maximal achievable efficiency of a wind turbine is found to be given by the Betz number $B = 16/27$. Derivation of the classical Betz limit could be followed as given by [14] and [15].

The question of the maximum wind kinetic energy that can be utilized by a wind turbine, which is of fundamental importance for employment of wind energy, was reconsidered in [16]. According to their study, the researchers observed that in previous studies, an answer to this question was obtained only for the case of an infinite number of turbine-rotor blades, in the framework of application of the one-dimensional theory of an ideal loaded disk without loss for friction and turbulence taken into account. This implies that for an ideal wind turbine, the maximum energy that can be extracted from the wind kinetic energy, or the power coefficient, does not exceed the Betz limit. Based on the exact calculation of the Goldstein function, the researchers determined the maximum power coefficient of an ideal wind turbine having a finite number of blades. As was expected, the maximum turned out to be always lower than the absolute Lanchester-Betz-Joukowski limit. According to their findings, with an increase in the number of blades, the power coefficient rises approaching the estimate of Glauert for a rotor with an infinite number of blades, only if by taking wake flow twisting into account [16].

In a different proposal, Cuerva and Sanz-Andre´s presented an extended formulation of the power coefficient of a wind turbine [17]. Their formulation was a generalization of the Betz-Lanchester expression for the power coefficient as a function of the axial deceleration of the wind speed provoked by the wind turbine in operation. The extended power coefficient took into account the benefits of the power produced and the cost associated to the production of this energy. By means of the proposed simple model, the researchers evidenced that the purely energetic optimum operation condition giving rise to the Betz-Lanchester limit (maximum energy produced) does not coincide with the global optimum operational condition (maximum benefit generated) if cost of energy and degradation of the wind turbine during operation is considered. The new extended power coefficient, according to the researchers claim, is a general parameter useful to define global optimum operation conditions for wind turbines, considering not only the energy production but also the maintenance cost and the economic cost associated to the life reduction of the machine [17].

Augmented Wind Turbines

It was suggested in [18] that one could extract more power from the wind by directing the wind by a diffuser that could be incorporated into the system. The benefit of such a device is to decrease the size of the system and thus decrease its cost [18].

According to [19], the performance of a diffuser-augmented wind turbine has been established by matching the forces acting on the blade element to overall momentum and energy balances. Good agreement with experimental data was obtained [19]. Based on computational fluid dynamics (CFD), an actuator disc CFD model of the flow through a wind turbine in a diffuser was developed and validated [20, 21]. Their research presumed a flow increase could be induced by a diffuser. They showed that from a one-dimensional analysis the Betz limit could be exceeded by a factor that is proportional to the relative increase in mass flow through the rotor. This result was verified by theoretical one-dimensional analysis by the CFD model [20, 21]. Supporting the same idea, [22] Sharpe stated that it is theoretically possible to exceed the Lanchester-Betz limit. His study presented a general momentum theory for an energy-extracting actuator disc that modeled a rotor with blades having radially uniform circulation. The study included the effects of wake rotation. Although the study reports that the general momentum theory is well known, the fall in pressure that is caused by the rotation of the wake that the theory predicts, is not usually recognized. Accounting for the wake rotational pressure drop changes some of the established conclusions of the momentum theory that appear in the literature. The conclusion from the study is that the theory establishes no loss of efficiency associated with the rotating wake [22]. Experimental and numerical investigations for flow fields of a small wind turbine with a flanged diffuser were carried out in [23].

The considered wind-turbine system gave a power coefficient higher than the Betz limit, which they attributed to the effect of the flanged diffuser. The experimental and numerical results gave useful information about the flow mechanism behind a wind turbine with a flanged diffuser. In particular, a considerable difference was seen in the destruction process of the tip vortex between the bare wind turbine and the wind turbine with a flanged diffuser [23]. According to the findings given in [24], suitable techniques to convert a country's wind availability (mostly in the low-speed regimes) as a renewable energy source must be scrutinized in order to achieve effective and efficient conversion. In this study, the researchers described efforts to step up the potential power augmentation offered by the Diffuser Augmented Wind Turbine (DAWT). Modification of the internal profile of the diffuser occurred by replacing the interior profile of the diffuser with an optimized airfoil shape as the interior profile of the diffuser. Additional velocity augmentation of approximately 66% could be achieved with the optimized profile when compared to a diffuser with an original flat interior [24]. As was pointed out in [25, 26], although there is an increase in maximum performance of a DAWT that is proportional to the mass flow of air, with application of simple momentum theory, the amount of energy extracted per unit of volume with a DAWT is the same as for an ordinary bare wind turbine [25, 26].

Modeling Controversy

Debate is ongoing around the issue of DAWT. In a recent study, a general momentum theory to study the behavior of the classical free vortex wake model of Joukowsky was used [27]. This model, as the researchers reported, has attained considerable attention as it shows the possibility of achieving a power performance that greatly exceeds the Lanchester-Betz limit for rotors running at low tip speed ratios. This behavior was confirmed even when including the effect of a center vortex, which, without any simplifying assumptions, allowed azimuthal velocities and the associated radial pressure gradient to be taken into account in the axial momentum balance. In addition, a refined model that remedies the problem of using the axial momentum theorem was proposed. Using this model the power coefficient never exceeds the Lanchester-Betz limit, but rather tends to zero at a zero tip speed ratio [27]. As asserted in [28], for reasons of energy and momentum conservation a conventional diffuser system, as it is commonly used in water turbines cannot augment the power of a wind turbine beyond the Betz limit. However, if the propeller of the turbine is embedded into an external flow of air from which by means of its static structure energy can be transferred to the internal flow through the propeller, the propeller can supersede the Betz limit with respect to this internal flow [28]. Features of such practical methods toward achieving such improvements in wind power are discussed in [29].

Blades Design

According to the report given by [30], the main problem of a wind turbine generator design project is the design of blades capable of satisfying, with optimum performance, the specific energy requirement of an electric system [30]. Simulations are very important to facilitate engineering and design of wind turbines for many reasons, especially those that concentrate upon reducing cost and saving human time. With regard to the designing the rotor blades, a CFD model for the evaluation of energy performance and aerodynamic forces acting on a straight-bladed vertical-axis Darrieus wind turbine was presented [31]. A modified blade element momentum theory for the counter-rotating wind turbine was developed [32]. This enabled the investigation of the effects of design parameters such as the combinations of the pitch angles, rotating speeds, and radii of the rotors on the aerodynamic performance of the counter-rotating wind turbine [32].

Wind Farms

Vermeer and others surveyed wind farms. The focus of this study was on standalone turbines and wind farm effects. The survey group suggested that when assembling many wind turbines together, several issues should be considered [33]. Other research studies discussed the issue of optimizing the placement of wind turbines in wind farms [34]. Factors considered included multidirectional winds and variable wind speeds, the effect of ambient turbulence in the wake recovery, the effect of ground, variable hub height of the wind turbines, and different terrains [34]. A review of the state of the art and present status of active aeroelastic rotor control research for wind turbines was presented in [35]. A wind farm controller was reported in [36]. That controller distributes power references among wind turbines while it reduces their structural loads.

In this study the effect of losses are considered and discussed for bare wind turbines and for shrouded wind turbines.

ONE-DIMENSIONAL FLUID DYNAMICS MODELS

In this section, one-dimensional fluid dynamics models are analyzed and formulated based on the extended Bernoulli equation [37], accounting for losses that are assumed to be proportional to the square of the velocity of the air crossing the rotor blades of the wind turbine. The performance characteristics of the wind turbine are given by the power and thrust coefficients. The efficiency of the wind turbine is addressed and its relation to the power coefficient is discussed.

While developing a model to describe the performance of a wind turbine, common assumptions regarding the fluid flow are as follows [18]:

1. The entire field is one-dimensional.
2. The fluid considered is not compressible.
3. The flow field in the proximity of the turbine is a pure axial flow.

Other assumptions are given for the specific models.

Bare Wind Turbine

Consider a wind turbine that intercepts the flow of air moving with velocity *V0*. The different quantities involved in the physics of the moving air are the pressure, *p*, the velocity, *V*, the cross section area, *A* (at different locations along the stream lines), and the thrust on the blades of wind turbine, *T*. The upstream condition is identified with a subscript of zero, the downstream condition is identified with a subscript of 3, and the turbine locations are identified by subscripts of 1, 2, and t (see Figure 1).

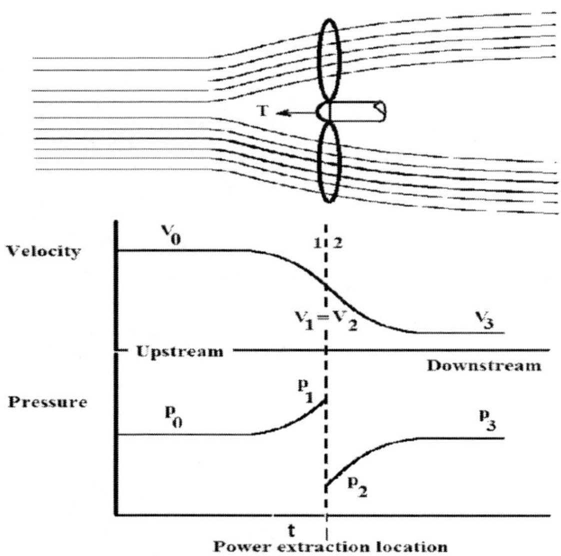

Figure 1. Schematics of the bare wind turbine. In the upper part, airstream lines are shown crossing the turbine's rotor. The velocity of the air at the rotor is the

same based on the mass flow rate: $V_1=V_2=V_t$. The pressure drop is determined by $\Delta p=p_1-p_2$.

The steady state mass flow rate is given by:

$$\dot{m} = \rho VA = const. \tag{1}$$

The modified Bernoulli equation (with reference to the turbine head), which describes the energy balance through the wind turbine, is written between the entrance and exit sections, and is given by:

$$\frac{p_0}{\gamma} + \frac{1}{2}\frac{V_0^2}{g} = \frac{p_0}{\gamma} + \frac{1}{2}\frac{V_3^2}{g} + h_t + h_{loss} \tag{2}$$

In this equation h_t is the head of the turbine (related to the amount of power extraction), h_{loss}represents the head losses, and γ is the specific weight. The loss term in the energy equation accounts for friction (mechanical and fluid) and is assumed proportional to the kinetic energy of the rotor blades, expressed as follows:

$$h_{loss} = C_{loss}\frac{V_t^2}{2g} \tag{3}$$

Equations (2) and (3) are rearranged and the head of the turbine is given by:

$$h_t = \frac{1}{2}\frac{V_0^2}{g} - \frac{1}{2}\frac{V_3^2}{g} - C_{loss}\frac{V_t^2}{2g} \tag{4}$$

The power output from the turbine, P, is given by:

$$P = \gamma V_t A_t h_t = \frac{1}{2}\rho V_0^3 A_t \frac{V_t}{V_0}\left(1 - \left(\frac{V_3}{V_0}\right)^2 - C_{loss}\left(\frac{V_t}{V_0}\right)^2\right) \tag{5}$$

From equation (5) we can determine the power coefficient, C_P, by:

$$C_P = \frac{P}{\frac{1}{2}\rho V_0^3 A_t} = \frac{V_t}{V_0}\left(1 - \left(\frac{V_3}{V_0}\right)^2 - C_{loss}\left(\frac{V_t}{V_0}\right)^2\right) \tag{6}$$

The developed thrust, T, on the turbine blades is given by the linear moment equation (see [15] for more details), as follows:

$$T = \rho V_t A_t (V_0 - V_3) \tag{7}$$

The thrust coefficient, C_T based on equation (7), is given by:

$$C_T = \frac{T}{\frac{1}{2}\rho V_0^2 A_t} = 2\frac{V_t}{V_0}\left(1 - \frac{V_3}{V_0}\right) \tag{8}$$

Equations (5) and (7) are governed by the following relationship:

$$P = TV_t \tag{9}$$

By observation, the velocity of air decreases toward the downstream. Therefore, we can simplify calculations by introducing the parameter, a, to express the velocity at the cross section of the turbine, and the upstream velocity can be determined by:

$$V_t = (1 - a)V_0 \qquad 0 < a < 1 \tag{10}$$

Equating equations (5) and (9) gives the velocity of the air, V_3, at the downstream, and after some algebraic manipulation, it could be shown to be given by:

$$V_3 = \left(1 - a - \sqrt{\left(a^2 - C_{loss}^{*}(1-a)^2\right)}\right)V_0 \tag{11}$$

The normalized loss coefficient is defined by:

$$C_{loss}^{*} = C_{loss}V_0^2 \tag{12}$$

Equations (11) and (12) are useful to calculate the power coefficient (equation (6)) and the thrust coefficient (equation (8)). For the case where the losses are negligible, the known results in the literature are reproduced and given by the following equations:

The downstream velocity is given by:

$$V_3 = 1 - 2a \tag{13}$$

The power coefficient is given by:

$$C_P = 4a(1-a)^2 \tag{14}$$

The thrust coefficient is given by:

$$C_T = 4a(1-a) \tag{15}$$

Inversing the relation given in equation (15), the parameter, *a,* is given as a function of C_T by:

$$a = \frac{1-\sqrt{1-C_T}}{2} \tag{16}$$

Finally, the power coefficient as a function of the thrust coefficient is given by:

$$C_P = \frac{C_T + C_T\sqrt{1-C_t}}{2} \tag{17}$$

Augmented Wind Turbine

In order to exploit wind power as economically as possible, it was suggested that the wind turbine should be enclosed inside a specifically designed shroud [38, 39]. Several models were reported in the literature to analyze wind turbine rotors surrounded by a device (shroud), which was usually a diffuser [18, 25, and 26]. Others suggested different approaches [28].

In this section, the extended Bernoulli equation and mass and momentum balance equations are used to analyze the augmented wind turbine. The power coefficient and the thrust coefficients are derived, accounting for losses in the same manner as was done for the bare turbine case. The efficiency of the wind turbine could be defined as the ratio of the net power output to the energy input to the system. The efficiency based on this definition agrees with the Betz limit.

The schematics of the shrouded wind turbine are shown in Figure 2.

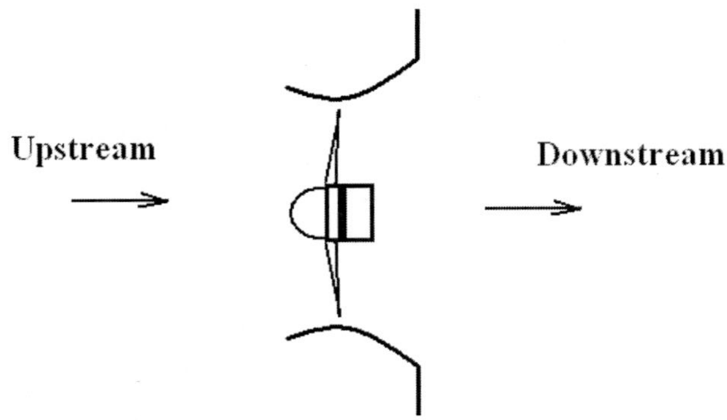

Figure 2. Schematics of the shrouded wind turbine. There is a vertical element at the exit of the wind turbine. This element contributes to reducing the power at the downstream side of the turbine, an effect that extracts more air through the wind turbine. (Idea reproduced similar to the description given by Ohya [40].).

This type of design has been recently reported [40], and it was shown that the power coefficient is about 2-5 times greater when compared to the performance of the bare wind turbine. The vertical part at the exit of the shroud reduces the pressure and therefore, the wind turbine draws more mass.

The balance equations are followed in the same manner as for the bare wind turbine. The modified Bernoulli equation differs by the pressure at the exit and is given by:

$$\frac{p_0}{\gamma} + \frac{1}{2}\frac{V_0^2}{g} = \frac{p_3}{\gamma} + \frac{1}{2}\frac{V_3^2}{g} + h_t + h_{loss} \tag{18}$$

The pressure drop between inlet and outlet (*p0-p3*) is rewritten as proportional (with C_F proportionality coefficient) to the difference in kinetic energies and it is given by:

$$\Delta p = p_0 - p_3 = \frac{1}{2}\rho\left(V_0^2 - V_3^2\right)C_F \tag{19}$$

The power coefficient for the shrouded wind turbine is given by:

$$C_P = \frac{P}{\frac{1}{2}\rho V_0^3 A_t} = \frac{V_t}{V_0}\left((C_F+1)\left(1-\left(\frac{V_3}{V_0}\right)^2\right) - C_{loss}\left(\frac{V_t}{V_0}\right)^2\right) \tag{20}$$

The thrust coefficient is given by:

$$C_T = \frac{T}{\frac{1}{2}\rho V_0^2 A_t} = 2(C_F+1)\frac{V_t}{V_0}\left(1-\frac{V_3}{V_0}\right) \tag{21}$$

Manipulating equations (9)-(12) makes it possible to produce sample plots to consider in the next section.

Wind Turbine Efficiency

Usually, efficiency is defined as the ratio between two terms: the amount of net work, w, to the input, q_{in}, energy to the device. Efficiency can be alternatively defined as the ratio between the derived power, P_{out}, and the rate of energy flowing to the system, P_{in}. Based on these definitions, the efficiency of power generating machine is given by:

$$\eta = \frac{w_{net}}{q_{in}} = \frac{P_{out}}{P_{in}} \tag{22}$$

As was observed by Betz, the maximal achievable efficiency of the bare wind turbine is given by the Betz number $B = 16/27$. In section 2.1, the power coefficient of the bare turbine was considered under the assumption of frictional losses. In this case, the power coefficient can also be identified as the efficiency of the wind turbine in this case. However, the power coefficient for the shrouded wind turbine as considered in section 2.2 is not efficiency. Based on the definition of efficiency, one could observe that if we divide the power coefficient by the factor (CF + 1) (taking into account the increased mass flow to the system due to pressure drop), a similar expression of the shrouded wind turbine could be given by equation (6). Thus, according to the modeling assumptions and with special care in treatment of the loss coefficient, one could conclude that the efficiency of the wind turbine could not exceed the Betz limit, although the power coefficient in general could exceed the Betz limit, as was observed previously by others.

Maximum Windmill Efficiency in Finite Time

In a different approach, a model to estimate the efficiency of a wind turbine was introduced [41] and the efficiency at maximum power output η_{map} was derived. Although the power developed in a wind turbine derives from kinetic energy rather than from heat, it was possible to view the basic model of the wind turbine in a schematic way, which is similar to the heat engine picture. After the wind turbine accepts energy input in its upstream side, it extracts power at the turbine blades and ejects energy at the downstream. Details of this approach are given elsewhere [41].

The derived value for the efficiency at maximum power operation was shown to be a function of the Betz number, *B,* and is given by the following formula:

$$\eta_{mp} = 1 - \sqrt{1 - B} \qquad (23)$$

This value is 36.2%, which agrees well with those for actually operating wind turbines. With an algebraic manipulation this expression could be approximated by:

$$\eta_{mp} = \frac{B}{2 - \frac{B}{2}} \qquad (24)$$

The expression given in equation (24) differs by only a small percentage when compared to equation (23). With the aid of equation (24), one could estimate the efficiency as 8/23.

As compared to the efficiency of heat engines, the system efficiency could be defined as the mean value between the maximum efficiency (Carnot efficiency) and the efficiency at maximum power point (Curzon-Ahlborn efficiency). Thus, the efficiency of the wind turbine system η_{ts}could be given by:

$$\eta_{ts} = \frac{\left(B + 1 - \sqrt{1 - B}\right)}{2} \qquad (25)$$

If equation (23) is used while neglecting the contribution of the Number $B/2$ compared to 2, the turbine efficiency given by equation (24) would be approximated by:

$$\eta_{ts} = \frac{3}{4}B \tag{26}$$

Wind Turbine Efficiency and the Golden Section

The golden section has been considered in different disciplines as a measure of beauty [42-49]. The schematics of the golden section are given in Figure 3

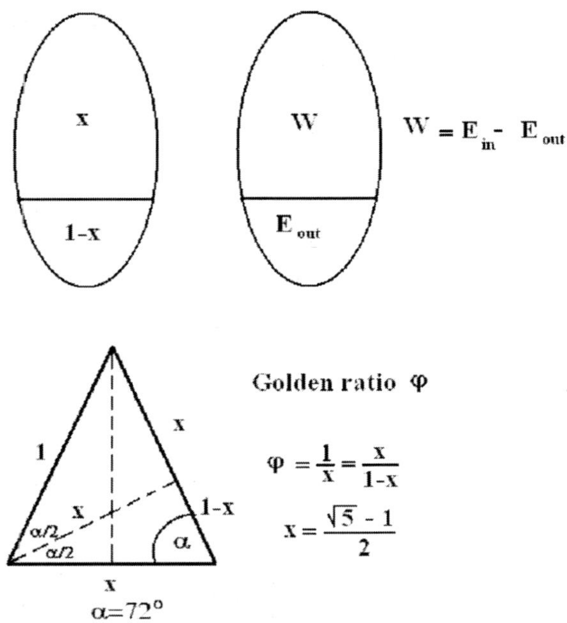

Figure 3. Schematics of the golden ratio. In the upper part to the left, the oval shape is divided into two parts, x and 1-x. in the upper part to the right, the same oval is divided into two parts with the following engine parameters: W represents the net work output; E_{out} represents the energy left the machine. E_{in} represents the energy that was put into the machine. In the lower part, a construction of the golden section is depicted by the isosceles triangle with a base angle of 72°. Comparing the different parts, x is defined as W/E_{in} or E_{out}/E_{in}, depending upon parts that have been produced and rejected.

In the upper part of the figure, the oval shape is divided into two parts, x and 1-x (as can be seen in the left side of the figure). In the right side, the same oval shape depicts the relation to quantities considered in engine machines. In the lower part of Figure 3, a golden section construction is depicted using the isosceles triangle with sides of unity and base triangle of 72°. If we apply the result of the golden section ratio (the golden section ratio is related to the sine of the angle of 18°, thus 2sin(18)=√5−1/2) to wind power efficiency, which is usually defined as the work gained divided by the energy input to the system, we can observe that the Betz limit agrees very well to the golden section (0.593 compared to 0.618). We can also see that the practical efficiencies of wind turbines in the proximity of 40% are comparable with the smaller part of the golden section 38.2%.

The golden section beauty can be related to wind turbines if we recall that the kinetic energy of the wind usually splits into two parts: useful and rejected. According to the Betz limit, about 60% of the energy is used to produce electric power and the rest is rejected. On the other hand, the real wind turbines extract about 40% and the rest is wasted. These findings are in match with the golden section division (61.8% and 38.2%). One could conclude by asking: Is this just a fortuitous result or is there something more deep and inherent in the beauty of nature?

Factors That Affect the Efficiency of the Wind Turbine

In terms of wind turbine efficiency, it is possible to highlight different parts of the turbine when estimating its value. Most of the studies discussed above considered extracting power from the kinetic energy of the wind. This could be defined as kinetic energy efficiencyη_{KE}. Other considerations should be accounted for, such as the mechanical efficiency, η_{me}(when power is decreased due to mechanical friction), the conversion to electricity efficiency,η_{con}, and the blockage efficiency, η_{bl}(which is defined as the amount of air blocked by the turbine blades as depicted by Figure 4). The overall turbine efficiency is given by:

$$\eta_{net} = \eta_{bl}\eta_{con}\eta_{me}\eta_{KE} \tag{27}$$

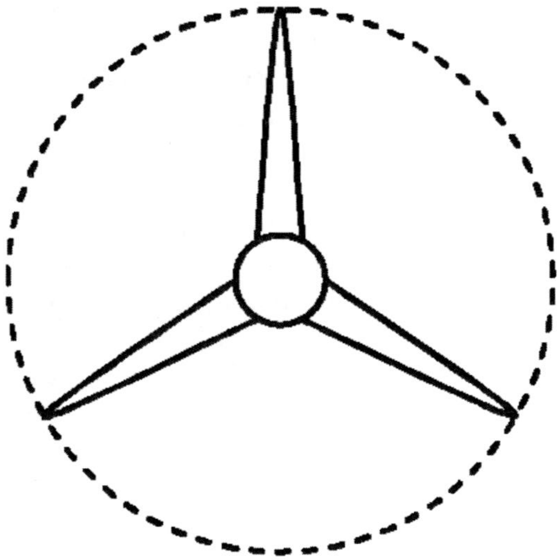

Figure 4. Schematic of the cross section of the rotor blades. The cross section illustrates the fact that the physical body of the rotor blades blocks some of the air particles, reducing potential power production from the air.

The blocking efficiency can be defined as $\eta_{bl}=1-\frac{A_{rotor}}{A_{cross}}$. In this equation, A_{rotor} is the projected cross section of the rotor blades and A_{cross} is the cross section without the rotor blades.

NUMERICAL CONSIDERATIONS

In this section sample, plots of the results are considered.

The Ideal Bare Wind Turbine Model

The one-dimensional bare wind turbine model without losses has been treated extensively and is well documented in textbooks [15]. The velocity of the air crossing the wind turbine velocity is assumed to be a fraction of the upstream air velocity. This fraction is introduced as a parameter, *a*, which expresses the ratio between the latter and the former. Parameter *a* is assigned a value in the range of numbers between zero and one. It is important to note that the physical range of this parameter is limited, for example, $0<a<0.5$, otherwise the velocity at the downstream becomes negative. To help clarify this point, calculations were performed covering the full range of parameter *a*. Equations (6) and (8) give the power coefficient and the thrust coefficient, respectively. Figure 5 shows these coefficients as functions of parameter a (a very well-known result in the professional literature in the field).

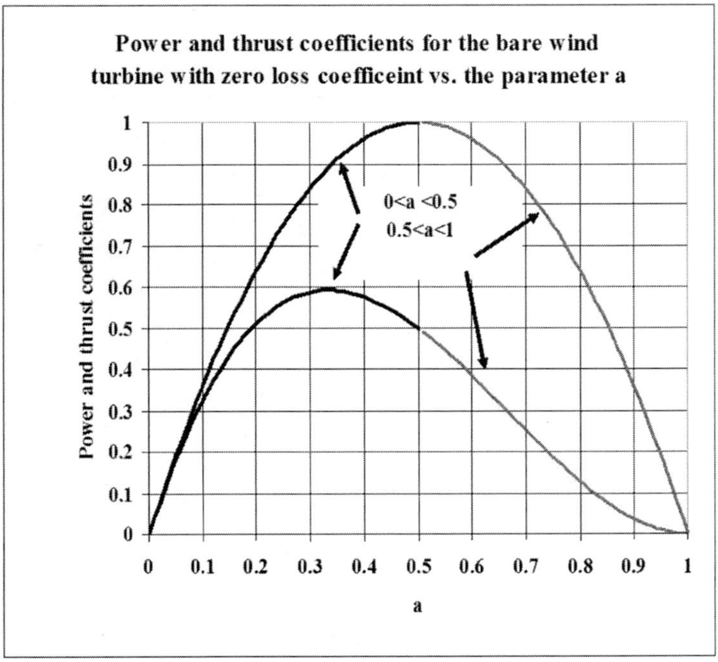

Figure 5. Power and thrust coefficients for the ideal bare wind turbine as a function of the parameter a (the ratio between the air velocity crossing the turbine blades and the upstream velocity of the air). The plot is reproduced similar to what is known in the literature, but highlighting the physical region ($0<a<0.5$) with thicker black color and the non-physical region ($0.5<a<1$) with thinner red color.

For explicit presentation, the physical range of parameter a (up to the value of 1/2) was drawn in a thick black color, while the rest of the plot was prepared using a thinner red color. It is clear from the figure that the coefficients vanish at the zero and one values of the parameter. In between, the maximum thrust coefficient (with a value of unity) occurs at the value of $a=1/2$. On the other hand, the maximum value of the power coefficient occurs at the value of $a=1/3$, for which the Betz limit is given ($C_P = B = 16/27$). If we plot the power coefficient as a function of the thrust coefficient, as Figure 6 shows, a loop shape would be produced. Again, the physical range was highlighted using a thick black color.

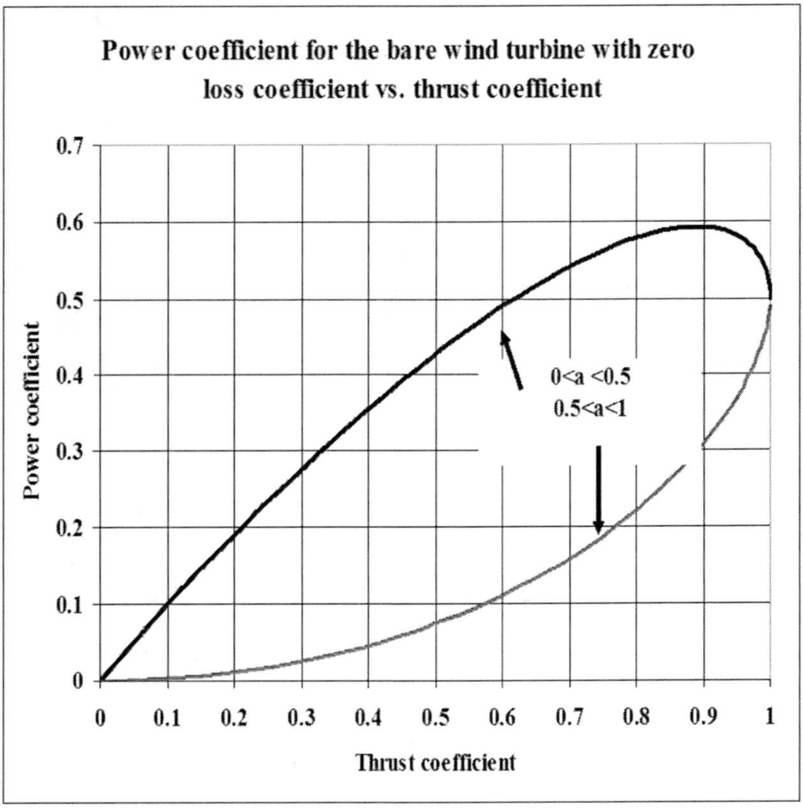

Figure 6. Power coefficient for the ideal bare wind turbine as a function of the thrust coefficient. The plot has been extended to include the non-physical region as was done in Figure 5 for reasons of consistency.

Two important points must be noted on such a plot: the maximum power coefficient (the Betz limit) for which the thrust coefficient receives a value of 8/9; and the maximum thrust (with the value of unity), for which the power coefficient gets a value of 1/2. These relations can be checked using equation (9), or more explicitly by using equation (17).

The Bare Wind Turbine with Losses

In this section, sample plots are given to demonstrate the effect of the losses as modeled in section 2.2. The losses are due to friction and are modeled as proportional to the velocity of the square of the velocity of the air flowing through the wind turbine. The plots are prepared for three different values of the non-dimensional loss coefficient C^*_{loss}: 0, 0.05, and 0.1. Figure 7 shows a plot of the power coefficient and of the thrust coefficient as a function of parameter *a,* covering the physical range while accounting for losses.

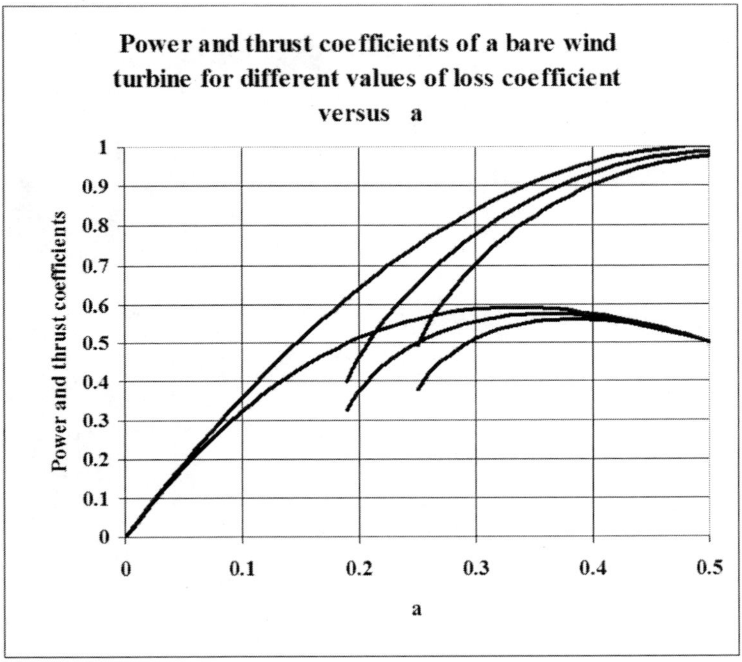

Figure 7. Power and thrust coefficients for the bare wind turbine as a function of parameter *a*, accounting for frictional losses, with different values.

It is clear from the figure that there is degradation in both coefficients in the order of a few percentage points. The two points (maximum power coefficient and maximum thrust coefficient) can be better visualized as illustrated by Figure 8.

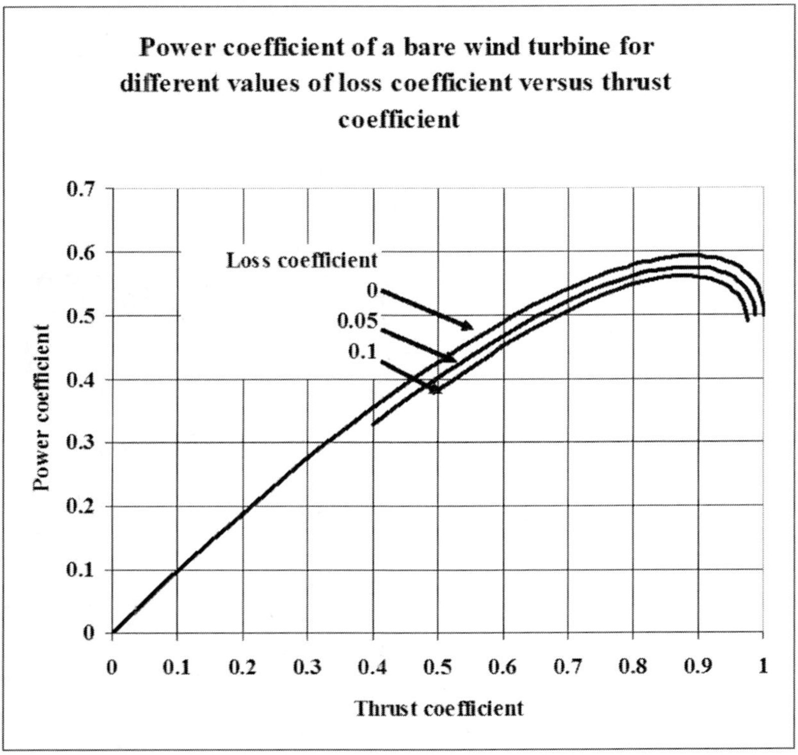

Figure 8. Power coefficient for the bare wind turbine as a function of the thrust coefficient, accounting for frictional losses, with different values.

The Shrouded Wind Turbine

The shrouded wind turbine was analyzed based on the extended Bernoulli equation, while accounting for frictional losses in the same manner as was done for the bare wind turbine. The increased air mass flow due to the larger drop in pressure was modeled as proportional to the kinetic energy difference, using the coefficient C_F (pressure drop coefficient—see equation (17)). As can be seen from Figure 9and Figure 10 (equations (18) and (19) respectively), the power coefficient and the thrust coefficient are increased proportionally to the pressure coefficient, which is in agreement with the findings in the literature.

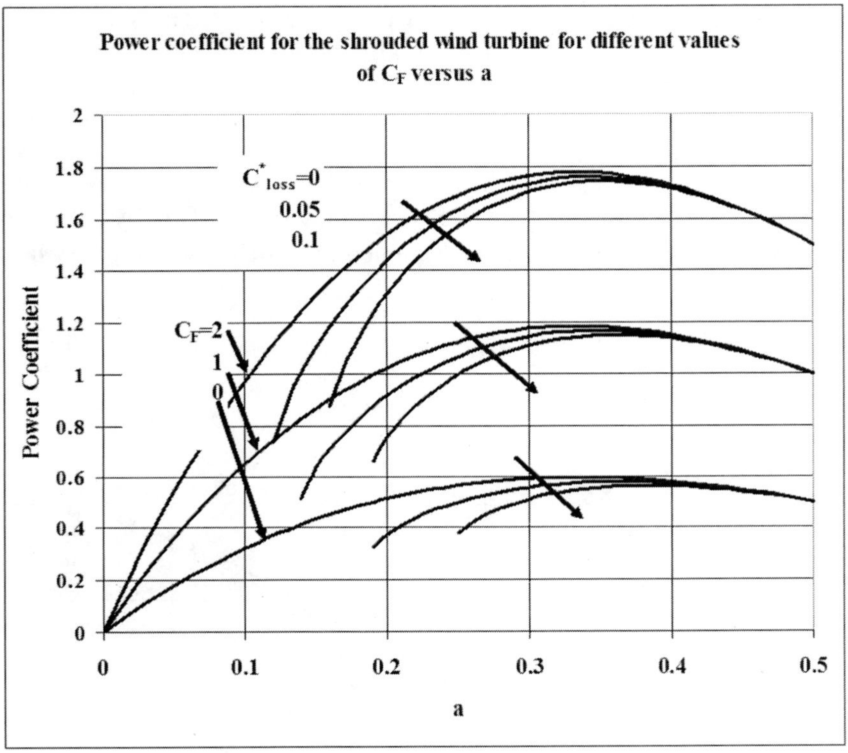

Figure 9. Power coefficient for the shrouded wind turbine as a function of parameter a, accounting for frictional losses and for augmentation coefficient C_F.

The maximum power coefficient and the maximum thrust points are illustrated in Figure 11.

By consulting Figure 11, one can observe that both maximum points are degraded by an increasing loss coefficient.

Efficiency of the Wind Turbine

As was considered by Betz, the power coefficient as originally defined agrees with the definition of efficiency for a device that extracts work from a given amount of energy. Thus, for the bare wind turbine, the maximum efficiency that could be extracted is actually given by the Betz limit. The effect of friction on wind turbine efficiency, as was expressed through the power coefficient, decreases with friction. A similar observation could be stated for the shrouded wind turbine if we use the definition as given by equation (22). Accordingly, the Betz limit is exceeded, that is, the shrouded wind turbine produces more power, but the amount of energy extracted per unit of volume with a shroud is the same as for an ordinary bare wind turbine. These results were found to be in agreement with results observed in [Van Bussel, 2007]. The efficiency of maximum power output that was observed by the finite time analysis was approximately 36%, which is comparable to experimental findings [14]. When compared to heat engines, the efficiency of the wind turbine could be expressed in terms of the Betz number by using equation (23). If the Betz number is substituted, the efficiency could be approximated as 47%, but if the other factors are taken into account, the practical efficiency could reach much lower values.

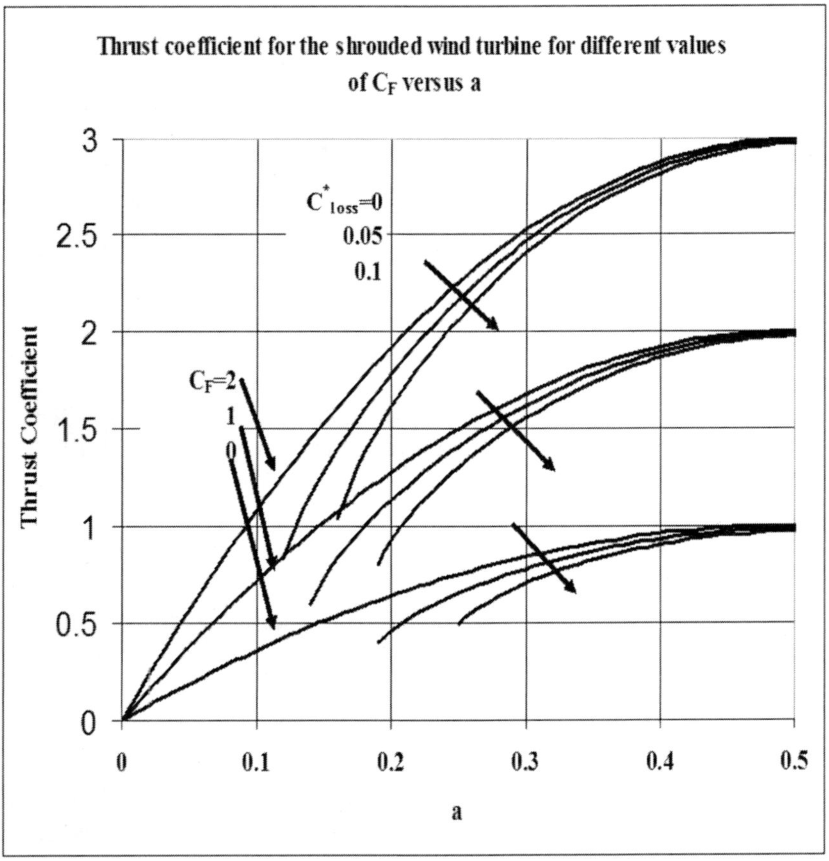

Figure 10. Thrust coefficient for the shrouded wind turbine as a function of parameter a, accounting for frictional losses and for augmentation coefficient C_F, with different values.

WIND TURBINE ARRANGEMENTS

One could suggest ideas to increase power extraction from the wind, thus decreasing the overall cost. One suggestion is the bottoming wind turbine; another is the *flower leaves* arrangement of wind turbines. Both are discussed in the following sections.

Bottoming Wind Turbines

According to Betz, the maximal power extraction efficiency is 16/27. If it is possible to extract energy from the downstream expelled air, assuming the same limit exists, one could estimate an extra amount of 11/27*16/27 which is approximately 24%. This estimate suggests adding a smaller bottoming rotor behind the main larger rotor. The idea of the bottoming wind turbine is depicted schematically in Figure 12.

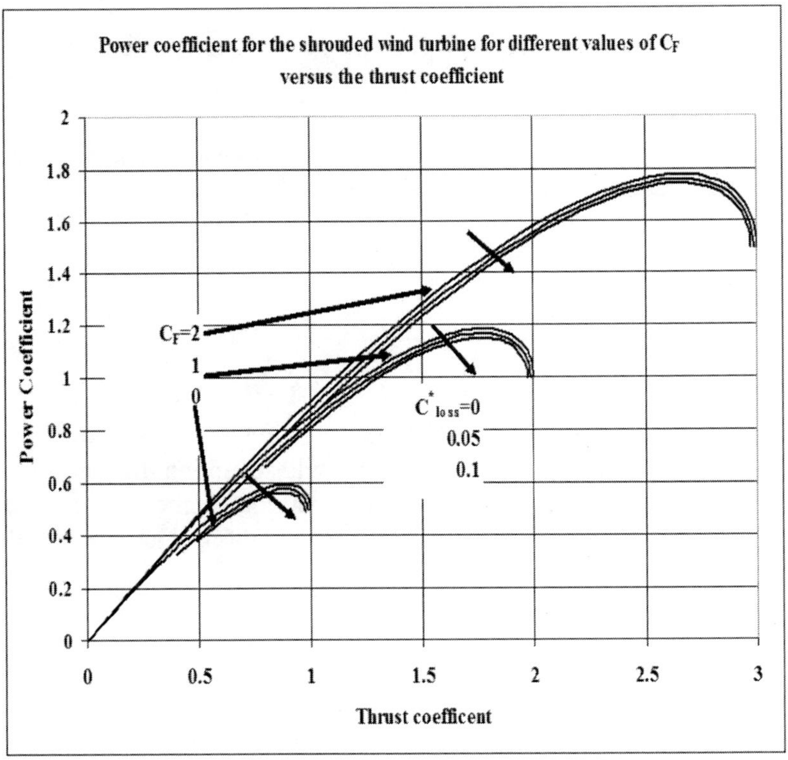

Figure 11. Power and coefficients for the shrouded wind turbine as a function of the thrust coefficient, accounting for frictional losses and for augmentation coefficient C_F, with different values.

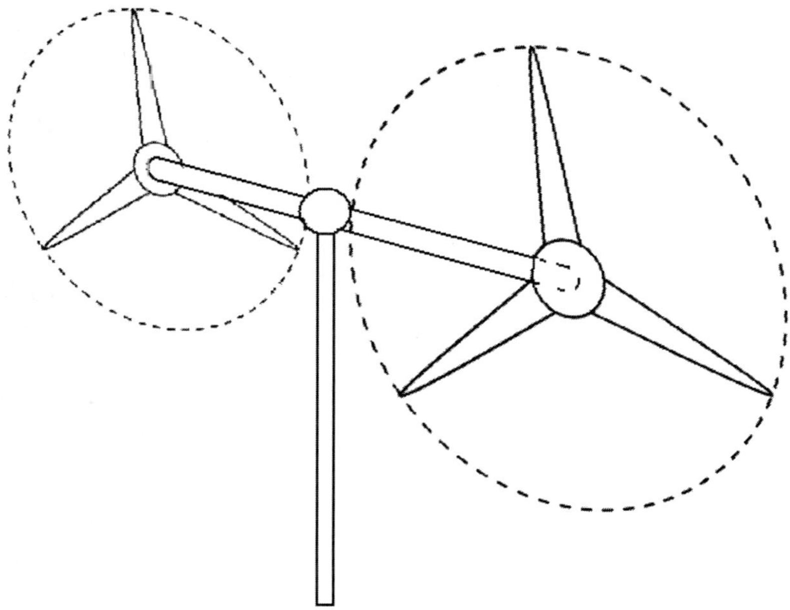

Figure 12. Schematics of the bottoming wind turbine idea. The main rotor is the first to intercept the airflow. Outlet air is then directed to the secondary rotor. The attractiveness of this idea is to gain more output with the same tower installation, reducing the inherently larger cost of erecting multiple towers.

Flower Leaves Arrangement of Wind Turbines

Considering the shrouded wind turbine as being a relatively small device is given (sometimes called flower power while searching the web), one could suggest installing different devices on the same tower. Such an arrangement reduces the cost of the installation. The idea is depicted schematically in Figure 13.

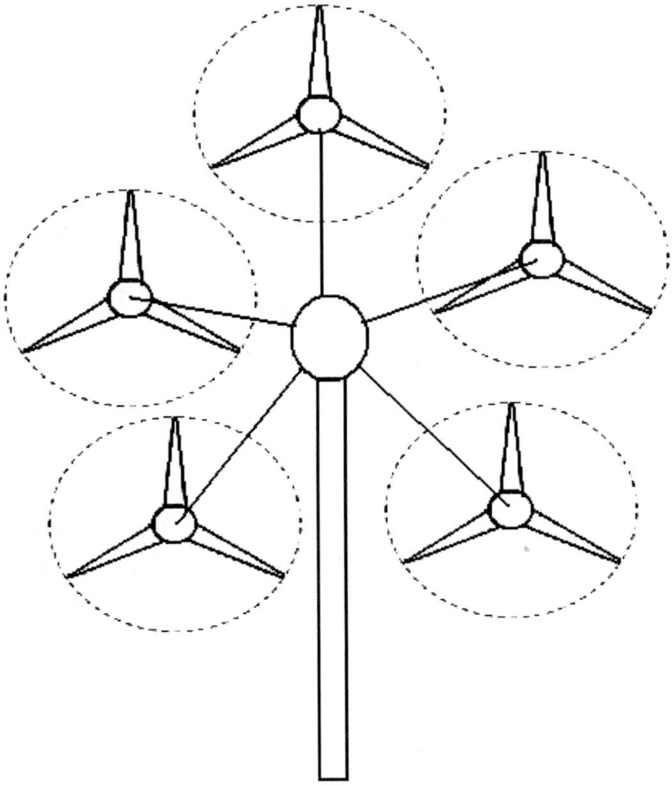

Figure 13. Schematics of *flower leaves* arrangement of wind turbines. Because they are small devices, it should be possible to install more of these shrouded systems on a single tower. While the concept of reducing installation costs is similar to that of the bottoming wind turbine, the attractiveness of the flower leaves configuration is that many smaller turbines can be accommodated on a single tower, with significant cost reductions.

SUMMARY AND CONCLUSIONS

In this study, wind turbine power was reconsidered. At the beginning, a literature review was given with relation to the potentiality of wind power, worldwide applications of wind power, and different factors that affect the performance of wind turbines, especially those related to one-dimensional modeling of the flow through the wind turbine. Later, different models were addressed, taking into account the effect of friction, which is usually neglected in the literature. In this study, friction is modeled to be proportional to the square of the velocity of the air crossing the wind turbine blades. The bare wind turbine model and the shrouded wind turbine model were analyzed based on the following balance equations: the mass balance equation, the momentum balance equation, and the energy balance equation that is exposed in the form of an extended (modified) Bernoulli equation. Through the analysis it was observed that both the power coefficient and the thrust coefficient degrade with friction. As was noticed in previous studies, the power coefficient given by the Betz number is the efficiency of the wind turbine (originally derived for the bare wind turbine). Following the same type of definition, a similar expression for the shrouded wind turbine could be derived. In a different approach, the wind turbine could be analyzed using finite time methods, as was given by [14]. In this study, the results were briefly summarized.

The well-known golden ratio usually is considered as a measure of beauty. It is interesting to notice that the Betz number differs from the golden ratio by only 4% (0.618 compared to 16/27=0.593).

In an effort to explain the discrepancy between theoretical efficiency and practical or measured efficiencies, different factors that affect the extraction of wind power are considered. These include mechanical friction, conversion efficiency to electricity, and blockage efficiency, which accounts for the blocked amount of air (usually is not mentioned in the literature), thus reducing the power output.

Finally, plots were given to suggest ways of assembling wind turbines to gain more of wind power for each tower installation.

REFERENCES

1. Sahin A. D., "Progress and recent trends in wind energy," Progress in Energy and Combustion Science 30, pp. 501–543, 2004.
2. Şen Z., "Modified wind power formulation and its comparison with Betz limits," Int. J. Energy Res., pp. unknown, 2012, http://onlinelibrary.wiley.com/doi/10.1002/er.2900/full, last visited on 22/6/2012.
3. Sorensen J. N., "Aerodynamic aspects of wind energy conversion," Annu. Rev. Fluid Mech. 43, pp. 427–48, 2011.
4. Villanueva D., Feijoo A., "Wind power distributions: A review of their applications," Renewable and Sustainable Energy Reviews 14, pp. 1490–1495, 2010.
5. Hubbard H. H., and Shepherd K. P., "Wind turbine acoustics," NASA technical paper 3057, DOE/NASA/2030-77, pp. 1-54, 1990.

6. Rojas-Sola J. I., Juan Manuel A., "Southern Spanish windmills: Technological aspects," Renewable Energy, 30, 1943–1953, 2005.

7. Rojas-Sola J. I., Gonzalez M., Martın E. P.," Computer-aided design and engineering: A study of windmills in la Mancha (Spain)," Renewable Energy, 31, 10, pp. 1471-1656, 2006.

8. Al-Mohamad A., Karmeh H., "Wind energy potential in Syria," Renewable Energy 28, pp. 1039–1046, 2003.

9. Islam M.R., Saidur R., Rahim N.A., "Assessment of wind energy potentiality at Kudat and Labuan, Malaysia using Weibull distribution function," Energy 36, pp. 985-992, 2011.

10. James P. A. B., Sissons M.F., Bradford J., Myers L. E., Bahaj A.S., Anwar A., Green S., "Implications of the UK field trial of building mounted horizontal axis micro-wind turbines," Energy Policy, 38, pp. 6130–6144, 2010.

11. Abdeen M. O., "On the wind energy resources of Sudan," Renewable and Sustainable Energy Reviews, 12, pp. 2117–2139, 2008.

12. Mostafaeipour A., Sedaghat A., Dehghan A.A. N., Kalantar V., "Wind energy feasibility study for city of Shahrbabak in Iran," Renewable and Sustainable Energy Reviews 15, pp. 2545– 2556, 2011.

13. Okulov V. L., Sørensen J. N., "An Ideal Wind Turbine with a Finite Number of Blades", Doklady Physics, Vol. 53, No. 6, pp. 337–342. 2008.

14. Rauh A., Seelert W., "The Betz optimum efficiency for windmills," Applied Energy, 17 pp. 15-23, 1984.

15. Shames, I. H. Shames, Mechanics of Fluids, 2nd ed., pp. A26 –A31,McGraw-Hill, New York, 1982.

16. Okulov V. L., Sørensen J. N., "An Ideal Wind Turbine with a Finite Number of Blades," Doklady Physics, Vol. 53, No. 6, pp. 337–342. 2008.

17. Cuerva A., Sanz-Andre´s A., "The extended Betz-Lanchester limit," Renewable Energy 30, pp. 783–794, 2005.

18. Igra, O., "Research and development for shrouded wind turbines," Energy Conv. & Mgmt., 21 pp. 13-48, 1981.

19. Fletcher C. A. J., "Diffuser-augmented wind turbine analysis," 7th Australasian hydraulics and fluid mechanics conference, Brisbane, 18-22 August 1980, pp. 435-438, 1980.

20. Philips D. G., Richards P. J. Mallinson G. D. and Flay R. G. J., "Computational modeling of diffuser for a diffuser augmented wind turbine," 13thAustralAsian fluid mechanics conference, Monash university, Melbourne Australia, 13-18 December 1998, pp. 207-210, 1998.

21. Hansen M. O. L., Sørensen N. N., Flay R. G. J., "Effect of Placing a Diffuser around a Wind Turbine," Wind Energy, 3, pp. 207–213, 2000.

22. Sharpe D. J., "A general momentum theory applied to an energy-extracting actuator disc," Wind Energy, 7, pp. 177–188, 2004.

23. Abe K., Nishida M., Sakurai A., Ohya Y., Kihara H., Wada E., Sato K., " Experimental and numerical investigations of flow fields behind a small wind turbine with a flanged diffuser," J. Wind Eng. Ind. Aerodyn. 93, pp. 951–970, 2005.

24. Nasutiona A., Purwantoa D. W., "Optimized Curvature Interior Profile for Diffuser Augmented Wind Turbine (DAWT) to increase its energy-conversion performance," 2011 IEEE First Conference on Clean Energy and Technology CET, pp. unknown, 2011.

25. Van Bussel, G.J.W, “An Assessment of the Performance of Diffuser Augmented Wind Turbines (DAWT’s),” 3rd ASME/JSME Fluid Engineering Conference FEDSM99-7830, San Francisco, USA, 1999.

26. Van Bussel, G. J. W., "The science of making more torque from wind: Diffuser experiments and theory revisited," J. Phys.: Conf. Ser. 75 012010, pp. 1-12, 2007.

27. Sørensen J. N., Van Kuik G. A. M., "General momentum theory for wind turbines at low tip speed ratios," Wind Energy, 14, pp. 821–839, 2011.

28. Bet F., Grassmann H., "Upgrading conventional wind turbines," Renewable Energy 28, pp. 71–78, 2003.

29. Grassmann H., Bet F., Ceschia M., Ganis M.L., "On the physics of partially static turbines," Renewable Energy, 29, pp. 491–499, 2003.

30. Vitale A.J., Rossi A.P., "Software tool for horizontal-axis wind turbine simulation," International Journal of Hydrogen Energy 33, pp. 3460-3465, 2008.

31. Castelli M. R., Englaro A., Benini E., "The Darrieus wind turbine: Proposal for a new performance prediction model based on CFD," Energy 36, pp. 4919-4934, 2011.

32. Seungmin L., Hogeon K., Eunkuk S., Soogab L., "Effects of design parameters on aerodynamic performance of a counter-rotating wind turbine," Renewable Energy 42, pp. 140-144, 2012.

33. Vermeer L.J., Sorensen J.N., Crespo A., "Wind turbine wake aerodynamics," Progress in Aerospace Sciences 39, 467–510, 2003.

34. Carlos M. Ituarte-Villarreal, Espiritu J. F., "Optimization of wind turbine placement using a viral based optimization algorithm," Procedia Computer Science 6, pp. 469–474, 2011.

35. Barlas T.K., van Kuik G.A.M., "Review of state of the art in smart rotor control research for wind turbines," Progress in Aerospace Sciences 46, pp. 1–27, 2010.

36. Maryam Soleimanzadeh, Rafael Wisniewski, "Controller design for a wind farm, considering both power and load aspects," Mechatronics 21, pp. 720–727, 2011.

37. Sabersky, R. H., Acosta, A. J., Hauptmann, E. G., Fluid flow, 2nd ed., 1971, Macmillan Publishing Co, Inc., Printed in the USA.

38. Igra O., "Compact shrouds for wind turbines," Energy conversion, 16, pp. 149-157, 1977.

39. Igra O., "The shrouded aerogenerator," Energy, 2, pp. 429-439, 1977.

40. Ohya Y., Karasudani, T., "A shrouded wind turbine generating high output power with wind-lens technology," Energies, 3, pp. 634-649, 2010.

41. Huleihil M., "Maximum windmill efficiency in finite time," J. Appl.Phys., 105, Issue 10, pp. 104908,1-4, 2009.

42. BEJAN, A., "The golden ratio predicted: Vision, cognition and locomotion as a single design in nature," Int. J. of Design & Nature and Ecodynamics. Vol. 4, No. 2, 97–104, 2009.

43. Markowsky G., "Book – review: the golden ratio," Notices of the AMS, vol. 52, 3, pp. 344-347, 2005.

44. Kaplan A., Tortumlu N. and Hizarci S., “A Simple Construction of the Golden Ratio,” World Applied Sciences Journal 7 (7): 833-833, 2009.

45. Shechtman, "Crystals of golden proportions," The Nobel prize in chemistry, the royal Sweedish academy of sciences, 2011.

46. Xu, L., Zhong, T., "Golden Ratio in Quantum Mechanics," Nonlinear Science Letters B: Chaos, Fractal and Synchronization, pp. 24, 2011.

47. Nikolic, S. T., Cosic I., Pecujlija, M., Miletic A., "The effect of the 'golden ratio' on consumer behavior", African Journal of Business Management Vol. 5 (20), pp. 8347-8360, 2011.

48. Seebregts, A., J., "Gas-Fired power," ©IEA ETSAP - Technology Brief E02, pp. 1-5, April 2010.

49. Michel Spira, "On the golden ratio," 12th International Congress on Mathematical Education Program Name XX-YY-zz (pp. abcde-fghij) 8 July – 15 July, 2012, COEX, Seoul, Korea.

CITATION

Mahmoud Huleihil and GedalyaMazor (2012). Wind Turbine Power: The Betz Limit and Beyond, Advances in Wind Power, Dr. Rupp Carriveau (Ed.), ISBN: 978-953-51-0863-4, InTech, DOI: 10.5772/52580.

CHAPTER 6

Power Sharing Algorithm for Vector Controlled Six-Phase AC Motor with Four Customary Three-Phase Voltage Source Inverter Drive

Sanjeevikumar Padmanaban[1], Gabriele Grandi[1], Frede Blaabjerg[2], Joseph Olorunfemi Ojo[3], Patrick William Wheeler[4]

[1]Department of Electrical, Electronic, and Information Engineering, Alma Mater Studiorum 1088AD, University of Bologna, 40136 Bologna, Italy
[2]Department of Energy Technology, Aalborg University, Pontoppidanstraede 101, 9220 Aalborg, Denmark
[3]Center for Energy System Research, Department of Electrical & Computer Engineering, Tennessee Technological University, 38505 Cookeville, TN, USA
[4]Power Electronics, Machines and Control Group, Department of Electrical & Electronics Engineering, Nottingham University, NG7 2RD Nottingham, UK

ABSTRACT

This paper considered a six-phase (asymmetrical) induction motor, kept 30° phase displacement between two set of three-phase open-end stator windings configuration. The drive system consists of four classical three-phase voltage inverters (VSIs) and all four dc sources are deliberately kept isolated. Therefore, zero-sequence/homopolar current components cannot flow. The original and effective power sharing algorithm is proposed in this paper with three variables (degree of freedom) based on synchronous field oriented control (FOC). A standard three-level space vector pulse width modulation (SVPWM) by nearest three vectors (NTVs) approach is adopted to regulate each couple of VSIs. The proposed power sharing algorithm is verified by complete numerical simulation modeling (Matlab/Simulink-PLECS software) of whole ac drive system by observing the dynamic behaviors in different designed condition. Set of results are provided in this paper, which confirms a good agreement with theoretical development.

INTRODUCTION

For the past 30 years the technology of multi-phase machines had proved that they are predominant solution over their classical three-phase ac machines counterpart [1],[2] and [3]. In specific by increased reliability, redundant structure, reduced torque ripple, increased torque density, fault tolerability, and reduced per-phase of inverter rating [2]. Furthermore, for any m-phase machines ($m > 3$) only two currents are sufficient to control the torque/flux independently, even if with reduced performance [2]. Multi-phase configurations, in particular six-phase machines, are the foremost choice for high power (low voltage/high current) applications, where two adjacent windings are spatially shifted by 60° (symmetrical type) [3] or by 30° (asymmetrical or dual three-phase type) [4], [5],[6], [7], [8], [9], [10] and [11].

Literatures are focused on modeling/controlling of six-phase (asymmetrical/symmetrical) machines with indirect/direct, synchronous/stationary field oriented control (FOC), direct torque control (DTC) [6], [7], [8] and [9], pulse width modulations (PWMs) techniques[12], [13], [14], [15], [16], [17], [18] and [19], parallel/series operation of multi-phase machines with single inverter drive [1], [2] and [4], and fault tolerant strategies [19], [20], [21], [22], [23], [24], [25], [26] and [27]. But kept starving for proper and effective power sharing algorithm for six-phase ac machine driven by multi-phase ac drive (VSIs).

This paper investigates a six-phase (asymmetrical or dual three-phase) induction motor with open-end stator windings configuration, as shown by Fig. 1. Priorities in selecting such machines over symmetrical version are described in Refs. [5], [10], [11] and [12]. The drive circuit consists of four classical three-phase voltage source inverters (VSIs) and each inverter is connected across the open-end of two three-phase stator windings. The resultant configuration is a multi-phase inverter motor drive, as shown in Fig. 2. To be noted, all the four dc sources are completely isolated and the entire circuit is absolutely free of zero-sequence/homopolar components.

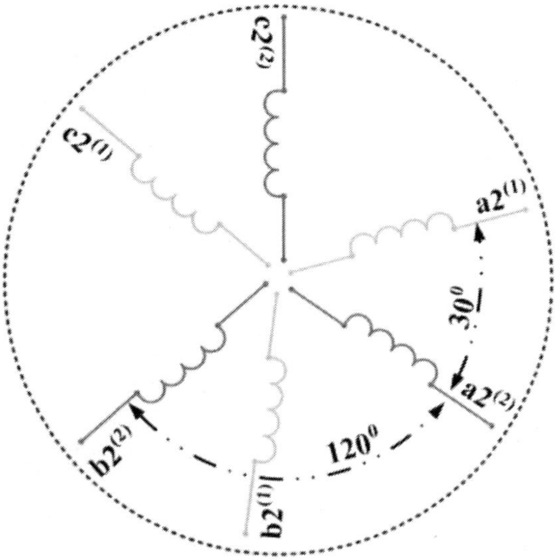

Figure 1. Six-phase asymmetrical (quasi) induction motor open-end stator windings.

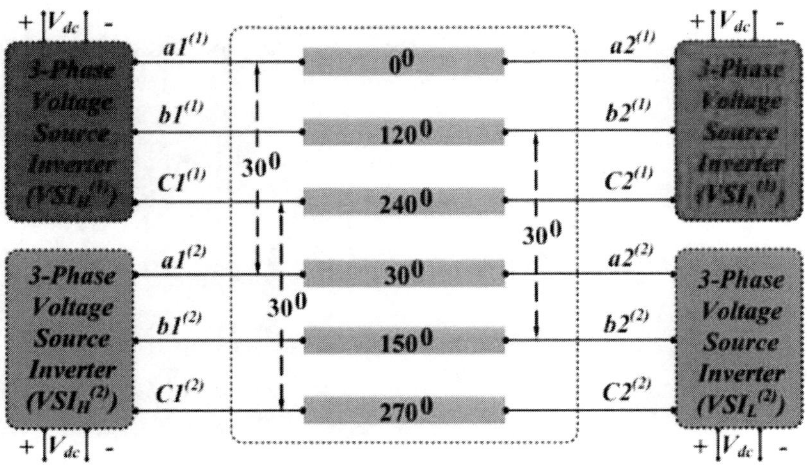

Figure 2. Configuration of a six-phase asymmetrical (quasi) open-end stator winding motor driven by four customary voltage source inverters (VSIs), one fed on each open-end.

Further, an original power sharing algorithm based on synchronous field oriented control (FOC) was proposed in this paper with three variables (three degree of freedom)[25] and [26]. First variable provides the proper currents sharing between two three-phase open-end stator windings. Other two provides the proper voltages sharing between each two inverters of two three-phase open-end stator windings. Investigation are focused in particular to dynamic behaviors of the six-phase induction motor direct–quadrature (d–q) axis components, subjected to balanced/unbalanced conditions to verify the effectiveness of the theoretical developments.

SIX-PHASE ORTHOGONAL SPACE VECTOR TRANSFORMATION

Six-phase orthogonal space vector transformation in sinusoidal rotating frame is given by[12], [16], [17], [18], [28] and [29]:

$$\overline{x}_h = \frac{1}{3}\left[x_1 + x_2\alpha^{h} + x_3\alpha^{4h} + x_4\alpha^{5h} + x_5\alpha^{8h} + x_6\alpha^{9h}\right], \tag{1}$$

where $h = 1,\ 3,\ 5$ and $X_1 \ldots X_6$ are variables of voltage or current and $\alpha = \exp(j\pi/6)$. Now substituting index h in Eq. (1), leads to:

$$\begin{cases} \overline{x}_1 = \frac{1}{3}\left[x_1 + x_2\alpha + x_3\alpha^4 + x_4\alpha^5 + x_5\alpha^8 + x_6\alpha^9\right] \\ \overline{x}_3 = \frac{1}{3}\left[(x_1 + x_3 + x_5) + j(x_2 + x_4 + x_6)\right] \\ \overline{x}_5 = \frac{1}{3}\left[x_1 + x_2\alpha^5 + x_3\alpha^8 + x_4\alpha + x_5\alpha^4 + x_6\alpha^9\right]. \end{cases} \tag{2}$$

Inverse of Eq. (2) is written as:

$$\begin{cases} x_1 = \overline{x}_3 \cdot 1 + \left(\overline{x}_1 + \overline{x}_5^*\right) \cdot 1 \\ x_3 = \overline{x}_3 \cdot 1 + \left(\overline{x}_1 + \overline{x}_5^*\right) \cdot \alpha^4, \\ x_5 = \overline{x}_3 \cdot 1 + \left(\overline{x}_1 + \overline{x}_5^*\right) \cdot \alpha^8 \end{cases} \begin{cases} x_2 = \overline{x}_3 \cdot j + \left(\overline{x}_1 - \overline{x}_5^*\right) \cdot \alpha \\ x_4 = \overline{x}_3 \cdot j + \left(\overline{x}_1 - \overline{x}_5^*\right) \cdot \alpha^5 \\ x_6 = \overline{x}_3 \cdot j + \left(\overline{x}_1 - \overline{x}_5^*\right) \cdot \alpha^9, \end{cases}$$

(3)

where "*" and "·" denote complex conjugate and scalar product, respectively. The space vectors $\overline{x}_1$, $\overline{x}_3$, $\overline{x}_5$ lie in the planes d_1–q_1, d_3–q_3, d_5–q_5 respectively.

Equivalent Two Three-phase Space Vector Transformation

For the purpose of simplification in analysis, space vector transformations given by Eqs.(2) and (3) are split and given as below:

$$\begin{cases} x_1^{(1)} = x_1 \\ x_2^{(1)} = x_3, \\ x_3^{(1)} = x_5 \end{cases} \begin{cases} x_1^{(2)} = x_2 \\ x_2^{(2)} = x_4 \\ x_3^{(2)} = x_6. \end{cases}$$

(4)

The two three-phase space vectors $\overline{x}^{(1)}$, $\overline{x}^{(2)}$ and zero-sequence components $x_0^{(1)}$, $x_0^{(2)}$ of six-phase machine would be written as:

$$\begin{cases} \overline{x}^{(1)} = \frac{2}{3}\left[x_1^{(1)} + x_2^{(1)}\alpha^4 + x_3^{(1)}\alpha^8\right] \\ x_0^{(1)} = \frac{1}{3}\left[x_1^{(1)} + x_2^{(1)} + x_3^{(1)}\right] \end{cases}, \begin{cases} \overline{x}^{(2)} = \frac{2}{3}\left[x_1^{(2)} + x_2^{(2)}\alpha^4 + x_3^{(2)}\alpha^8\right] \\ x_0^{(2)} = \frac{1}{3}\left[x_1^{(2)} + x_2^{(2)} + x_3^{(2)}\right]. \end{cases}$$

(5)

The relationship between six-phase space vectors and two three-phase space vectors is obtained by applying Eqs. (4) and (5) to Eq. (1) and given by:

$$\begin{cases} \overline{x}_1 = \frac{1}{2}\left[\overline{x}^{(1)} + \alpha\overline{x}^{(2)}\right] \\ \overline{x}_5^* = \frac{1}{2}\left[\overline{x}^{(1)} - \alpha\overline{x}^{(2)}\right] \end{cases}, \quad \overline{x}_3 = x_0^{(1)} + jx_0^{(2)}. \tag{6}$$

Inverse transformation of Eq. (6) given by:

$$\begin{cases} \overline{x}^{(1)} = \overline{x}_1 + \overline{x}_5^* \\ x_0^{(1)} = \overline{x}_3 \cdot 1 \end{cases}, \quad \begin{cases} \overline{x}^{(2)} = \alpha^{-1}\left(\overline{x}_1 - \overline{x}_5^*\right) \\ x_0^{(2)} = \overline{x}_3 \cdot j. \end{cases} \tag{7}$$

By Eqs. (6) and (7), it is simplified to represent six-phase (asymmetrical or dual three-phase) open-end stator windings in two three-phase dimensions [5]. It should be considered that:

- The fundamental components with the harmonics of the order $k = 6 \cdot n \pm 1$, $(n = 0, 2, 4\ldots)$ are mapped in $\overline{x}_1$ (first sub-space d_1–q_1).
- $k = 6 \cdot n \pm 1$, $(n = 1, 3, 5\ldots)$ order harmonics are mapped in $\overline{x}_5^*$ (fifth sub-spaced_5–q_5).
- $k = 3 \cdot n$, $(n = 0, 1, 2, 3\ldots)$ order harmonics (homopolar components or zero-sequences components) are mapped in the third sub-space $\overline{x}_3$ (third sub-spaced_3–q_3).

The multiple space vectors therefore, $\overline{x}_1$, $\overline{x}_3$, $\overline{x}_5^*$ are orthogonal to each other in space, i.e. three vectors are independent of each other, hence first sub-space free of odd harmonics and zero-sequence components. Therefore, the complete behavior of the six-phase machine is described by two space vectors $\overline{x}_1$ and $\overline{x}_5^*$ moving arbitrarily in space and controlled independently [10], [11] and [12].

3. Six-phase open-end induction motor drive

In Fig. 1, the sinusoidally distributed stator windings behavior of the six-phase induction machine is described in stationary reference frame (Eqs. (5) and (6)) by the following space vectors equations:

$$\overline{v}_{S1} = R_S \overline{i}_{S1} + \frac{d\overline{\varphi}_{S1}}{dt}, \ \overline{\varphi}_{S1} = L_{S1} \overline{i}_{S1} + M_1 \overline{i}_{R1}, \tag{8}$$

$$0 = R_R \overline{i}_{R1} - jp\omega_m \overline{\varphi}_{R1} + \frac{d\overline{\varphi}_{R1}}{dt}, \ \overline{\varphi}_{R1} = M_1 \overline{i}_{S1} + L_{R1} \overline{i}_{R1}, \tag{9}$$

$$\overline{v}_{S5} = R_S \overline{i}_{S5} + \frac{d\overline{\varphi}_{S5}}{dt}, \ \overline{\varphi}_{S5} = L_{S5} \overline{i}_{S5}, \tag{10}$$

$$T = 3pM_1 \overline{i}_{S1} \cdot j\overline{i}_{R1}, \tag{11}$$

where p represents the pole pair, ωm the rotor speed and the subscripts S, R represents the stator and rotor reference components. The space vectors currents $\overline{i}_{S1}$ and $\overline{i}_{R1}$ are responsible for the magnetic field in the air gap and $\overline{i}_{S5}$ does not contribute any air gap field and produce only leakage flux [5], [6], [7], [8] and [12].

PROPOSED SYNCHRONOUS REFERENCE FOC POWER SHARING ALGORITHM

Fig. 3 depicts the proposed effective power sharing algorithm actually derived from standard synchronous FOC. The total power is shared among the four dc sources by three variables (degree of freedom) in two developed control strategies. In first one, the variables $k_v^{(1)}$, $k_v^{(2)}$ share the voltages between two inverters (VSI_H and VSI_L) of two three-phase open-end stator windings. In second one, the variable ki share the currents between two three-phase open-end stator windings.

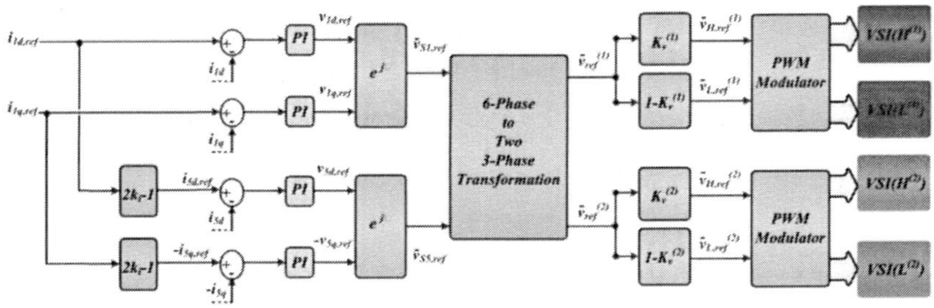

Figure 3. Complete effective power sharing algorithm with three realizable sharing variables in synchronous field oriented control (FOC).

First control strategy deals with voltage sharing, the reference output voltage of first three-phase open-end stator windings $(\overline{v}^{(1)})$ and second three-phase open-end stator windings $(\overline{v}^{(2)})$ are determined by the space vectors $\overline{v}_{S1,ref}$ and $\overline{v}^*$ given by Eq. (6). The reference voltage $\overline{v}^{(1)}$ is produced by the sum of $\overline{v}_{H}^{(1)}$ and $\overline{v}_{L}^{(1)}$ reference output voltages by inverters $VSI_{H}^{(1)}$ and $VSI_{L}^{(1)}$ for first three-phase open-end windings. Similarly, the reference voltage $\overline{v}^{(2)}$ is produced by the sum of $\overline{v}_{H}^{(2)}$ and $\overline{v}_{L}^{(2)}$ reference output voltages by inverters $VSI_{H}^{(2)}$ and $VSI_{L}^{(2)}$ for second three-phase open-end windings, then:

$$\overline{v}^{(1)} = \overline{v}_{H}^{(1)} + \overline{v}_{L}^{(1)}, \tag{12}$$

$$\overline{v}^{(2)} = \overline{v}_{H}^{(2)} + \overline{v}_{L}^{(2)}. \tag{13}$$

By introducing first and second voltage sharing variables say $k_v^{(1)}$ to Eq. (12), $k_v^{(2)}$ to Eq.(13), then output voltage references are predicted as

$$\begin{cases} \overline{v}_{H}^{(1)} = k_v^{(1)}\overline{v}^{(1)} \\ \overline{v}_{L}^{(1)} = \left(1 - k_v^{(1)}\right)\overline{v}^{(1)} \end{cases}, \quad \begin{cases} \overline{v}_{H}^{(2)} = k_v^{(2)}\overline{v}^{(2)} \\ \overline{v}_{L}^{(2)} = \left(1 - k_v^{(2)}\right)\overline{v}^{(2)} \end{cases}. \tag{14}$$

Eq. (14) allows their maximum dc bus voltage utilization by the inverters, the total power developed by the first and second three-phase open-end stator windings is defined as:

$$P^{(1)} = P_{\mathrm{H}}^{(1)} + P_{\mathrm{L}}^{(1)} = \frac{3}{2}\overline{v}^{(1)} \cdot \overline{i}^{(1)};\ P^{(2)} = P_{\mathrm{H}}^{(2)} + P_{\mathrm{L}}^{(2)} = \frac{3}{2}\overline{v}^{(2)} \cdot \overline{i}^{(2)}, \tag{15}$$

where $P_{\mathrm{H}}^{(1)}$, $P_{\mathrm{L}}^{(1)}$ and $P_{\mathrm{H}}^{(2)}$, $P_{\mathrm{L}}^{(2)}$ are the individual powers of inverters $\mathbf{VSI}_{\mathrm{H}}^{(1)}$, $\mathbf{VSI}_{\mathrm{L}}^{(1)}$ and $\mathbf{VSI}_{\mathrm{H}}^{(2)}$, $\mathbf{VSI}_{\mathrm{L}}^{(2)}$.

By comparing Eqs. (14) and (15), the individual power generated by each inverters $\mathbf{VSI}_{\mathrm{H}}^{(1)}$, $\mathbf{VSI}_{\mathrm{L}}^{(1)}$ and $\mathbf{VSI}_{\mathrm{H}}^{(2)}$, $\mathbf{VSI}_{\mathrm{L}}^{(2)}$ is derived as

$$\begin{cases} P_{\mathrm{H}}^{(1)} = k_{\mathrm{v}}^{(1)} P^{(1)} \\ P_{\mathrm{L}}^{(1)} = \left(1 - k_{\mathrm{v}}^{(1)}\right) P^{(1)} \end{cases};\ \begin{cases} P_{\mathrm{H}}^{(2)} = k_{\mathrm{v}}^{(2)} P^{(2)} \\ P_{\mathrm{L}}^{(2)} = \left(1 - k_{\mathrm{v}}^{(2)}\right) P^{(2)}. \end{cases} \tag{16}$$

Therefore, the total power of two three-phase open-end stator windings written in terms of two voltage sharing variables $k_{\mathrm{v}}^{(1)}$ and $k_{\mathrm{v}}^{(2)}$ from Eqs. (15) and (16) can be summarized as [27], [28] and [29]:

$$\begin{aligned} P &= P^{(1)} + P^{(2)} = \left[P_{\mathrm{H}}^{(1)} + P_{\mathrm{L}}^{(1)}\right] + \left[P_{\mathrm{H}}^{(2)} + P_{\mathrm{L}}^{(2)}\right] = \frac{3}{2}\left[\overline{v}^{(1)} \cdot \overline{i}^{(1)} + \overline{v}^{(2)} \cdot \overline{i}^{(2)}\right] \\ P &= \frac{3}{2}\left[\left(\overline{v}_{\mathrm{H}}^{(1)} + \overline{v}_{\mathrm{L}}^{(1)}\right) \cdot \overline{i}^{(1)} + \left(\overline{v}_{\mathrm{H}}^{(2)} + \overline{v}_{\mathrm{L}}^{(2)}\right) \cdot \overline{i}^{(2)}\right] \\ P &= \frac{3}{2}\left[\left(k_{\mathrm{v}}^{(1)}\overline{v}_{\mathrm{H}}^{(1)} + \left(1 - k_{\mathrm{v}}^{(1)}\right)\overline{v}_{\mathrm{L}}^{(1)}\right) \cdot \overline{i}^{(1)} + \left(k_{\mathrm{v}}^{(2)}\overline{v}_{\mathrm{H}}^{(2)} + \left(1 - k_{\mathrm{v}}^{(2)}\right)\overline{v}_{\mathrm{L}}^{(2)}\right) \cdot \overline{i}^{(2)}\right]. \end{aligned} \tag{17}$$

Both the references ($\bar{v}_H^{(1)}$ and $\bar{v}_L^{(1)}$), ($\bar{v}_H^{(2)}$ and $\bar{v}_L^{(2)}$) should be within the range of obtainable output voltages and they depend solely on their dc bus voltages. The total reference voltages $\bar{v}^{(1)}$ and $\bar{v}^{(2)}$ should satisfy if one inverter voltage saturates, the second inverter should generate the missing quantity of voltage in the corresponding two three-phase open-end stator windings.

Second control strategy deals with proper current sharing between two three-phase open-end stator windings. The space vector $\bar{i}_{S1}$ stator current which is determined on the basis of torque and flux commands by Eq. (6) as:

$$\bar{i}_{S1} = \frac{1}{2}\left(\bar{i}^{(1)} + \alpha\bar{i}^{(2)}\right). \tag{18}$$

Introducing the third variable *ki* in Eq. (18), leads to:

$$\begin{cases} \frac{1}{2}\bar{i}^{(1)} = k_i\bar{i}_{S1} \\ \frac{1}{2}\alpha\bar{i}^{(2)} = (1-k_i)\bar{i}_{S1}. \end{cases} \tag{19}$$

Now, from the above expression, the first and second three-phase open-end stator windings currents $\bar{i}^{(1)}$ and $\bar{i}^{(2)}$ are obtained as:

$$\begin{cases} \bar{i}^{(1)} = 2k_i\bar{i}_{S1} \\ \bar{i}^{(2)} = 2(1-k_i)\alpha^{-1}\bar{i}_{S1}. \end{cases} \tag{20}$$

By introducing Eq. (20) in Eq. (7), the space vector stator current $\bar{i}_{S5}^{*}$ which produces the leakage fluxes is determined as:

$$\bar{i}_{S5}^{*} = \frac{1}{2}\left(\bar{i}^{(1)} - \alpha\bar{i}^{(2)}\right) = (2k_i - 1)\bar{i}_{S1}. \tag{21}$$

Also, Eq. (21) can be rewritten in equivalent *d–q* components as

$$\begin{cases} i_{Sd} = (2k_i - 1)i_{1d} \\ i_{Sq} = -(2k_i - 1)i_{1q}. \end{cases} \tag{22}$$

The total motor power can be determined as [27], [28] and [29]:

$$P = P^{(1)} + P^{(2)} = 3\overline{v}_{S1} \cdot \overline{i}_{S1} + 3\overline{v}_{S5} \cdot \overline{i}_{S5}. \tag{23}$$

On the basis of Eq. (7), individual space vectors of two three-phase open-end stator winding currents and voltages are calculated as:

$$\begin{cases} \overline{i}^{(1)} = \overline{i}_{S1} + \overline{i}^{*}_{S5} \\ \overline{i}^{(2)} = \alpha^{-1}\left(\overline{i}_{S1} - \overline{i}^{*}_{S5}\right) \end{cases}, \quad \begin{cases} \overline{v}^{(1)} = \overline{v}_{S1} + \overline{v}^{*}_{S5} \\ \overline{v}^{(2)} = \alpha^{-1}\left(\overline{v}_{S1} - \overline{v}^{*}_{S5}\right). \end{cases} \tag{24}$$

By substituting Eq. (24) in Eq. (15) by considering Eq. (21), leads to predict individual power delivered by the two three-phase open-end stator windings as:

$$\begin{aligned} P^{(1)} &= 3k_i\overline{v}_{S1} \cdot \overline{i}_{S1} + 3k_i\overline{v}^{*}_{S5} \cdot \overline{i}_{S1} \\ P^{(2)} &= 3(1 - k_i)\overline{v}_{S1} \cdot \overline{i}_{S1} - 3(1 - k_i)\overline{v}^{*}_{S5} \cdot \overline{i}_{S1}. \end{aligned} \tag{25}$$

In the characteristic equation of motor given by Eqs. (8) and (10), the space vector $\overline{v}^{*}_{S5}$(generating the leakage flux in the air-gap only) is negligible in comparison to $\overline{v}_{S1}$(generating the magnetic flux in the air-gap). Then Eq. (25) can be approximates as:

$$P^{(1)} \cong k_i P; \;\; P^{(2)} \cong (1 - k_i)P. \tag{26}$$

It has been proved that the third variable *ki* introduced in the power equations is good approximation for current sharing between two open-end stator windings.

Finally from Eqs. (16) and (26), the expression for individual power of each inverters and the total power of the motor with three power sharing variables is given by:

$$\begin{aligned} P_{\mathrm{H}}^{(1)} &\cong k_i \cdot k_v^{(1)} P^{(1)}; \quad P_{\mathrm{H}}^{(2)} \cong k_i \cdot k_v^{(2)} P^{(2)} \\ P_{\mathrm{L}}^{(1)} &\cong k_i \cdot \left(1 - k_v^{(1)}\right) P^{(1)}; \quad P_{\mathrm{L}}^{(2)} \cong k_i \cdot \left(1 - k_v^{(2)}\right) P^{(2)} \\ P &\cong k_i \cdot \left[k_v^{(1)} P_H^{(1)} + \left(1 - k_v^{(1)}\right) P_L^{(1)}\right] + (1 - k_i) \\ &\cdot \left[k_v^{(2)} P_{\mathrm{H}}^{(2)} + \left(1 - k_v^{(2)}\right) P_{\mathrm{L}}^{(2)}\right]. \end{aligned} \tag{27}$$

Fig. 3 elaborates the complete power sharing algorithm based on synchronous FOC. The d_1–q_1 components of the stator currents which produce the flux and torque in the machine, where *d*-axis aligned with the rotor flux, displaced by angle ϑ (rotor flux angle). The reference values in synchronous rotating frame of flux $i_1 d_{,\mathrm{ref}}$ and torque $i_1 q_{,\mathrm{ref}}$ producing currents are derived from the flux and torque commands. Then, d_1–q_1 components of the reference current are directly determined on the basis of the corresponding d_1–q_1 components current ratio. The stator voltage space vectors whose reference values in stationary reference frames $\overline{v}_{S1,\mathrm{ref}}$ and $\overline{v}_{S5,\mathrm{ref}}$ are determined from P–I (proportional–integral) controller output rotated in rotor flux angle.

Further, standard three-level space vector pulse width modulation (SVPWM) is adopted to regulate each couple of inverter ($\mathrm{VSI_H}$ and $\mathrm{VSI_L}$) with two voltage sharing variable to regulate the output voltages as shown in the same Fig. 3[13], [16], [17], [18] and [19]. It is defined $\overline{v}_{S5,\mathrm{ref}} = 0$, the balanced power sharing by the two three-phase open-end stator windings, i.e. quadrupling the power capability of a single VSI with given voltage and current rating.

VERIFICATION WITH SIMULATION TESTS

To verify the proposed power sharing FOC algorithm, the complete model of ac motor is implemented in numerical simulation software (Matlab/Simulink-PLECS) [30] with the parameters taken from Table 1.

Table 1. Parameters taken for simulation software examination of the ac motor.

$P_{rated} = 8$ kW	$R_S = 0.51\ \Omega$
$I_{S,rated} = 16\ A_{rms}$	$R_R = 0.42\ \Omega$
$V_{S,rated} = 125\ V_{rms}$	$L_{S1} = 58.2$ mH
$\omega_{S,rated} = 2\pi 50$ rad/s	$L_R = 58.2$ mH
$P = 2$ (pairs)	$M_1 = 56$ mH
DC bus voltage = 155 V	
Switching frequency = 5 kHz	

First simulation test deals with balanced condition, obtained by setting the three power sharing variable as $k_P^{(1)} = k_P^{(2)} = \mathbf{1/2}$, $ki = \mathbf{1/2}$, electro-magnetic torque $T = 53.4$ N m (rated) and rotor speed fixed at $\omega m = 1440$ rpm (rated). Fig. 4(a) shows the step reference torque command (red color) set to increasing/decreasing between (50–100-–50)% of rated value. It is observed that actual torque (blue color) follows the step variations exactly in every 30 ms.

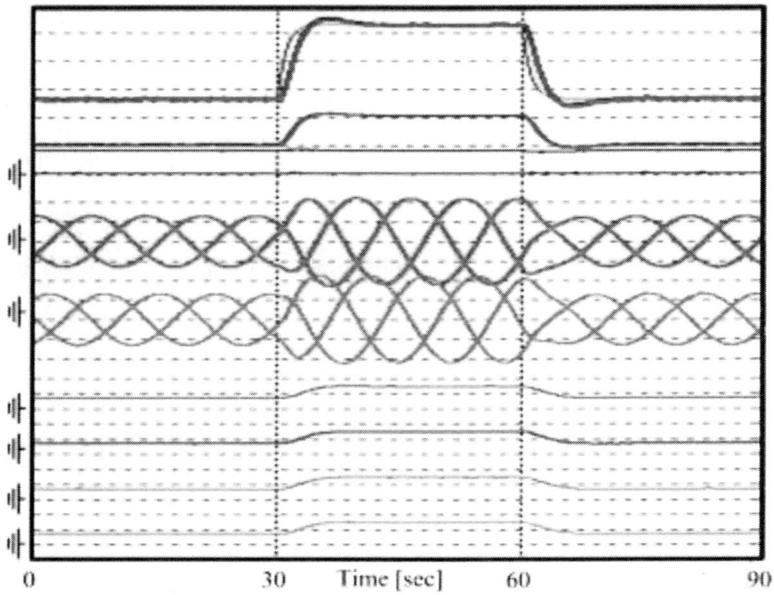

Figure 4. Behavior of six-phase asymmetrical (quasi) open-end stator windings induction motor in balanced condition. Characteristic waveform in sequential order: a. Torque response with respect to step changes (blue-torque, red-step command). [*Y*-axis scale: 10 N m/div]. b. Torque producing current (first-space) component i_1q. [*Y*-axis scale: 10 A/div]. c. Flux producing current (first-space) component i_1d. [*Y*-axis scale: 10 A/div]. d. Leakage current (fifth-space) components i_5q, i_5d. [*Y*-axis scale: 10 A/div]. e. Three-phase current of first open-end stator windings. [*Y*-axis scale: 10 A/div]. f. Three-phase current of second open-end stator windings. [*Y* -axis scale: 10 A/div]. g. DC link current of inverter $VSI_H^{(1)}$. [*Y* -axis scale: 10 A/div]. h. DC link current of inverter $VSI_L^{(1)}$. [*Y* -axis scale: 10 A/div]. i. DC link current of inverter $VSI_H^{(2)}$. [*Y* -axis scale: 10 A/div]. j. DC link current of inverter $VSI_L^{(2)}$. [*Y*-axis scale: 10 A/div].

As expected the synchronous reference frame (first space) currents iq_1 (blue color) andid_1 (violet color), follow the reference torque command and the reference rotor flux command, respectively, as shown by Fig. 4(b) and (c). As determined by Eq. (22)($ki = 1/2$), the synchronous reference frame (fifth space) current iq_5 and id_5 (leakage components) are null as shown by Fig. 4(d), which confirms the balanced operation of the ac motor drive.

First (purple color) and second (turquoise color) three-phase open-end stator windings current have a sinusoidal waveform, as shown in Fig. 4(e) and (f); they change according to the torque demand in every 30 ms. It is confirmed that both the two three-phase (six-phase) stator windings currents have the same amplitude in every 30 ms according to torque demand, and the expected 30° phase displacement between them is also noticeable. Hence, the balanced currents operation of ac motor drive in the entire testing period is proved.

Fig. 4(g)–(j) shows the DC link currents of the four inverters (low-pass filtered by time constant $\tau = 2$ ms) $VSI_H^{(1)}$ (green color), $VSI_L^{(1)}$ (red color), $VSI_H^{(2)}$ (gray color), $VSI_L^{(2)}$(orange color). DC link currents of four VSIs vary equally in amplitude during every 30 ms, following the behavior given by Eqs. (16), (26) and (27), and ensuring balanced power operation among the VSIs.

The trajectories of stationary reference frame currents are shown in Fig. 5 for an additional verification. Where, Fig. 5(a) confirms the sinusoidal nature of first phase current components i_sd_1 and i_sq_1, which actually produces the electromagnetic field in the air-gap and its amplitude depends on torque demands. Fig. 5(b) shows that the fifth space current components i_sd_5 and i_sq_5 are null, confirming the balanced operation of the ac motor drive according to Eq. (21) ($ki = 1/2$). Fig. 5(c) shows that the third space current components i_sd_3 and i_sq_3 are practically zero, confirming that the ac motor drive system is absolutely free of zero-sequence components.

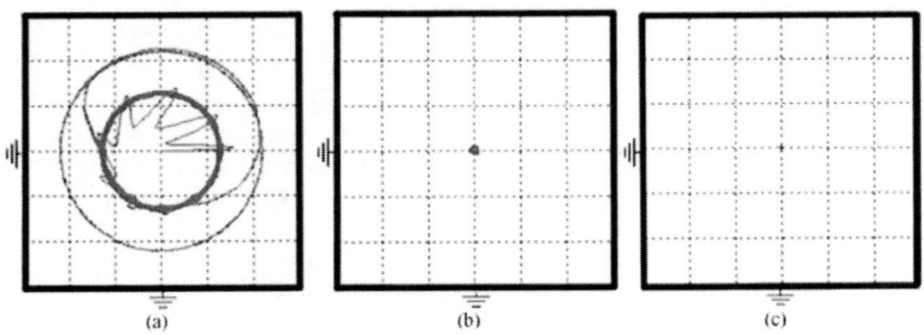

Figure5. Response trajectories in stationary reference frame current components. a. i_{s1} vs i_{s1}, b. i_{s5} vs i_{s5}, c. i_{s3} vsi_{s3}. [X–Y-axis scale: +/−10 A/div].

Second simulation test deals with the verification of first control strategy, where the stator voltages are effective shared between each couple of inverters (VSI_H and VSI_L) of two three-phase open-end stator windings. The three power sharing variables are set to $k_P^{(1)} = \mathbf{1/2{-}1/3}$, $k_P^{(2)} = \mathbf{1/2{-}1/3}$, and $ki = \mathbf{1/2}$, electromagnetic torque is $T = 53.4$ N m (rated) and rotor speed is set to $\omega m = 1100$ rpm. To avoid inverters saturation (speed directly proportional to voltage), the rotor speed ωm is reduced from 1440 rpm (rated) to 1100 rpm. From Fig. 6(a) it is observed that actual torque (blue color) exactly follows the set reference torque command (red color). It also clearly indicates that torque is unaffected by the transient introduced in the two power sharingvariable $(k_V^{(1)}, k_V^{(2)})$.

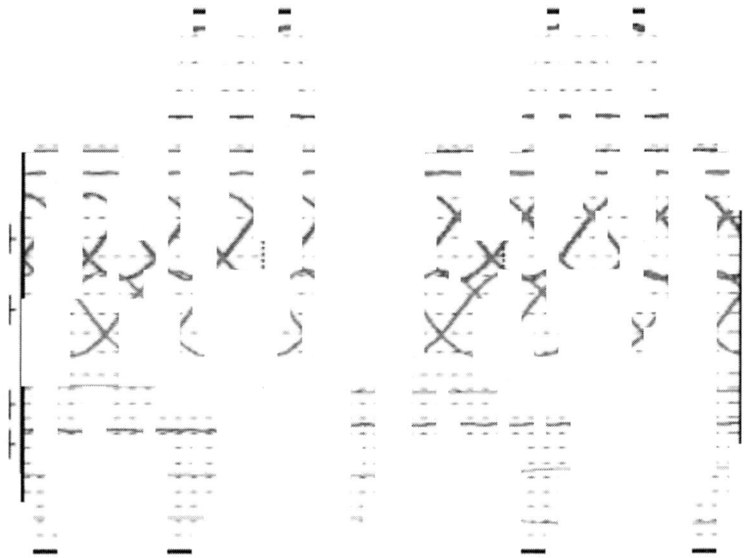

Figure 6. Behavior of six-phase asymmetrical (quasi) open-end stator windings induction motor in first prediction when unbalanced voltage sharing between each couple of inverters (VSI_H and VSI_L) condition. Characteristic waveform in sequential order: a. Torque response with respect to step changes (blue-torque, red-step command). [*Y*-axis scale: 10 N m/div]. b. Torque producing current (first-space) component i_1q. [*Y*-axis scale: 10 A/div]. c. Flux producing current (first-space) component i_1d. [*Y*-axis scale: 10 A/div]. d. Leakage current (fifth-space) components i_5q, i_5d. [*Y*-axis scale: 10 A/div]. e. Three-phase current of first open-end stator windings. [*Y*-axis scale: 10 A/div]. f. Three-phase current of second open-end stator windings. [*Y* -axis scale: 10 A/div]. g. DC link current of inverter $VSI_H^{(1)}$. [*Y* -axis scale: 10 A/div]. h. DC link current of inverter $VSI_L^{(1)}$. [*Y* -axis scale: 10 A/div]. i. DC link current of inverter $VSI_H^{(2)}$. [*Y* -axis scale: 10 A/div]. j. DC link current of inverter $VSI_L^{(2)}$. [*Y*-axis scale: 10 A/div].

As expected, the synchronous reference frame (first space) currents iq_1 (blue color) andid_1 (violet color), follow the reference torque command and the reference rotor flux command, respectively, as shown by Fig. 6(b) and (c). Also in this test, the synchronous reference frame (fifth space) current iq_5 and id_5 (leakage components) are null as shown by Fig. 6(d), as determined by Eq. (22) ($ki = 1/2$).

First (purple color) and second (turquoise color) three-phase open-end stator windings current (purple color) have a sinusoidal waveform as shown by Fig. 6(e) and (f). Six-phase currents maintain equal amplitude in the whole period confirming balanced current operation according to constant torque; a correct 30° phase displacement between them is observed. It is noted that currents are unaffected by the transient created by the two voltage sharing variables when subjected to a change in their values.

Fig. 6(g)–(j) shows that the DC link currents of four VSIs vary its values according to Eq.(16). Therefore, it is confirmed that sum of power delivered by inverters ($VSI_H^{(1)}$ and $VSI_L^{(1)}$)is equal to $P^{(1)}$. In detail, when $k_r^{(1)} = 1/2-1/3$ changes its value at 30 ms, $VSI_H^{(1)}$ provides only 1/3 of $P^{(1)}$ whereas inverter $VSI_L^{(1)}$ provides 2/3 of $P^{(1)}$. Their sum correctly maintains the constant power delivered to the first three-phase open-end stator winding. Similarly, the sum of power delivered by inverters ($VSI_H^{(2)}$ and $VSI_L^{(2)}$) is equal to $P^{(2)}$. In particular when $k_r^{(2)} = 1/2-2/3$ changes its value at 60 ms, $VSI_H^{(2)}$ provides only 2/3 of $P^{(2)}$ whereas inverter $VSI_L^{(2)}$ provides 1/3 of $P^{(2)}$. Their sum correctly maintains the constant power delivered to the second three-phase open-end stator winding. Hence the dc currents of VSIs confirm the effectiveness of first control strategy, sharing the voltages between inverters (VSI_H and VSI_L) of two three-phase open-end stator windings (according to Eqs. (16) and (27)).

The trajectories of stationary reference frame currents are shown in Fig. 7. Fig. 7(a) shows that the first phase current components i_sd_1 and i_sq_1, are sinusoidal with steady-state constant amplitude, depending on the given torque demand. This again confirms that both torque and motor currents are unaffected from the transient created by the two voltage sharing variables. Fig. 7(b) shows that the fifth space current components $i_sd_5$and i_sq_5 are zero, according to Eq. (21) ($ki = 1/2$). Fig. 7(c) shows that the third space current components i_sd_3 and i_sq_3 are practically zero, confirming that the ac motor drive system is absolutely free of zero-sequence components.

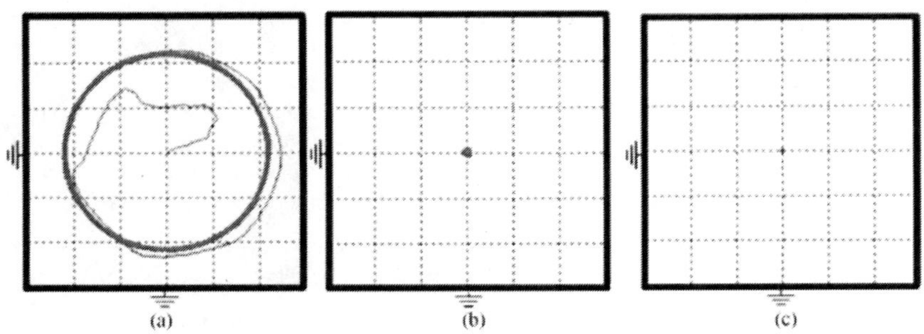

Figure 7. Response trajectories in stationary reference frame current components. a. i_{s1} vs i_{s1}, b. i_{s5} vs i_{s5}, c. i_{s3} vsi_{s3}. [X–Y-axis scale: +/−10 A/div].

Third simulation test deals with the verification of second control strategy, where the currents are effectively shared between two three-phase open-end stator windings. The three power sharing variables are set to $k_r^{(1)} = k_r^{(2)} = 1/2$, *ki* = **1/2–1/3–2/3**, electromagnetic torque is $T = 38$ N m, and rotor speed is set to $\omega m = 1430$ rpm (rated). To avoid over-current delivery from inverters w.r.t. rated power (torque directly proportional to current), the torque reference is reduced from 53.4 N m (rated) to 38 N m. From Fig. 8(a), it is observed that actual torque (blue color) exactly follows the set reference torque command (red color). It also indicates clearly that torque is unaffected by the transient created by current sharing variable *ki* when subjected to a change in its value.

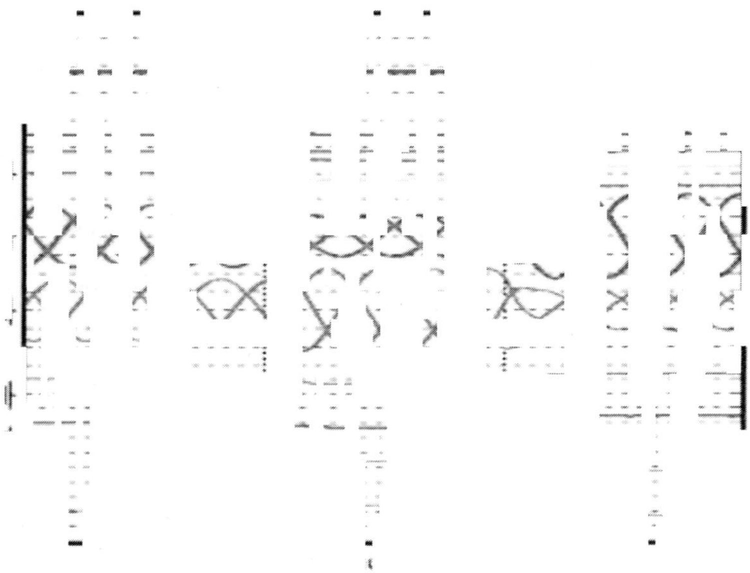

Figure 8. Behavior of six-phase asymmetrical (quasi) open-end stator windings induction motor in second prediction when unbalanced current sharing between two open-end stator windings. Characteristic waveform in sequential order: a. Torque response with respect to step changes (blue-torque, red-step command). [*Y*-axis scale: 10 N m/div]. b. Torque producing current (first-space) component i_{1q}. [*Y*-axis scale: 10 A/div]. c. Flux producing current (first-space) component i_{1d}. [*Y*-axis scale: 10 A/div]. d. Leakage current (fifth-space) components i_{5q}, i_{5d}. [*Y*-axis scale: 10 A/div]. e. Three-phase current of first open-end stator windings. [*Y*-axis scale: 10 A/div]. f. Three-phase current of second open-end stator windings. [*Y* -axis scale: 10 A/div]. g. DC link current of inverter $VSI_H^{(1)}$. [*Y* -axis scale: 10 A/div]. h. DC link current of inverter $VSI_L^{(1)}$. [*Y* -axis scale: 10 A/div]. i. DC link current of inverter $VSI_H^{(2)}$. [*Y* -axis scale: 10 A/div]. j. DC link current of inverter $VSI_L^{(2)}$. [*Y*-axis scale: 10 A/div].

As expected the synchronous reference frame (first space) currents iq_1 (blue color) andid_1 (violet color), follows the reference torque command and the reference rotor flux command as shown by Fig. 8(b) and (c). Now, the synchronous reference frame (fifth space) currents iq_5 and id_5 are null up to first 30 ms, proving the balanced operation, as shown in Fig. 8(d). Furthermore, the currents are generated according to Eq. (21)(ki = **1/2–1/3–2/3**) at 30 ms and 60 ms, ensuring the proper currents sharing between two three-phase stator windings, as shown in the same Fig. 8(d).

First (purple color) and second (turquoise color) three-phase open-end stator windings current (purple color) have a sinusoidal waveform as shown by Fig. 8(e) and (f). It is observed that until 30 ms both three-phase stator winding currents have the same amplitude, confirming the balanced operation. Then six-phase currents share 1/3 and 2/3 of the total current (power) demand at 30 ms and 60 ms for the reference torque requirement, according to Eq. (20). Furthermore, it is noticed that the correct 30° phase displacement is maintained between the two three-phase stator currents, and the torque is unaffected during the transients of the current sharing variable.

Fig. 8(g)–(j) shows that DC link current of the four VSIs changes its value according to Eq. (26), confirming the current sharing between the two three-phase stator windings. In particular, DC currents have same amplitude until 30 ms; once the current sharing variable changes its value ki = 1/2–1/3–2/3 at 30 ms and 60 ms, it observed that DC currents drops in inverters $VSI_H^{(1)}$ and $VSI_L^{(1)}$ are compensated by increasing DC currents in inverters $VSI_H^{(2)}$ and $VSI_L^{(2)}$, and vice-versa. It is concluded that the total current remains always constant as per torque demand.

The trajectories of stationary reference frame currents are shown in Fig. 9. In particular,Fig. 9(a) shows that the first phase current components i_sd_1 and i_sq_1 have a sinusoidal waveform, with constant amplitude at steady-state, depending on the torque demand. This again confirms that the torque is unaffected by the transient created by the current sharing change. Fig. 9(b) shows the fifth space current components i_sd_5 and i_sq_5, generated according to Eq. (22) (ki = 1/2–1/3–2/3). This further confirms that the fifth space is responsible for the current sharing between two stator open-end windings. Fig. 9(c) shows that the third space current components i_sd_3 and i_sq_3 is practically zero, confirming that the ac drive system is absolutely free of zero-sequence components.

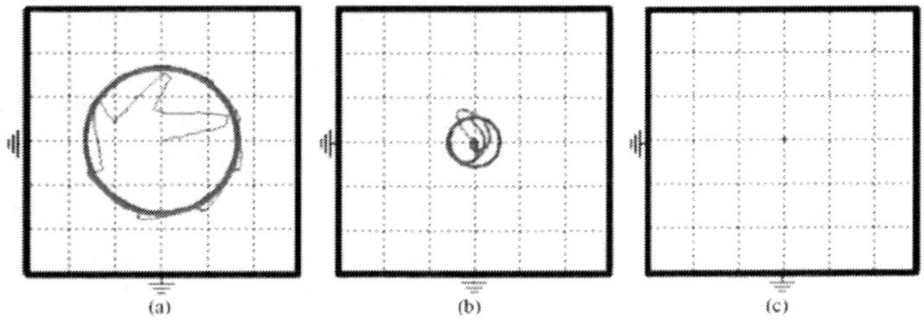

Figure9. Response trajectories in stationary reference frame current components. a. i_{s1} vs i_{s1}, b. i_{s5} vs i_{s5}, c. i_{s3} vsi_{s3}. [X–Y-axis scale: +/−10 A/div].

Finally, it is confirmed from all this numerical simulation tests that the proposed power sharing algorithm (FOC) is effective, maintaining the constant power delivered to motor irrespective of sharing voltage and/or currents among the two three-phase stator open-end windings.

CONCLUSION

This paper presented an original and effective power sharing algorithm based on field oriented control (FOC) in synchronous reference frame for six-phase (asymmetrical) open-end stator windings motor. Power driver circuit consists of four customary three-phase voltage source inverters (VSIs), each one connected across the open-ends of the stator windings to framed multi-phase inverter drive configuration. The standard three-level space vector modulation is adopted for each VSI to behave as three-level output voltage generator. The proposed power sharing algorithm consists of three variables corresponding to three degrees of freedom in control. Two variables can share the two voltages between two inverters of each two three-phase open-end stator windings. The third variable can share the current between two open-end stator windings. Detailed investigation are carried out by the implementation of the complete ac drive model with power sharing (FOC) algorithm in numerical simulation software, and all the proposed balanced/unbalanced theoretical developments successfully are verified.

ACKNOWLEDGEMENT

We authors would like to acknowledge Ministry of University Research (**MIUR**), Government of Italy, for providing research fund for carry out this research activity with Power Electronics Laboratory, Department of Electrical, Electronics and Information Engineering, University of Bologna, Viale Risorgimento 2, Bologna-40136, Italy.

REFERENCES

1. E. Levi, Multiphase electric machines for variable-speed applications, IEEE Trans. Ind. Electron. 55 (5) (2008) 1893e1909.
2. E. Levi, R. Bojoi, F. Profumo, H.A. Toliyat, S. Williamson, Multiphase induction motor drives e a technology status review, IET Electr. Power Appl. J. 1 (4) (2007) 489e516.
3. W. Cao, B.C. Mecrow, G.J. Atkinson, J.W. Bennett, D.J. Atkinson, Overview of electric motor technologies used for more electric aircraft (MEA), IEEE Trans. Ind. Electron. 59 (9) (2012) 3523e3531.
4. M. Jones, E. Levi, Series connected quasi-six-phase two-motor drives with independent control, J. Math. Comput. Simul. 71 (2006) 41e424. Elsevier Publications.
5. Y. Zhao, T.A. Lipo, Space vector PWM control of dual three-phase induction machine using vector space decomposition, IEEE Trans. Ind. Appl. 31 (5) (1995) 1100e1109.
6. R. Bojoi, F. Farina, F. Profumo, A. Tenconi, Dual three-phase induction machine drives controlda survey, IEE J. Trans. Ind. Appl. 126 (4) (2006).
7. R. Bojoi, M. Lazzari, F. Profumo, A. Tenconi, Digital field oriented control of dual three-phase induction motor drives, IEEE Trans. Ind. Appl. 39 (2003) 75e760.
8. R. Bojoi, F. Farina, G. Griva, F. Profumo, A. Tenconi, Direct torque control for dual three-phase induction drives, IEEE Trans. Ind. Appl. 41 (6) (2005) 1627e1636.
9. G.K. Singh, K. Nam, S.K. Lim, A simple indirect field-oriented control scheme for multiphase induction machine, IEEE Trans. Ind. Electron. 52 (4) (2005) 1177e1184.
10. A. Tessarolo, C. Bassi, Stator harmonic currents in VSI-fed synchronous motors with multiple three-phase armature windings, IEEE Trans. Energy Convers. 25 (4) (2010) 974e982.
11. R.H. Nelson, P.C. Krause, Induction machine analysis for arbitrary displacement between multiple winding sets, IEEE Trans. Power Apparatus Syst. (1974) 841e848.

12. G. Grandi, A. Tani, G. Serra, Space vector modulation of six-phase VSI based on three-phase decomposition, in: Proceeding IEEE 19th International Conference Symposium on Power Electronics, Electrical Drives, SPEEDAM'08, Taormina, Italy, 2008, pp. 674e679.
13. G. Grandi, C. Rossi, A. Lega, D. Casadei, Multilevel operation and input power balancing for a dual two-level inverter with insulated dc sources, IEEE Trans. Ind. Appl. 44 (6) (2008) 1815e1824.
14. N. Bodo, E. Levi, M. Jones, Investigation of carrier-based PWM techniques for a five-phase open-end winding drive topology, IEEE Trans. Ind. Electron. 60 (5) (2013) 2054e2065.
15. J. Prieto, M. Jones, F. Barrero, E. Levi, S. Toral, Comparative analysis of discontinuous and continuous PWM techniques in VSI-fed five-phase induction motor, IEEE Trans. Ind. Electron. 58 (12) (2011) 5324e5335.
16. O. Dordevic, E. Levi, M. Jones, A vector space decomposition based space vector PWM algorithm for a three-level seven-phase voltage source inverter, IEEE Trans. Power Electron. 28 (2) (2013) 637e649.
17. M. Jones, N. Satiawan, N. Bodo, E. Levi, A dual five-phase space vector modulation algorithm based on the decomposition method, IEEE Trans. Ind. Appl. 48 (6) (2012) 2110e2120.
18. K. Marouani, L. Baghli, D. Hadiouche, A. Kheloui, A. Rezzoug, A new PWM strategy based on a 24-sector vector space decomposition for a six-phase VSIfed dual stator induction motor, IEEE Trans. Ind. Electron. 55 (5) (2008) 1910e1920.
19. O. Lopez, D. Dujic, M. Jones, F. Freijedo, J. Doval-Gandoy, E. Levi, Multidimensional two-level multiphase space vector PWM algorithm and its comparison with multi frequency space vector PWM method, IEEE Trans. Ind. Electron. 58 (2) (2011) 465e475.
20. L. Alberti, N. Bianchi, Experimental tests of dual three-phase induction motor under faulty operating condition, IEEE Trans. Ind. Electron. 59 (5) (2012) 2041e2048.
21. F. Meinguet, P. Sandulescu, X. Kestelyn, E. Semail, A method for fault detection and isolation based on the processing of multiple diagnostic indices: application to inverter faults in AC drives, IEEE Trans. Vehic. Technol. 62 (3) (2013) 995e1009.
22. J.M. Apsley, De-rating of multiphase induction machines due to supply imbalance, IEEE Trans. Ind. Appl. 46 (2) (2010) 798e805.
23. I.P. Georgakopoulos, E.D. Mitronikas, A.N. Safacas, Detection of induction motor faults in inverter drives using inverter input current analysis, IEEE Trans. Ind. Electron. 58 (9) (2011) 4365e4373.
24. F. Betin, G. Capolino, Shaft positioning for six-phase induction machines with open phases using variable structure control, IEEE Trans. Ind. Electron. 59 (6) (2012) 2612e2620.

25. G. Grandi, A. Tani, P. Sanjeevikumar, D. Ostojic, Multi-phase multi-level ac motor drive based on four three-phase two-level inverters, in: Proceeding IEEE 20th International Conference Symposium on Power Electronics, Electrical Drives, SPEEDAM'10, Pisa, Italy, 2010, pp. 1768e1775.
26. G. Grandi, P. Sanjeevikumar, Y. Gritli, F. Filippetti, Experimental investigation of fault-tolerant control strategies for quad-inverter converters, in: Proceeding IEEE International Conference on Electrical System for Aircraft, Railway and Ship Propulsion, ESARS'12, Bologna, Italy, 2012.
27. A. Tani, M. Mengoni, L. Zarri, G. Serra, D. Casadei, Control of multi-phase induction motors with an odd number of phases under open circuit faults, IEEE Trans. Power Electron. 27 (2) (2012) 565e577.
28. G. Grandi, G. Serra, A. Tani, General analysis of multiphase systems based on space vector approach, in: Proceeding International Power Electronics and Motion Control Conference, EPE-PEMC'06, Portoroz, Slovenia, 2006, pp. 834e840.
29. A. Tessarolo, G. Zocco, C. Tonello, Design and testing of a 45-MW100-Hz quadruple-star synchronous motor for a liquefied natural gas turbocompressor drive, IEEE Trans. Ind. Appl. 47 (3) (2011) 1210e1219.
30. H. Abu-Rub, A. Iqbal, J. Guzinski, High Performance Control of AC Drives with Matlab/Simulink Models, first ed., John Wiley & Sons Ltd., Publications, UK, 2012.

CITATION

Sanjeevikumar Padmanaban, Gabriele Grandi, Frede Blaabjerg, Joseph Olorunfemi Ojo, Patrick William Wheeler, Power sharing algorithm for vector controlled six-phase AC motor with four customary three-phase voltage source inverter drive, Engineering Science and Technology, an International Journal, Volume 18, Issue 3, September 2015, Pages 408-415, ISSN 2215-0986, http://dx.doi.org/10.1016/j.jestch.2015.02.002.

CHAPTER 7

Control Strategy and Hardware Implementation for DC–DC Boost Power Circuit Based on Proportional–Integral Compensator for High Voltage Application

Sanjeevikumar Padmanaban[1], Ersan Kabalci[2], Atif Iqbal[3], Haitham Abu-Rub[4], Olorunfemi Ojo[5]

[1]School of Electrical & Electronics Engineering, Dublin Institute of Technology, Dublin 8, Ireland
[2]Department of Electrical & Electronics Engineering, Nevsehir University, Nevsehir 50300, Turkey
[3]Department of Electrical Engineering, Qatar University, Post Box Number. 2713, Doha, Qatar
[4]Department of Electrical & Computer Engineering, Texas A&M University, Post Box Number. 23874, Doha, Qatar
[5]Center for Energy System Research, Department of Electrical & Computer Engineering, Tennessee Technical University, Cookeville, TN 38505, USA

ABSTRACT

For high-voltage (HV) applications, the designers mostly prefer the classical DC–DC boost converter. However, it lacks due to the limitation of the output voltage by the gain transfer ratio, decreased efficiency and its requirement of two sensors for feedback signals, which creates complex control scheme with increased overall cost. Furthermore, the output voltage and efficiency are reduced due to the self-parasitic behavior of power circuit components. To overcome these drawbacks, this manuscript provides, the theoretical development and hardware implementation of DC–DC step-up (boost) power converter circuit for obtaining extra output-voltage high-performance. The proposed circuit substantially improves the high output-voltage by voltage-lift technology with a closed loop proportional–integral controller. This complete numerical model of the converter circuit including closed loop P-I controller is developed in simulation (Matlab/Simulink) software and the hardware prototype model is implemented with digital signal processor (DSP) TMS320F2812. A detailed performance analysis was carried out under both line and load regulation conditions. Numerical simulation and its verification results provided in this paper, prove the good agreement of the circuit with theoretical background.

INTRODUCTION

The classical DC–DC step-up (boost) power converter drastically suffers the restricted output-voltage due to the transfer gain ratio. Both output-voltage and efficiency are further reduced due to the self-parasitic behavior of the power circuit components. Moreover, two sensors (voltage and current) which are essential for control algorithm, increases the complexity and overall cost of the system when subjected to high-voltage applications [1], [2], [3] and [4]. Voltage-lift technology (i.e. additional passive (inductor/capacitor) components inclusion), implemented in power circuit design of electronic components is integrated to DC–DC power circuit system recently [1], [2], [3],[4], [5], [6], [7] and [8]. This provides an opportunity to design extra high-voltage (EHV) DC–DC power circuit, which actually derived from classical DC–DC buck power converter configuration [9], [10] and [11].

Alternate solutions are also found in literatures to increase the output voltage, particularly on isolating the load side in order to ensure high-voltage operation of the DC–DC converter [12], [13] and [14], while others were focused on [15], [16], [17] and [18].

- The boot-strap capacitors
- Power-conditioning circuits
- High-frequency transformers

The main drawback of these HV power converters are reduced efficiency and increased ripple in output waveforms. Although the isolated DC–DC power converters operate safely in HV applications, yet they are costlier and bulky in size due to inclusion of transformer. Furthermore, in case of utilization of the dynamically slow systems such as battery and fuel cell, the overall efficiency and the response of the system will be decreased in comparison to transformer-less configuration.

The performance of the proposed EHV power circuit is advantageous in comparison to classical DC–DC configuration [1] and [2] as:

- Increased voltage transfer (k) gain ratio
- Wide range of control and reduced ripple components at the outputs
- Increased power density as well as efficiency
- Required closed loop P–I controller with one sufficient sensor (voltage feedback)

The numerical simulation software (Matlab/Simulink) model of the entire converter system based on governing equations is used for the operational modes of EHV power circuit unit and the analytical development of the entire system. The hardware prototype is implemented with complete controller algorithm written in TMS320F2812 processor. Analyses are carried out on both line and load variations to determine the performance of the whole EHV DC–DC converter system. Numerical simulations with experimental verification results provided in this paper matches closely with the analytical predictions.

EHV BOOST POWER CONVERTER CONFIGURATION

EHV DC–DC boost power circuit diagram shown in Fig. 1(a), represents that the converter consists of N-channel MOSFET as static switch S, diodes (D, D_1, ... , D_{13}),voltage-lift circuit with an inclusion of additional capacitors (C, C_1, ... , C_4) and inductors (L, L_0, ... , L_3). The voltage-lift gain is obtained by the capacitors (C_2, C_3& C_4). The capacitor voltage (V_C) is built-up by the gain transfer ratio with the four times of the battery input voltage V_S ($V_{C0} = V_C1$). Fig. 1(a) indicates the voltage and current flow directions in the circuit. When the circuit operates in the continuous conduction mode, it is assumed that all components are ideal and capacitors are large enough. The load voltage (V_0), load current (I_0), supply voltage (V_I) and supply current (i_I) of the power circuit are depicted in the same Fig. 1(a) and its complete analytical predictions are provided in the next section.

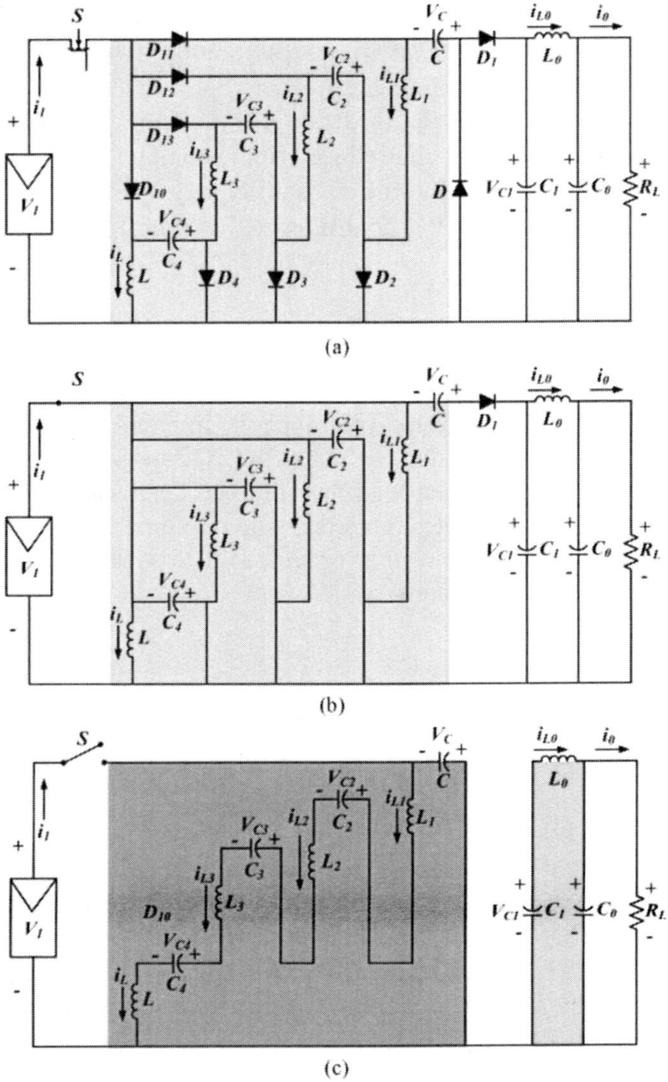

Figure 1. (a) Investigated extra high voltage (HV) DC–DC boost power converter configuration; (b) Equivalent power converter circuit configuration during switch turned ON period; (c) Equivalent power converter circuit configuration during switch turned OFF period.

ANALYSIS OF CONVERTER OPERATING MODES

Fig. 1(b) explains and provides the details about the instant at which the switch *S* is turned ON and its corresponding equivalent circuit. At this moment, the instantaneous input current ($i1$-source current) gets equal to sum of all capacitor and inductor currents except i_{C0} and i_{C1}. The load current (I_O-load) flows through the circuit depending on the summation of the battery supply voltage V_1 and the capacitor voltage V_C. In this circumstance, the capacitors C_2, C_3 and C_4 are charged by the input voltage and increase all inductor currents.

Fig. 1(c) represents about the instant at which the switch S is turned OFF and its corresponding equivalent circuit. The instantaneous input current ($i1$-source current) equals to zero at this interval. The inductors present in the voltage-lift part which is located on the left-hand side of capacitor C (L_1, L_2, L_3 and L) store the energy, while the capacitor C_2, C_3 and C_4 discharges the stored energy. Therefore, corresponding directions that are leading to charge the capacitor C are shown in the Fig. 1(c). At this situation, the current (i_{L0}) flows through the load over the inductor and decreases the current with all the inductors. In steady–state condition, the average inductor voltage will exist over a period as given below:

$$V_{C0} = V_0 \tag{1}$$

When switch $S$is turned ON, the circuit voltages appear as:

$$V_{C2} = V_{C3} = V_{C4} = V_1 \tag{2}$$

that also yields:

$$V_0 = V_{C1} = V_C + V_1 \tag{3}$$

It should be noted that, the inductor current IL is increased during the period when the switch S is turned ON and is decreased during switch S is turned OFF. Therefore, the voltage components V_1 as well $-V_{L\text{-}OFF}$, which actually predicts the voltages across the inductor L as:

$$kTV_1 = (1 - k)\cdot TV_{L\text{-}OFF} \tag{4}$$

$$V_{L\text{-}OFF} = [k/(1 - k)]\cdot V_1 \tag{5}$$

Whereas, the voltages across inductors (L_1, L_2 and L_3) are determined as:

$$V_{L1\text{-}OFF} = [k/(1 - k)]\cdot V_1 \tag{6}$$

$$V_{L2\text{-}OFF} = [k/(1 - k)]\cdot V_1 \tag{7}$$

$$V_{L3\text{-}OFF} = [k/(1 - k)]\cdot V_1 \tag{8}$$

The capacitor voltage VC and output voltage $V0$ are analytically predicted from Fig. 1(c) as:

$$V_C = V_{C\text{-}OFF} = V_{L\text{-}OFF} + V_{L1\text{-}OFF} + V_{L2\text{-}OFF} + V_{L3\text{-}OFF} + V_{C2} + V_{C3} + V_{C4} \tag{9}$$

$$V_C = 4k/(1 - k)\cdot V_1 + 3V_1 \tag{10}$$

$$V_0 = V_C + V_1 \tag{11}$$

$$V_0 = [4/(1 - k)]\cdot V_1 \tag{12}$$

Gain transfer ratio of output current and output voltage are predicted as:

$$i_0 = [(1 - k)/4]\cdot i_1, \tag{13}$$

$$M_0 = 4/(1 - k). \tag{14}$$

Finally, the average voltages and currents are described as:

$$V_C = [(3 + k)/(1 - k)] \cdot V_1, \tag{15}$$

$$V_{C1} = V_0, \tag{16}$$

$$V_{C2} = V_{C3} = V_{C4} = V_1. \tag{17}$$

$$i_{L0} = i_0, \tag{18}$$

$$i_L = [k/(1 - k)] \cdot i_0, \tag{19}$$

$$i_{L1} = i_{L2} = i_{L3} = i_L + i_{L0} = [1/(1 - k)] i_0. \tag{20}$$

Table 1 explains the comparison between the EHV DC–DC boost proposed converter with the classical DC–DC boost converter with respected to voltage-life transfer gain k for better clarification. In addition, Table 2 proves the efficiency of the proposed power circuit in comparison with the classical DC–DC boost converter. It is noted that the classical one varies its output voltage from 1.11 V to 90 V with the set duty ratio k variation ranges from 0.1 to 0.9 and battery input voltage is 10 V respectively, while the proposed power circuit varies its output voltage from 44.44 V to 400 V.

Table 1. Comparative performances emphasis the proposed DC–DC boost power converter with classical boost power converter.

Converter type	Output voltage (V_O) (Volts)	Output current (i_O) (Amps)
Classical converter	$V_O = [k/(1 - k)] \cdot V_1$	$i_O = [(1 - k)/k] \cdot i_1$
Proposed converter	$V_O = [4/(1 - k)] \cdot V_1$	$i_O = [(1 - k)/4] \cdot i_1$

Table 2. Comparative performances emphasis the simulation result of the proposed power circuit with classical power circuit unit (steady-state rated condition).

Duty ratio (k)	Output voltage (V_O) Classical converter	Output voltage (V_O) Proposed topology
0.1	1.11	44.44
0.2	2.5	50
0.3	4.28	57.14
0.4	6.66	66.66
0.5	10	80
0.6	15	100
2/3	19.999	120
0.7	23.33	133.33
0.8	40	200
0.9	90	400

Fig. 2 shows the closed loop P-I controller scheme for the power circuit and notably consists of one voltage sensor feedback without having any additional sensing unit for current. The feedback signal obtained from the DC load voltage is then compared with the set reference voltage V_{dc}. Hence the obtained error between the set reference and feedback signal is applied to the P-I controller to compensate the available error. The manipulated signal provided by the P-I controller defines the set duty ratio k for pulse-width modulation (PWM) generator. The achieved set duty ratio k is then compared with the high frequency ramp-signal to provide controlled pulses for the static switch S. The controller parameters (P-proportional gain, I-integral gain) are fine-tuned to get the set reference DC voltage under different perturbation conditions. The P-I regulator transfer function are given by $G(s) = \frac{0.065011(s+9000)}{s}$ (simulation) and $G(s) = \frac{2.8(s+12000)}{s}$ (hardware), parameter are obtained based on bode plot technique and design concepts provided by [10] and [20].

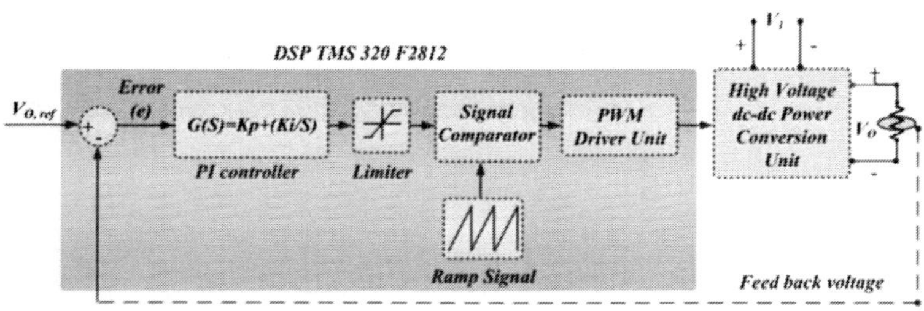

Figure 2. Schematic circuit of simplified closed loop control strategy for power conversion unit under line/load regulation.

Fig. 3 elaborates the hardware prototype model implementation based on TMS320F2812 processor with N-channel MOSFET (IRFPC60 version) EHV DC–DC power converter circuit. A set of numerical simulation as well as experimental results obtained are provided and discussed in detail in the next section.

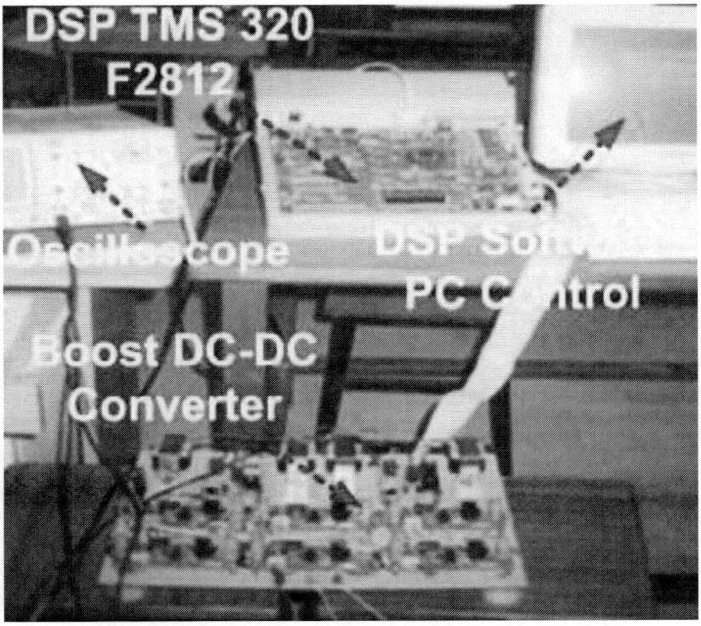

Figure 3. Hardware prototype module of extra HV DC–DC boost power converter with DSP TMS320F2812 processor in closed loop operation.

IMPLEMENTATION AND ANALYSIS OF PROPOSED DC–DC BOOST CONVERTER

a) Simulation and experimental results

Table 3, indicates the parameters taken into account for the numerical simulation test as well as experimental verifications. Inductance and capacitance values are determined with respect to the criteria of 5% (<2 mV) output ripple requirement as per IEEE standards [1] and [19].

Table 3. Parameters taken for simulation and hardware prototype for investigations.

Parameter		Value
Input voltage	(V_I)	10 V
Inductance	(L)	100 μH
Capacitance	(C)	5 μF
Load resistance	(R)	44 Ω
Duty ratio	(k)	2/3
Switching	(F_{SW})	50 KHz
Digital processor	DSP	TMS 320F2812
N-channel MOSFET	(IRFPC60)	$V_{DSS} = 600$ V, $R_{DS(ON)} = 0.40$ Ω, $I_D = 16$ A

Fig. 4 and Fig. 5 shows the simulated and experimental results at the rated condition of the set reference output voltage 120 V with the load resistance of 44Ω and duty ratio k = 2/3 (constant) of the proposed EHV DC–DC power converter. It is observed from the experimental result that the output voltage is slightly lower than the set reference and settles at 119.8 V with the negligible loss of 200 mV, which is caused by the self-parasitic effects of power circuit that is practically appreciable. This confirms that the output voltage closely matches (simulation/experimental) and verifies the prediction given by Eq. (12) at k = 2/3.

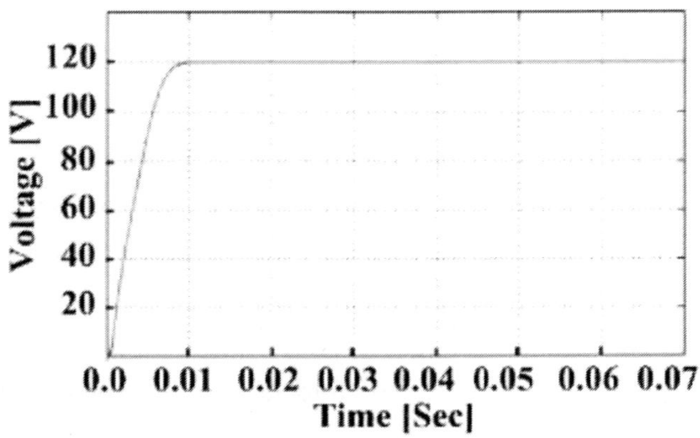

Figure 4. Output voltage of extra HV DC–DC boost power converter with transient behavior (Simulation).

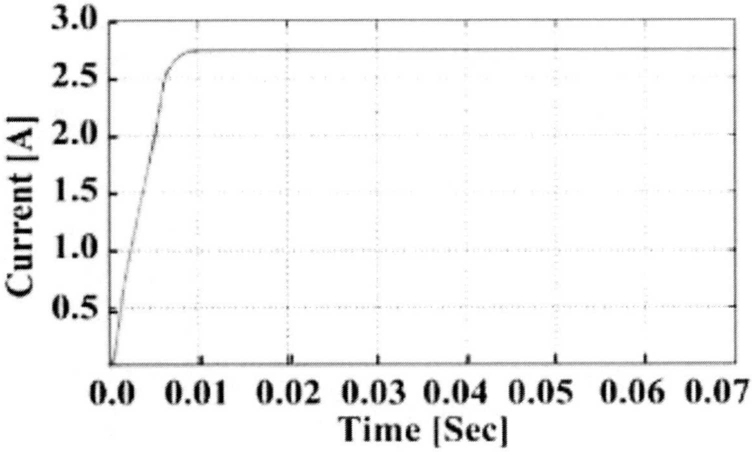

Figure 5. Output current of extra HV DC–DC boost power converter with transient behavior (Simulation).

Correspondingly, Fig. 6 and Fig. 7 shows the output currents such as the simulated (2.7272 A) and experimental (2.722 A) observed. Obviously a small deviation is noticeable with the experimental result, as explained above due to the parasitic effects, which verifies the prediction given by Eq. (13) and satisfies the power-circuit laws.

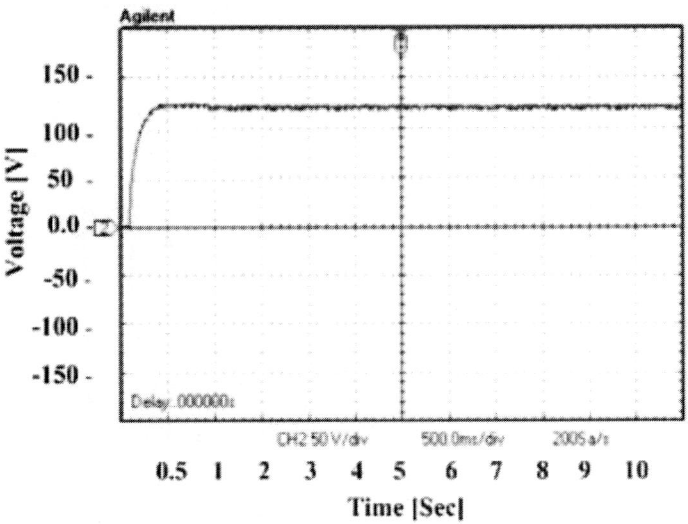

Figure 6. Output voltage of extra HV DC–DC boost power converter with transient behavior (Experimental).

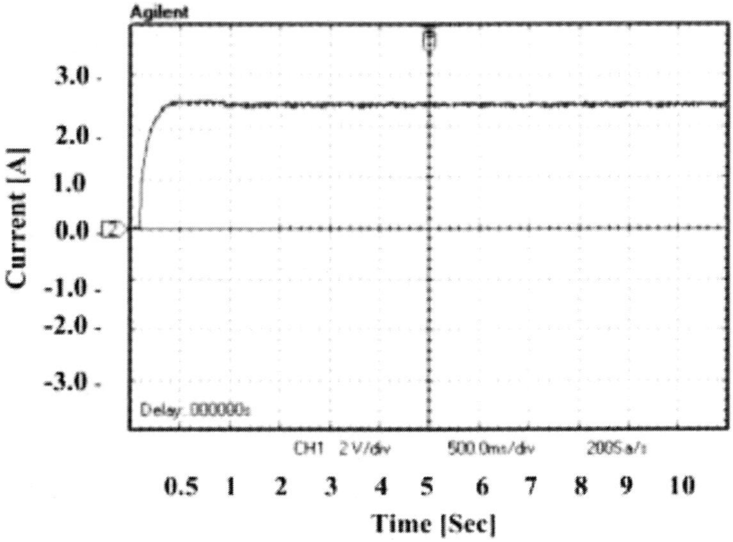

Figure 7. Output current of extra HV DC–DC boost power converter with transient behavior (Experimental).

Fig. 8 and Fig. 9 shows the output voltages where both the simulated (120 V) and experimental (119.8 V) results are obtained under same line and load perturbation condition. It is noticed that the output voltage stabilized at 0.005 s (simulated) and 0.625sec (experimental) by the closed-loop controller when the battery voltage represents a step variation from 10 V to 9 V.

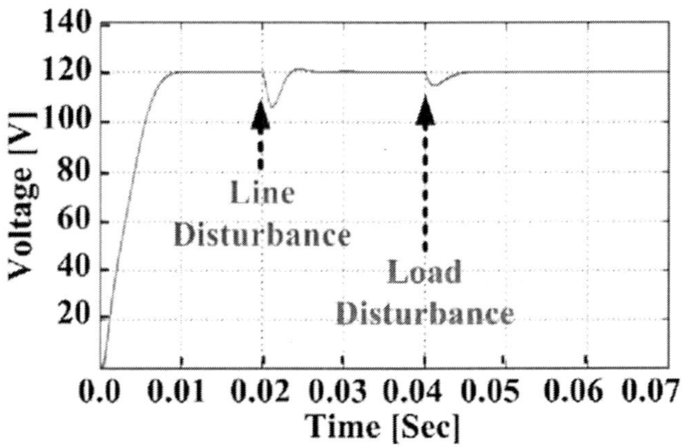

Figure 8. Output voltage of extra HV DC–DC boost power converter with line-load regulation behavior (Simulation).

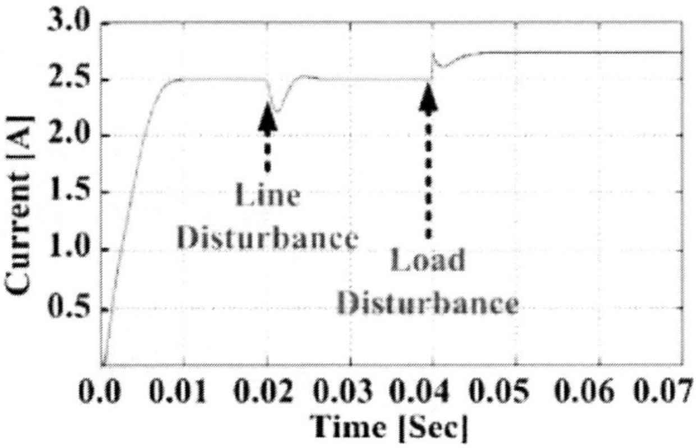

Figure 9. Output current of extra HV DC–DC boost power converter with line-load regulation behavior (Simulation).

Although the load resistance varies from 48 Ω to 44 Ω in the same analytical testing, the output voltage retains the same value of 120 V (set reference) in simulation and settles at 0.005 s, whereas the 119.8 V experimental result settles at 0.850 s as shown in Fig. 8 and Fig. 9 respectively. It is observable from experimental test, that the settling time required are little higher than simulated ones due to the parasitic effects. However, the output voltages of both the simulated and experimental results closely match and thereby verify the prediction given in Eq. (12).

Fig. 10 and Fig. 11 correspondingly depict both the simulated and experimental output currents under the same line and load perturbation conditions. As expected both the simulated (2.5 A) and experimental (2.495 A) output currents remain the same value irrespective of variation in DC input voltage (battery) from 10 V to 9 V which settles at 0.005 s (simulated) and 0.625 s (experimental) respectively. Furthermore, if the load resistance is varied from 48 Ω to 44 Ω, load current increases from 2.5 A to 2.7272 A which gets settle at 0.005 s in the simulation, where as in the experimental study it increases from 2.495 A to 2.722 A and gets settle at 0.850 s. It is observed that the settling time slightly higher in case of experimental result due to self-parasitic nature and thereby it is practically appreciable. In addition, in this case both the simulated and experimental results closely match and thereby verify the prediction given in Eq. (13) by satisfying the power-circuit laws.

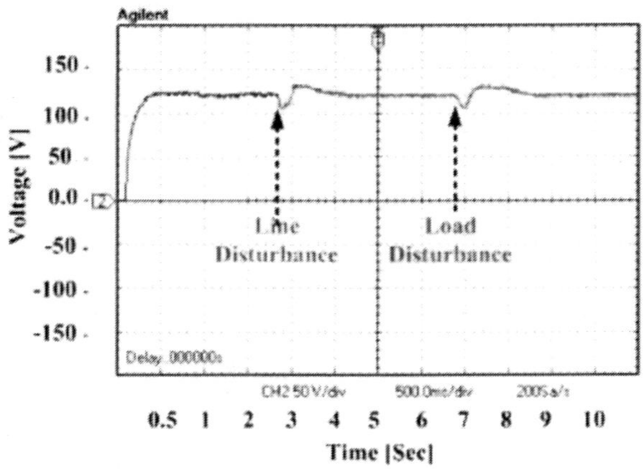

Figure 10. Output voltage of extra HV DC–DC boost power converter with line-load regulation behavior (Experimental).

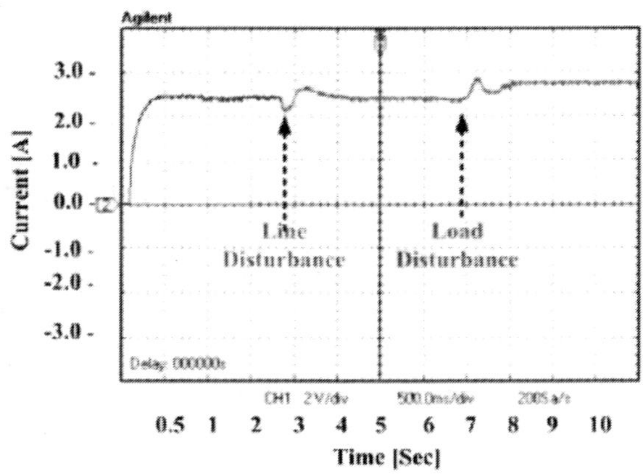

Figure 11. Output current of extra HV DC–DC boost power converter with line-load regulation behavior (Experimental).

Further, a complete set of performance outputs obtained from numerical simulation and experimental tests are summarized in Table 4. All operating conditions (duty cycle variation $k = 0.1–0.9$) the converter performed with efficiency of 100% in simulation and 99% in experimental test with reduced ripple content of 0.22% in simulation and 0.43% in experimental test at the outputs is observed. Finally, the both simulation and experimental test parameters, the output voltage (VO), output current (iO) and efficiency (η) from the Table 4 are shown in three-dimensional (radar plot) view given by Fig. 12, to confirms the results (simulation/experimental) obtained are very closely matched.

Table 4. Investigated output performance indices of numerical simulation and hardware prototype testing.

Duty ratio (k)	Output voltage (V_O) volt	Output current (i_O) amp	Output power (P_O) watt	Input voltage (V_I) volt	Input current (i_I) amp	Input power (P_I) watt	% Efficiency (η)	% Ripple
Simulation study performances								
0.1	44.44444	1.010101	44.89338	10	4.489338	44.89338	100	0.238
0.2	50	1.136364	56.81818	10	5.681818	56.81818	100	0.235
0.3	57.14286	1.298701	74.2115	10	7.42115	74.2115	100	0.233
0.4	66.66667	1.515152	101.0101	10	10.10101	101.0101	100	0.22
0.5	80	1.818182	145.4545	10	14.54545	145.4545	100	0.22
0.6	100	2.272727	227.2727	10	22.72727	227.2727	100	0.22
2/3	**120**	**2.7272**	**327.264**	**10**	**32.7264**	**327.264**	**100**	**0.22**
0.7	133.3333	3.030303	404.0404	10	40.40404	404.0404	100	0.22
0.8	200	4.545455	909.0909	10	90.90909	909.0909	100	0.22
0.9	400	9.090909	3636.364	10	363.6364	3636.364	100	0.22
Experimental study performances								
0.1	44.24444	1.004901	44.46129	10	4.466227	44.66227	99.55	0.486
0.2	49.8	1.131164	56.33195	10	5.655818	56.55818	99.6	0.478
0.3	56.94286	1.293501	73.65566	10	7.391436	73.91436	99.65	0.472
0.4	66.46667	1.509952	100.3614	10	10.06634	100.6634	99.7	0.466
0.5	79.8	1.812982	144.6759	10	14.50385	145.0385	99.75	0.441
0.6	99.8	2.267527	226.2992	10	22.67527	226.7527	99.8	0.432
2/3	**119.8**	**2.722**	**326.094**	**10**	**32.664**	**326.64**	**99.83**	**0.431**
0.7	133.1333	3.025103	402.7421	10	40.33471	403.3471	99.85	0.431
0.8	199.8	4.540255	907.1429	10	90.80509	908.0509	99.9	0.431
0.9	399.8	9.085709	3632.466	10	363.4284	3634.284	99.95	0.431

Bold values represent parameter taken for experimental and simulation task in closed operation.

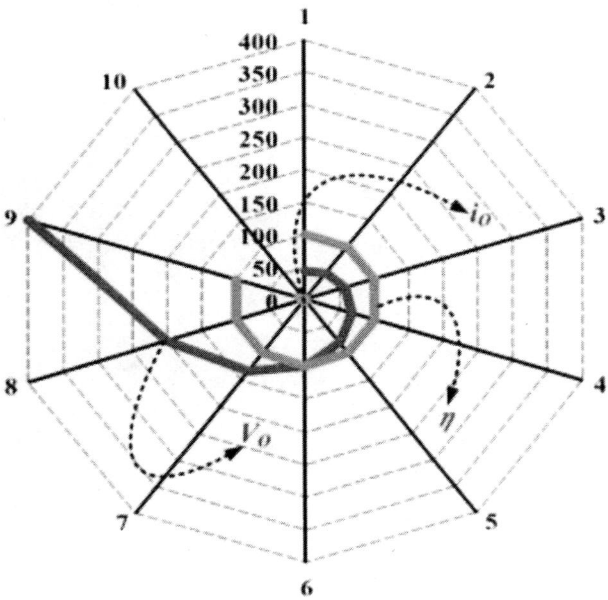

Figure 12. Output performances of simulation and experimental test in 3D (radar plot) response. [*X*-axis: Output voltage (*VO*) (50 v/div), *Y*-axis: Output current (*iO*) (50 amp/div), *Z*-axis: Efficiency (η) (50 amp/div)].

Complete investigation test confirms that the proposed converter produce high efficiency, due to the fact that inclusion of additional passive (L, C) components. This actually reduces the several parasitic effects (L,

R, C, and MOSFET switching, conduction etc.) [21], i.e. losses and increases the voltage transfer ratio gain *k*, which is actually proved in Eq. (12). It is verified by the generated outputs of hardware prototype having losses in the range of milli-amplitude which is practically appreciable. Also, DSP has its own sampling rate to interface with external hardware modules in real time, still the outputs settles less than 1 s in all investigated test.

b). Power Loss Analysis of Proposed DC–DC Converter

In literatures, the parasitic effects due to inductance are neglected in investigations of DC–DC converters. Hence, the RC time constant of converter is always comparable to the switching frequency to attain high efficiency. Therefore, the small parasitic inductance is considered, the results could differ and the inductance connected in the circuit, the current can be expressed by the following equation [21]:

$$i(t) = \frac{V_O}{R} e^{-\frac{t}{RC}} + I_{Load} \tag{21}$$

where, V_O is the initial voltage difference between the voltage source and the voltage across the capacitor. By considering the parasitic inductor, the input current is given by:

$$i(t) = I_{Load} e^{-\frac{R}{2L}t} \cos\left(\sqrt{\frac{1}{LC} + \left(\frac{R}{2L}\right)^2}\, t\right) + I_{Load} + \frac{V_O - \frac{R I_O}{2}}{\sqrt{\frac{L}{C} - \left(\frac{R}{2}\right)^2}} e^{-\frac{R}{2L}t} \sin\left(\sqrt{\frac{1}{LC} + \left(\frac{R}{2L}\right)^2}\, t\right) \tag{22}$$

Further,

$$i(t) = I_{Load} + f(I_O, V_O) e^{-\frac{R}{2L}t} \times \sin\left(\sqrt{\frac{1}{LC} + \left(\frac{R}{2L}\right)^2}\, t + \varphi(I_O, V_O)\right) \tag{23}$$

where, $I_{Load} + I_O$ and V_O are the initial current and initial voltage difference between the input voltage and capacitor. The conduction loss of the circuit is directly proportional to the square of the rms value. Based on the above equations, one can easily numerically calculate the ratio of rms current over the average current based on the different parameters, and from which the loss can be calculated. The conduction losses of the MOSFET can be calculated by:

$$P_{COND.} = I_{ON}^2 R_{DS,ON} \tag{24}$$

where, I_{ON} is the drain current of the MOSFET, when it is ON state, $R_{DS,ON}$ is the drain-source resistance of the MOSFET, when it is ON state. Therefore, to obtain the average value of the conduction loss, the above Eq. (24) simply multiply by the duty ratio (k) of the MOSFET.

The switching losses of the MOSFET are due to the non-zero product of the drain current (ID) and drain to source voltage ($VDSS$). If the MOSFET were an ideal switch, the rise- and fall time of the current and voltage would be zero and would not have any switching loss. Then, switching losses can be calculated by:

$$P_{SW} = \frac{T_{SW,ON} V_{OFF} I_{ON} F_{SW}}{2} + \frac{T_{SW,OFF} V_{OFF} I_{ON} F_{SW}}{2} \tag{25}$$

where, V_{OFF} is the drain source voltage of the MOSFET, when it is OFF state, $T_{SW,ON}$, $T_{SW,OFF}$ are the time to turn the MOSFET ON and OFF state, and F_{SW} is the switching frequency. From Table 3, N-channel MOSFET ON state resistance RDS(0.4Ω), C each capacitor (5 μF), L each inductor (100 μH), and Fs switching frequency (50 kHz) with duty ratio $k = 2/3$ (i.e. $T_{SW,ON} = 2/3*1/50$ KHz, $T_{SW,ON} = 1/3*1/50$ KHz) set for calculate the losses by Eqs. (24) and (25). Fig. 13 and Fig. 14 depicts the variation of normalized conduction and switching loss when parasitic L changes (simulation/experimental). It is observed, that the conduction losses are dramatically reduced with variation of inductance profile (at 100 μH). Relatively, the switching loss is reduced and minimal at the resonant point (at 100 μH) and further increased with variation of inductance profile. It is concluded that, the increasing parasitic effects of power converters can be overcome by inclusion of additional passive components (L and C i.e. voltage lift technique) within the power circuit without any additional external compensation network/ circuitry [1], [2], [3], [4], [5], [6], [7], [8], [9] and [21].

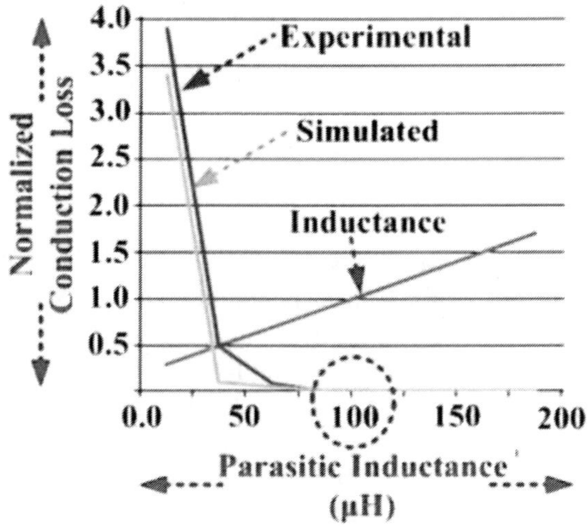

Figure 13. Normalized conduction loss versus inductance variation.

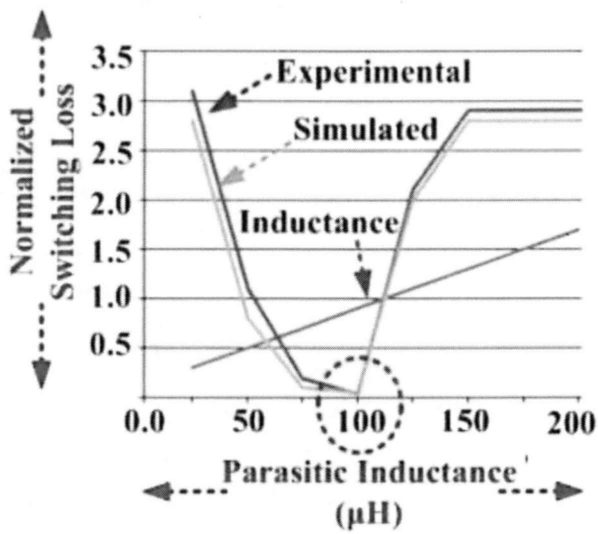

Figure 14. Normalized switching loss versus inductance variation.

Finally, the obtained performances higher output voltage and higher efficiency, reduced % ripple and faster settling time (Table 4), which verifies that the proposed DC–DC converter (hardware prototype) has better power density factor as per standard [22],[23] and [24]. Proving exact viability for parasitic compensation and suits the high voltage industrial needs.

CONCLUSION

Experimental implementation of hardware prototype EHV DC–DC boost power converter based on DSP TMS320F2812 process controller is described along with the relevant theoretical validations in this manuscript. The DC–DC power converter circuit integrated with the voltage-lift technology generates a high performance output-voltage compared to the conventional DC–DC boost converter whereas the duty ratio remains the same. This approach significantly overcomes the parasitic effects and reduces ripples at the output waveforms (voltage/current). A close loop controller with only one voltage sensor feedback, algorithm is developed to maintain the output voltage requirements under the line and load perturbation conditions. Simulation results provided in this manuscript are in accordance with the experimental results that are verified by the theoretical predictions. Hence the analyzed EHV DC–DC converter is suitable for the industrial applications, where the high-voltage becomes mandatory with reduced ripple at the output.

ACKNOWLEDGMENT

We authors would like to acknowledge and convey our gratitude for making this publication possible by NPRP grant # [NPRP X-033-2-007] from the Qatar National Research Fund (a member of Qatar Foundation). The statements made herein are solely the responsibility of the authors.

REFERENCES

1. P. Sanjeevikumar, K. Rajambal, Extra high voltage DCeDC boost converters with simplified control strategy, Int. J. Modeling and Simulation 2008 (2008) 8. Hindawi Publishing Corporation, US, Article ID 593042.
2. P. Sanjeevikumar, B. Geethalakshmi, DCeDC buck converter operated in boost mode for high voltage applications, in: Proc. National Level Power Engineering Conference (MEPPCON'06), MEPPCO Schlenek Engineering College, India, 2006, pp. 195e199.
3. K. Ramash Kumar, S. Jeevananthan, Design and implementation of reduced order sliding mode controller plus proportional double integral controller for negative output elementary super lift Luo-converter, Trans. IET Power Electronics 6 (5) (2013) 974e989.
4. K. Ramash Kumar, S. Jeevananthan, Modelling and implementation of fixed switching frequency sliding mode controller for negative output elementary super lift Luo-converter, Trans. IET Power Electronics 5 (8) (2012) 1593e1604.
5. K. Ramash Kumar, S. Jeevananthan, Analysis, design and implementation of hysteresis modulation sliding mode controller for negative output elementary boost converter, Int. J. Electric Power Comp. Systems (IJEPCS) 40 (2) (2012) 292e311. Taylor & Francis Publications.
6. Y. He, F.L. Luo, Analysis of Luo converter with voltage lift technique, Proc. IEE Conf Power Appl. 152 (5) (Sept. 2005).
7. F.L. Luo, Y. Hong, Positive output super-lift converters, Trans. IEEE Power Electronics 18 (1) (January 2003) 105e113.
8. Luo, F.L., Chen, X.F. "Self-Lift DC-DC Converters", Proc. IEEE Power Electronics drives and Energy Systems for Industrial Growth (IEEE-PEDES'98), vol. 1, pp.441e446, 1e3 December 1998.
9. Luo, F.L. "Luo-Converters, A Series of New DC-DC Step-up (Boost) Circuit" Proc. IEEE Power Electronics and drives Systems (IEEE-PEDS'97), Singapore, pp. 582e588, 26e29 May, 1997.
10. Katsuhiko Ogata, Modern Control Engineering, 2nd ed., Prentice-Hall of India Private Limited, New Delhi, 1996.
11. K.M. Medley, S. Cuk, Dynamics of one cycle controlled Cuk converter, Trans. IEEE Power Electronics 10 (6) (1995) 634e639.
12. K.-I. Hwu, T.-J. Peng, High-voltage-boosting converter with charge pump capacitor and coupling inductor combined with buck-boost converter, Trans. IET Power Electronics 7 (1) (January 2014) 177e188.
13. Changwoo Yoon, Joongeun Kim, Sewan Choi, Multiphase DCeDC converters using a boost-half-bridge cell for high-voltage and high-power applications, Trans. IEEE Power Electronics 26 (2) (February 2011) 381e388.

14. M. Nymand, M.A.E. Andersen, High-efficiency isolated boost DCeDC converter for high-power low-voltage fuel-cell applications, Trans. IEEE Industrial Electronics, 57 (2) (February 2010) 505e514.
15. C.S. Leu, Pin-Yu Huang, Ming-Hui Li, A novel dual-inductor boost converter with ripple cancellation for high-voltage-gain applications, Trans. IEEE Industrial Electronics 58 (4) (April 2011) 1268e1273.
16. L.H.S.C. Barreto, P. Peixoto Praca, D.S. Oliveira, R.N.A.L. Silva, High-voltage gain boost converter based on three-state commutation cell for battery charging using PV panels in a single circuit stage, Trans. IEEE Power Electronics 29 (1) (January 2014) 150e158.
17. C.S. Leu, Ming-Hui Li, A novel current-fed boost converter with ripple reduction for high-voltage circuit applications, Trans. IEEE Industrial Electronics 57 (6) (June 2010) 2018e2023.
18. K.I. Hwu, C.F. Chuang, W.C. Tu, High voltage-boosting converters based on bootstrap capacitors and boost inductors, Trans. IEEE Industrial Electronics 60 (6) (June 2013) 2178e2193.
19. M.H. Rashid, Power Electronics Circuits, Devices and Applications, second ed., Prentice-Hall, USA, 1993.
20. Darko Ostoji, A Multilevel Converter Structure for Grid-Connected PV Plants
21. Ph.D. theses., in: XXII Cycle, University of Bologna, Italy, March 2010. Available online: http://www.die.ing.unibo.it/dottorato_it/Ostojic/Tesi_Ostojic.pdf.
22. Miaosen Shen, Fang Zheng Peng, Leon M. Tolbert, Multilevel DCeDC power conversion system with multiple DC sources, Trans. IEEE Power Electronics, 23 (1) (2008) 420e426.
23. Johann Kolar, Uwe Drofenik, Juergen Biel, Marcelo Heldwein, Hans Ertl, Thomas Friedli, et al., PWM converter power density barriers, IEE J. Trans. Ind. Appl. 128 (4) (2008) 1e14.
24. A. Matthew Porter, An Investigation of DCeDC Converter Power Density Using Si and SiC MOSFETS, U.S.N.A., Trident Scholar Project Report, no. 391, United States Naval Academy, Annapolis, Maryland, 2010.
25. Jeff Ham. "Thermal considerations: assuring performance of Vicor's maxi, mini, micro series high-density DCeDC converter modules", The Power Bench, Published by Vicor Corporation, Andover, MA 01810, pp. 1e3.

CITATION

Sanjeevikumar Padmanaban, Ersan Kabalci, Atif Iqbal, Haitham Abu-Rub, Olorunfemi Ojo, Control strategy and hardware implementation for DC–DC boost power circuit based on proportional–integral compensator for high voltage application, Engineering Science and Technology, an International Journal, Volume 18, Issue 2, June 2015, Pages 163-170, ISSN 2215-0986, http://dx.doi.org/10.1016/j.jestch.2014.11.005.

CHAPTER 8

Overview of Current Development in Electrical Energy Storage Technologies and the Application Potential in Power System Operation

Xing Luo, Jihong Wang, Mark Dooner, Jonathan Clarke

School of Engineering, The University of Warwick, Coventry CV4 7AL, UK

ABSTRACT

Electrical power generation is changing dramatically across the world because of the need to reduce greenhouse gas emissions and to introduce mixed energy sources. The power network faces great challenges in transmission and distribution to meet demand with unpredictable daily and seasonal variations. Electrical Energy Storage (EES) is recognized as underpinning technologies to have great potential in meeting these challenges, whereby energy is stored in a certain state, according to the technology used, and is converted to electrical energy when needed. However, the wide variety of options and complex characteristic matrices make it difficult to appraise a specific EES technology for a particular application. This paper intends to mitigate this problem by providing a comprehensive and clear picture of the state-of-the-art technologies available, and where they would be suited for integration into a power generation and distribution system. The paper starts with an overview of the operation principles, technical and economic performance features and the current research and development of important EES technologies, sorted into six main categories based on the types of energy stored.Following this, a comprehensive comparison and an application potential analysis of the reviewed technologies are presented.

INTRODUCTION

Global electricity generation has grown rapidly over the last decade. As of 2012, the annual gross production of electricity reached approximately 22,200 TW h, of which fossil fuels (including coal/peat, natural gas and oil) contribute around 70% of global electricity generation [1], [2] and [3]. To maintain the power network stability, the load balance has mainly been managed through fossil fuel power plants. To achieve the target of reducing CO_2 emissions, future electricity generation will progress with diminishing reliance on fossil fuels, growing use of renewable energy sources and with a greater respect for the environment [3]. However, most renewable energy sources are intermittent in their nature, which presents a great challenge in energy generation and load balance maintenance to ensure power network stability and reliability. Great efforts have been made in searching for viable solutions, including Electrical Energy Storage (EES), load shifting through demand management, interconnection with external grids, etc. Amongst all the possible solutions, EES has been recognized as one of the most promising approaches [4] and [5].

EES technology refers to the process of converting energy from one form (mainly electrical energy) to a storable form and reserving it in various mediums; then the stored energy can be converted back into electrical energy when needed [4] and [5]. EES can have multiple attractive value propositions (functions) to power network operation and load balancing, such as: (i) helping in meeting peak electrical load demands, (ii) providing time varying energy management, (iii) alleviating the intermittence of renewable source power generation, (iv) improving power quality/reliability, (v) meeting remote and vehicle load needs, (vi) supporting the realization of smart grids, (vii) helping with the management of distributed/standby power generation, (viii) reducing electrical energy import during peak demand periods.

In many scenarios, demand for EES and selection of appropriate EES technologies have been considered to be important and challenging in countries with a relatively small network size and inertia. For example, the UK electric power network currently has a capacity of Pumped Hydroelectric Storage (PHS) at 27.6 GW h [6]. Although PHS facilities have been built worldwide as a mature and commercially available technology, it is considered that the potential for further major PHS schemes is restricted in the UK [6]. Therefore, it is of great importance that suitable EES technologies in addition to PHS are explored. Derived from the study of recent publications, Fig. 1 illustrates various EES technologies with potentials to address the challenges faced by the UK energy systems[4], [6], [7], [8] and [9]. Many countries potentially need to address similar challenges which can be solved or improved by suitable EES technologies.

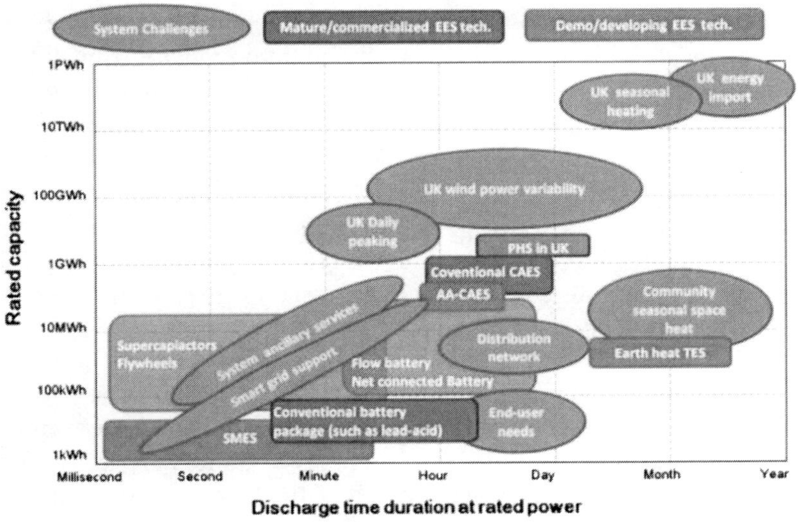

Figure 1. Electrical energy storage technologies with challenges to the UK energy systems [4], [6], [7], [8] and [9].

Due to the great potential and the multiple functions of EES, in the literature many authors have reviewed and summarized the EES research and development, demonstrations and industrial applications from different perspectives, particularly in recent years. The paper presented by Ibrahim et al. highlighted the need to store energy for improving power networks and maintaining load levels [10]. A group of characteristics of different EES technologies is given, which can help improve performance and cost estimates for storage systems. However relatively few references are cited in [10]. Chen et al. provided a well-organized and comprehensive critical review on progress in EES systems, which covered various types of EES technologies and their applications/deployment status [4]. The discussion on the selection of appropriate EES candidates for specific applications was relatively brief. Hall et al. also presented a review article concentrating on several EES technologies, i.e., batteries, supercapacitors, superconducting magnetic energy storage and flywheels [11]. Liu et al. provided an insightful review of the advanced materials for several EES technologies[12]. The strategies for developing high-performance hydrogen storage materials and electrochemical lithium-ion battery materials were discussed in detail [12]. The paper also highlighted the prospects in the future development of advanced materials for EES. With the rapid penetration of intermittent renewables, the review articles [13], [14],[15] and [16] have made effort to assess and summarize the EES options for increased renewable electricity applications. Díaz-González et al. [13] and Zhao et al. [15] focus on the review of EES technologies for wind power applications. A detailed discussion of existing EES applications in wind power is a highlight provided by the article [13], whilst the planning issues, the operation and control strategies of the ESS applications for wind power integration support are summarized by the paper [15]. Furthermore, from a novel viewpoint, Connolly et al. assessed available computer tools for analyzing the integration of renewable energy into various energy systems [17]. Researchers have also reviewed specific aspects of EES systems, such as in [18], [19], [20], [21] and [22]. For instance, Dunn et al. contributed a high quality review on battery energy storage for the grid applications, mainly focusing on commercially available sodium–sulfur batteries, relatively low cost redox-flow batteries and developing lithium-ion batteries, all with the aim to be used in grid storage [22]. The reviews of the developments and challenges in materials for electrochemical relevant energy storage are presented in [23], [24] and [25]. For example, Whittingham addressed the current challenges in the subject of electrochemical energy storage materials, which can be summarized as: reducing the cost and extending the lifetime of devices whilst improving their performance and making them more environmentally friendly [23]. In addition, some journals have published special issues dedicated to EES research and development, such as the special issue in 2013 from the Wiley journal Advanced Functional Materials: "Grand Challenges in Energy Storage".

A brief statistical study has been carried out to ascertain the trends in EES related research using the search engine 'Web of Science' and choosing 'Topic' as the search field. Fig. 2 shows the results detailing the number of research papers published in six EES related fields over the past ten years (2004–2013). The titles of the subfigures in Fig. 2 are the input keywords used in the search engine. The results indicate that research in EES in the past ten years has tended to increase, with rapid increases in 2012 and 2013. In particular, research into compressed air energy storage grew significantly in 2012 whilst, in contrast, research into superconducting magnetic energy storage has remained relatively stable. It can also be seen that there has been a large increase in the research into renewable and energy management with EES topics. The statistic figure in EES research literatures shows that the entire EES research and technology development have changed and advanced rapidly in the years since some previous reviews, such as[4], [10] and [11], were published. Thus it is timely to have a new critical overview of the current development in EES.

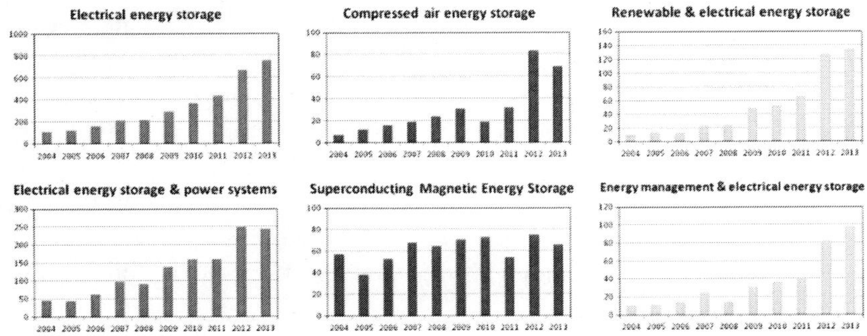

Figure 2. A brief statistical study to the trend in EES related research.

Although the potential benefits of EES installation to power system operation have been widely recognized, some significant challenges in the deployment of EES systems exist, mainly in: (1) how to choose the suitable EES technology to match the power system application requirements; (2) how to accurately evaluate the actual values of deployed EES facilities including technical and economic benefits; (3) how to bring the cost down to a realistically acceptable level for deployment, especially for newly developing EES technologies.

The objective of this paper is to provide a updated picture of the state-of-the-art EES technologies, in comparison with the previous review articles [4], [10] and [11]. The paper begins with an overview of the operation principles, technical and economic performance features and the current research and development of important EES technologies. Following this, a comprehensive comparison of EES technologies is conducted. The paper presents a detailed summarization and predication of the existing and promising EES technology options for different power system applications with their corresponding technical specifications. The overview will help address the challenges faced in deployment of EES and provide useful information and guidance in selecting suitable technologies for specific applications based on the nature of EES characteristics.

CLASSIFICATION OF ELECTRICAL ENERGY STORAGE TECHNOLOGIES

There are several suggested methods for categorization of various EES technologies, such as, in terms of their functions, response times, and suitable storage durations [4],[26] and [27]. One of the most widely used methods is based on the form of energy stored in the system [15] and [16] as shown in Fig. 3, which can be categorized into mechanical (pumped hydroelectric storage, compressed air energy storage and flywheels), electrochemical (conventional rechargeable batteries and flow batteries), electrical (capacitors, supercapacitors and superconducting magnetic energy storage), thermochemical (solar fuels), chemical (hydrogen storage with fuel cells) and thermal energy storage (sensible heat storage and latent heat storage). A detailed description and discussion of each type of EES technology will be given in the next section following the above order of category.

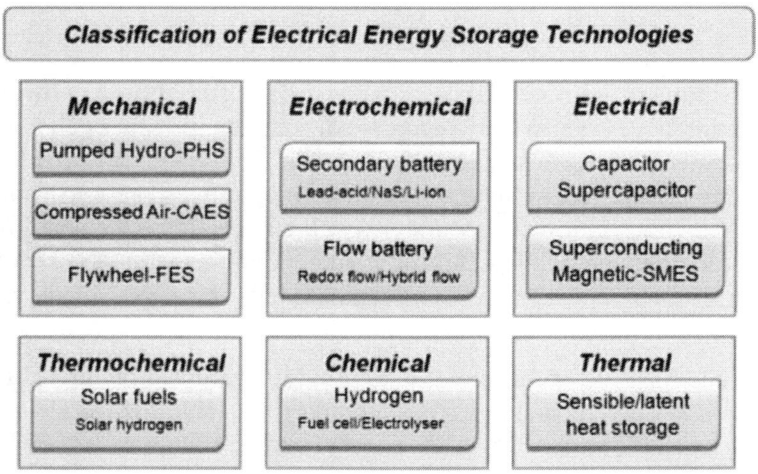

Figure 3. Classification of EES technologies by the form of stored energy.

DESCRIPTION OF ELECTRICAL ENERGY STORAGE TECHNOLOGY

Pumped Hydroelectric Storage (PHS)

PHS is an EES technology with a long history, high technical maturity and large energy capacity. With an installed capacity of 127–129 GW in 2012, PHS represents more than 99% of worldwide bulk storage capacity and contributes to about 3% of global generation[26], [28] and [29]. As shown in Fig. 4, a typical PHS plant uses two water reservoirs, separated vertically. During off-peak electricity demand hours, the water is pumped into the higher level reservoir; during peak hours, the water can be released back into the lower level reservoir. In the process, the water powers turbine units which drive the electrical machines to generate electricity. The amount of energy stored depends on the height difference between the two reservoirs and the total volume of water stored. The rated power of PHS plants depends on the water pressure and flow rate through the turbines and rated power of the pump/turbine and generator/motor units, and [30].

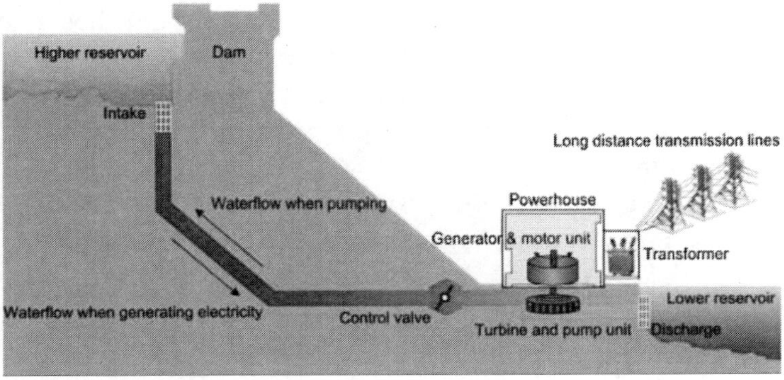

Figure 4. A pumped hydroelectric storage plant layout.

Various PHS plants exist with power ratings ranging from 1 MW to 3003 MW, with approximately 70–85% cycle efficiency and more than 40 years lifetime [4], [29],[31] and [32]. Some PHS facilities along with their features are listed in Table 1. The nature of the operation of PHS systems means that their applications mainly involve energy management in the fields of time shifting, frequency control, non-spinning reserve and supply reserve. However, with the restriction of site selection, PHS plants suffer long construction time and high capital investment.

Table 1. Selected pumped hydroelectric storage plants [4], [26], [31], [32] and [33].

Plant name	Country	Power rating	Features
Rocky river PHS plant	US	32 MW	The world's first large-scale commercial PHS plant
Bath County PHS plant	US	3003 MW	The world's largest power rated PHS plant
Okinawa Yanbaru PHS	Japan	~30 MW	Only commercial seawater PHS plant
Hawaiian Elec. Co. PHS facility	US	—	Claimed 87% relatively high cycle efficiency
HPS of Ikaria Island	Greece	2.655 MW	One of the first wind-PHS plants (under construction)

Recently, with the advance of technology, some PHS plants using flooded mine shafts, underground caves and oceans as reservoirs have been planned or are in operation , such as the Okinawa Yanbaru in Japan, a 300 MW seawater-based PHS plant in Hawaii, the Summit project in Ohio and the Mount Hope project in New Jersey [28], [34] and [35]. In addition, wind or solar power generation coupled with PHS is now being developed. This could help the adoption of renewable energy in isolated or distributed networks[36] and [37]. For instance, the Ikaria Island power station will integrate a 3 × 900 kW wind farms with a PHS facility [28]. The development trend of PHS facilities consists of building the hydroelectric set with higher speed and larger capacity compared to the current technical level, installing centralized monitoring and using intelligent control systems [26], [28] and [38].

Compressed Air Energy Storage (CAES)

In addition to PHS, CAES is another type of commercialized EES technology which can provide power output of over 100 MW with a single unit. A schematic diagram of a CAES plant is shown in Fig. 5. During the periods of low power demand, the surplus electricity drives a reversible motor/generator unit in turn to run a chain of compressors for injecting air into a storage vessel, which is either an underground cavern or over ground tanks. The energy is stored in the form of high pressure air. When the power generation cannot meet the load demand, the stored compressed air is released and heated by a heat source which can be from the combustion of fossil fuel or the heat recovered from the compression process. The compressed air energy is finally captured by the turbines. The waste heat from the exhaust can be recycled by a recuperator unit (Fig. 5).

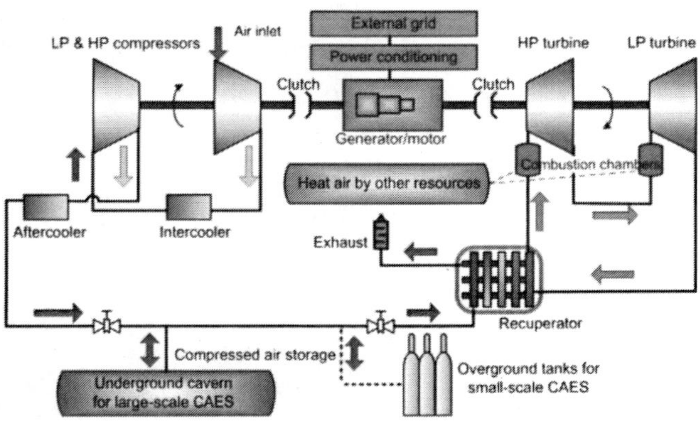

Figure 5. Schematic diagram of a CAES plant/facility.

The world's first utility-scale CAES plant, the Huntorf power plant, was installed in Germany in 1978 [39], [40] and [41]. It uses two salt domes as the storage caverns and it runs on a daily cycle with 8 h of compressed air charging and 2 h of operation at a rated power of 290 MW [39]. This plant provides black-start power to nuclear units, back-up to local power systems and extra electrical power to fill the gap between the electricity generation and demand. Another commercial CAES plant started operation in McIntosh, the US, in 1991 [39], [40] and [41]. The 110 MW McIntosh plant can operate for up to 26 h at full power. The compressed air is stored in a salt cavern. A recuperator is operated to reuse the exhaust heat energy. This reduces the fuel consumption by 22–25% and improves the cycle efficiency from ~42% to ~54%, in comparison with the Huntorf plant[4] and [42]. These two CAES plants have consistently shown good performances with 91.2–99.5% starting and running reliabilities [38] and [39].

CAES system can be built to have small to large scale of capacities; CAES technology can provide the moderate speed of responses and good partial-load performance. The practical uses of large-scale CAES plants involve grid applications for load shifting, peak shaving, and frequency and voltage control. CAES can work with intermittent renewable energy applications, especially in wind power, to smooth the power output, which have attracted much attentions from academic researchers and industrial sectors as described in [40], [43], [44] and [45]. The major barrier to implementing large-scale CAES plants is identifying appropriate geographical locations which will decide the main investment cost of the plant. Relative low round trip efficiency is another barrier for CAES compared to PHS and battery technologies.

In addition, the developing Liquid Air Energy Storage (LAES) has many components which are the same or similar as those used for CAES, such as compressors, turbines, electric machines and heat exchangers. Considering the type of energy stored, LAES can be classified into thermal energy storage, which will be introduced in Section 3.10.

Currently, the newly developing Advanced Adiabatic CAES (AA-CAES) is attracting attention. AA-CAES technology is normally integrated with a thermal energy storage subsystem, which has no fuel combustion involved in the expansion mode [39], [42],[43] and [46]. The world's first AA-CAES demonstration plant – ADELE – is in the development stage, at Saxony-Anhalt in Germany. The plant will have a storage capacity of 360 MWh and an electric output of 90 MW, aiming for ~70% cycle efficiency [43]. Because its compression mode will be powered by wind energy, the ADELE plant emits no CO_2 in a full cycle. The US based LightSail Energy Ltd. is also developing the AA-CAES facilities by using reversible reciprocating piston machines [46]. In 2007 Luminant and Shell-Wind Energy proposed wind farm projects involving

CAES in Texas, intending to evaluate the potential of incorporating CAES facilities in conjunction with the wind farm; after a long wait, in 2013 the project got underway and hosting 317 MW of CAES has been set as the current target [39] and [44]. In addition, a comparison between different adiabatic CAES plant configurations was recently published in [47].

Recently, apart from using salt caverns, researchers have attempted to study other geological structures for use in underground CAES technology. A 2 MW field test program has used a concrete-lined tunnel in an abandoned mine in Japan [39] and [48]. A test facility made by Electric Power Research Institute and others utilized a hard rock cavern with water compensation [49]. Italy's Enel operated a 25 MW porous rock based CAES facility in Sesta – the test was stopped due to a disturbed geothermal issue [39]. The Iowa Stored Energy Park project aimed to use porous sandstone aquifers to build an underground reservoir for constructing a 270 MW CAES plant; unfortunately, this project stopped in 2011 as the field test result indicated that the geological structure in Iowa cannot obtain a fast enough flow for large-scale CAES [50].

Over ground small-scale CAES has recently undergone rapid development. It can be used as an alternative to the battery for industrial applications, such as Uninterruptible Power Supplies (UPS) and back-up power systems. Compressed air battery systems developed by the UK based Flowbattery (previously named Pnu Power) were recently successfully commercialized [51]. It uses pre-prepared compressed air from air cylinders to drive a combination of a scroll expander and a generator to produce electricity[51] and [52]. In addition, the guideline study for the efficient design and sizing of small-scale CAES pressure vessels considering minimizing its cost was reported in [53]. Also, the feasibility on the direct mechanical coupling of a wind turbine and a scroll expander with small-scale CAES has been studied by the University of Warwick, and its on-site tests are on-going [52] and [54].

Flywheel Energy Storage (FES)

A modern FES system is composed of five primary components: a flywheel, a group of bearings, a reversible electrical motor/generator, a power electronic unit and a vacuum chamber [18]. Fig. 6 shows the simplified structure of a modern FES facility. FES systems use electricity to accelerate or decelerate the flywheel, that is, the stored energy is transferred to or from the flywheel through an integrated motor/generator. For reducing wind shear and energy loss from air resistance, the FES system can be placed in a high vacuum environment. The amount of energy stored is dependent on the rotating speed of flywheel and its inertia.

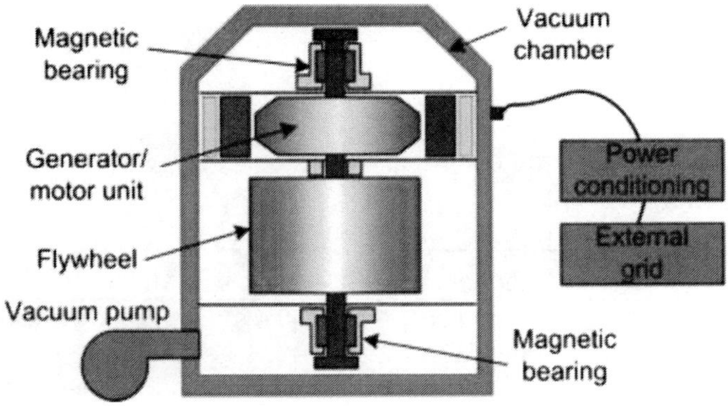

Figure 6. System description of a flywheel energy storage facility.

FES can be classified into two groups: (1) low speed FES: it uses steel as the flywheel material and rotates below 6×10^3 rpm; (2) high speed FES: it uses advanced composite materials for the flywheel, such as carbon-fiber, which can run up to $\sim 10^5$ rpm [55]. Low speed FES systems are typically used for short-term and medium/high power applications. High speed FES systems use non-contact magnetic bearings to mitigate the wear of bearings, thereby improving the efficiency. The application areas of high speed FES are continuously expanding, mainly in high power quality and ride-through power service in traction and the aerospace industry [56]. The specific energy of low speed flywheels is ~5 W h/kg, and the high speed composite rotor can achieve a specific energy of up to ~100 W h/kg [57]. The cost of high speed composite systems can be much higher than that of conventional metal flywheel systems. FES has some favorable characteristics, including high cycle efficiency (up to ~95% at rated power), relatively high power density, no depth-of-discharge effects and easy maintenance [18],[55] and [57].

Table 2 lists some selected FES facilities. In June 2011, a 20 MW modular plant built by Beacon Power was put into commercial operation in New York, which was the largest advanced EES facility operating in North America [58] and [59]. It employs 200 high speed flywheel systems to provide fast response frequency regulation services to the grid, providing ~10% of the whole state frequency regulation demand [58] and [59]. Normally, FES devices can supply sufficient power in a short time period with modest capacity. Thus it is not used as standalone backup power unless operated with other EES or power generation systems, such as batteries or fuel-fired generators. The main weakness of FES is that flywheel devices suffer from the idling losses during the time when the flywheel is on standby. This can lead to relatively high self-discharge, up to ~20% of stored capacity per hour [57].

Table 2. Selected flywheel energy storage facilities [18], [55], [57], [58], [59] and [60].

Firms/Institutes	Characteristics	Application area
Active Power Company	Clean Source series 100–2000 kW	Backup power supply, UPS systems
Beacon Power Company	100/150 kW a unit, 20 MW/5 MW h plant	Freq. regulation, power quality, voltage support
Boeing Phantom Works	100 kW/5 kW h, HT magnetic bearings	Power quality and peak shaving
Japan Atomic Energy Center	235 MVA, steel flywheel	High power supply to Nuclear fusion furnace
Piller power systems Ltd.	3600–1500 rpm, 2.4 MW for 8 s	Ride-through power and sources of backup power
NASA Glenn research center	2×10^4–6×10^4 rpm, 3.6 MW h	Supply on aerospace aviation & other transports

Currently, the research and development area of FES includes the material of the flywheel for increasing their rotation speed capabilities and power densities, high speed electrical machines, high carrying capacity of the bearings and the flywheel array technology. An advance in FES technology is the High Temperature Superconductor (HTS) bearings which is a promising option for improving bearing performance. The US Argonne National Laboratory developed a 2 kW h FES system using high-temperature superconductors and permanent magnets as passive bearings for a feasibility study [61]. A model-based power flow control strategy has been studied for improving flywheel performance in high power pulse systems [62]. The rail traction industry has tested FES devices for trackside voltage support [63]. Optimizing flywheels for relatively long-term operation (up to several hours) are being studied for use in vehicles and power plants[26].

Battery Energy Storage (BES)

The rechargeable battery is one of the most widely used EES technologies in industry and daily life. Fig. 7 shows the simplified operational principle of a typical BES system. A BES system consists of a number of electrochemical cells connected in series or parallel, which produce electricity with a desired voltage from an electrochemical reaction. Each cell contains two electrodes (one anode and one cathode) with an electrolyte which can be at solid, liquid or ropy/viscous states [64] and [65]. A cell can bi-directionally convert energy between electrical and chemical energy. During discharging, the electrochemical reactions occur at the anodes and the cathodes simultaneously. To the external circuit, electrons are provided from the anodes and are collected at the cathodes. During charging, the reverse reactions happen and the battery is recharged by applying an external voltage to the two electrodes (Fig. 7).

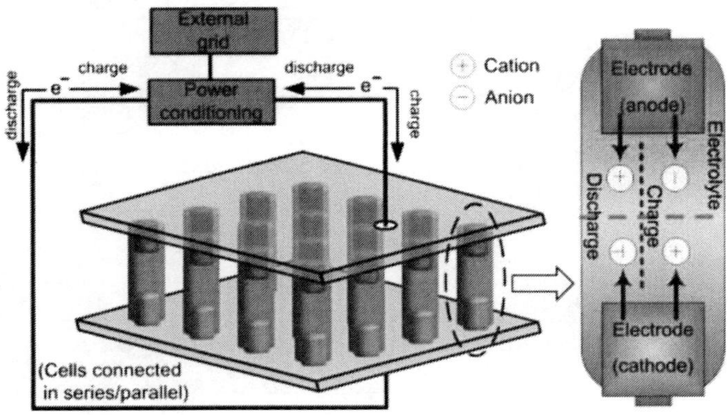

Figure 7. Schematic diagram of a battery energy storage system operation.

Batteries can be widely used in different applications, such as power quality, energy management, ride-through power and transportation systems. The construction of BES systems takes a relatively short time period (roughly within 12 months) [4] and [66]. The location for installation can be quite flexible, either housed inside a building or close to the facilities where needed. Currently, relatively low cycling times and high maintenance costs have been considered as the main barriers to implementing large-scale facilities. The disposal or recycling of dumped batteries must be considered if toxic chemical materials are used [9]. Furthermore, many types of battery cannot be completely discharged due to their lifetime depending on the cycle Depth-of-Discharge (DoD) [13]. A description of several important BES technologies will be presented in the following five subsections. The chemical reactions taking place in these battery types are listed in Table 3.

Table 3. Chemical reactions and single unit voltages of main batteries available to EES [4], [13], [67] and [68].

Battery type	Chemical reactions at anodes and cathodes	Unit voltage
Lead–acid	$Pb + SO_4^{2-} \Leftrightarrow PbSO_4 + 2e^-$ $PbO_2 + SO_4^{2-} + 4H^+ + 2e^- \Leftrightarrow PbSO_4 + 2H_2($	2.0 V
Lithium-ion	$C + nLi^+ + ne^- \Leftrightarrow Li_nC$ $LiXXO_2 \Leftrightarrow Li_{1-n}XXO_2 + nLi^+ + ne^-$	3.7 V
Sodium–sulfur	$2Na \Leftrightarrow 2Na^+ + 2e^-$ $\chi S + 2e^- \Leftrightarrow \chi S^{2-}$	~2.08 V
Nickel–cadmium	$Cd + 2OH^- \Leftrightarrow Cd(OH)_2 + 2e^-$ $2NiOOH + 2H_2O + 2e^- \Leftrightarrow 2Ni(OH)_2 + 2OH^-$	1.0–1.3 V
Nickel–metal hydride	$H_2O + e^- \Leftrightarrow 1/2H_2 + OH^-$ $Ni(OH)_2 + OH^- \Leftrightarrow NiOOH + H_2O + e^-$	1.0–1.3 V
Sodium nickel chloride	$2Na \Leftrightarrow 2Na^+ + 2e^-$ $NiCl_2 + 2e^- \Leftrightarrow Ni + 2Cl^-$	~2.58 V

Lead–acid Batteries

The most widely used rechargeable battery is the lead–acid battery [4] and [10]. The cathode is made of PbO_2, the anode is made of *Pb*, and the electrolyte is sulfuric acid. Lead–acid batteries have fast response times, small daily self-discharge rates (<0.3%), relatively high cycle efficiencies (~63-90%) and low capital costs (50–600 $/kW h) [4],[14], [57] and [69]. Some examples of EES facilities using lead–acid batteries are listed inTable 4. Lead–acid batteries can be used in stationary devices as back-up power supplies for data and telecommunication systems, and energy management applications. Also, they have been developed as power sources for hybrid or full electric vehicles. However, there are still limited installations around the world as utility-scale EES, mainly due to their relatively low cycling times (up to ~2000), energy density (50–90 W h/L) and specific energy (25–50 Wh/kg) [4], [70] and [71]. In addition, they may perform poorly at low temperatures so a thermal management system is normally required, which increases the cost [72].

Table 4. Selected lead–acid battery energy storage facilities [4], [13], [67], [75] and [76].

Battery type	Chemical reactions at anodes and cathodes	Unit voltage
Lead–acid	$Pb + SO_4^{2-} \Leftrightarrow PbSO_4 + 2e^-$	2.0 V
	$PbO_2 + SO_4^{2-} + 4H^+ + 2e^- \Leftrightarrow PbSO_4$ -	
Lithium-ion	$C+nLi^++ne^- \Leftrightarrow Li_nC$	3.7 V
	$LiXXO_2 \Leftrightarrow Li_{1-n}XXO_2+nLi^++ne^-$	
Sodium–sulfur	$2Na \Leftrightarrow 2Na^++2e^-$	~2.08 V
	$\chi S+2e^- \Leftrightarrow \chi S^{2-}$	
Nickel–cadmium	$Cd+2OH^- \Leftrightarrow Cd(OH)_2+2e^-$	1.0–1.3 V
	$2NiOOH+2H_2O+2e^- \Leftrightarrow 2Ni(OH)_2+2OH^-$	
Nickel–metal hydride	$H_2O+e^- \Leftrightarrow 1/2H_2+OH^-$	1.0–1.3 V
	$Ni(OH)_2+OH^- \Leftrightarrow NiOOH+H_2O+e^-$	
Sodium nickel chloride	$2Na \Leftrightarrow 2Na^++2e^-$	~2.58 V
	$NiCl_2+2e^- \Leftrightarrow Ni+2Cl^-$	
Name/locations	Characteristics	Application area
BEWAG, Berlin	8.5 MW/8.5 MW h	Spinning reserve, frequency control
Chino, California	10 MW/40 MW h	Spinning reserve, load leveling
PREPA, Puerto Rico	20MW/14 MW h	Spinning reserve, frequency control
Metlakatla, Alaska	1 MW/1.4 MW h	Enhancing stabilization of island grid
Kahuku Wind Farm, Hawaii	15 MW/3.75 MW h	Power management, load firming, grid integration
Notrees EES project, U.S.	36 MW/24 MW h	Solving intermittency issues of wind energy

Currently, the research and development of lead–acid batteries focuses on: (1) innovating materials for performance improvement, such as extending cycling times and enhancing the deep discharge capability; (2) implementing the battery technology for applications in the wind, photovoltaic power integration and automotive sectors. Several advanced lead–acid batteries that have fast responses comparable to flywheels and supercapacitors are being developed or are in the demonstration phase, such as Ecoult UltraBattery smart systems and Xtreme Power advanced lead–acid "Dry Cell"[73] and [74].

Lithium-ion (Li-ion) Batteries

In a Li-ion battery, the cathode is made of a lithium metal oxide, such as $LiCoO_2$ and $LiMO_2$, and the anode is made of graphitic carbon. The electrolyte is normally a non-aqueous organic liquid containing dissolved lithium salts, such as $LiClO_4$[13]. The Li-ion battery is considered as a good candidate for applications where the response time, small dimension and/or weight of equipment are important (milliseconds response time, ~1500–10,000 W/L, ~75–200 W h/kg, ~150–2000 W/kg) [4], [9], [26] and [57]. Li-ion batteries also have high cycle efficiencies, up to ~97% [4] and [26]. The main drawbacks are that the cycle DoD can affect the Li-ion battery's lifetime and the battery pack usually requires an on-board computer to manage its operation, which increases its overall cost.

The current research focuses for the Li-ion battery include: (1) increasing battery power capability with the use of nanoscale materials; (2) enhancing battery specific energy by developing advanced electrode materials and electrolyte solutions. Several companies have experience in using Li-ion batteries in the utility-scale energy market. The U.S. based AES Energy Storage has been commercially operating a Li-ion BES system (8 MW/2 MW h in 2010, enlarged 16 MW in 2011) in New York for supplying frequency regulation [8] and [77]. The AES also installed a 32 MW/8 MW h Li-ion BES system (Laurel Mountain) for supporting a 98 MW wind generation plant in 2011 [77] and [78]. Currently, the largest European Li-ion battery EES trial is underway in the UK. The project will deploy a 6 MW/10 MWh Li-ion battery at a primary substation to assess the cost effectiveness of EES as part of the UK's Carbon Plan [79]. The companies claimed that the storage could save more than $9 million compared to traditional system upgrades; the project can be used to balance the intermittency of wind and other renewables [79]. Also, in December 2013 Toshiba announced a project to install a 40 MW/20 MWh Li-ion battery project in Tohoku, which will help integrate renewables into the grid [80]. In addition, Li-ion batteries are now applied in Hybrid and full Electric Vehicles (HEVs and EVs), which use large-format cells and packs with capacities of 15–20 kW h for HEVs and up to 50 kW h for EVs [28].

Sodium–sulfur (NaS) Batteries

A NaS battery uses molten sodium and molten sulfur as the two electrodes, and employs beta alumina as the solid electrolyte. The reactions normally require a temperature of 574–624 K to ensure the electrodes are in liquid states, which leads to a high reactivity[8]. The desirable features of NaS batteries include relatively high energy densities (150–300 W h/L), almost zero daily self-discharge, higher rated capacity than other types of batteries (up to 244.8 MW h) and high pulse power capability [13], [26] and [81]. The battery uses inexpensive, non-toxic materials leading to high recyclability (~99%)[4] and [13]. However, the limitations are high annual operating cost (80 $/kW/year) and an extra system required to ensure its operating temperature [72].

The NaS battery is considered as one of the most promising candidates for high power EES applications. Table 5 lists some NaS battery facilities with their applications. The research and development focuses are mainly on enhancing the cell performance indices and decreasing/eliminating the high temperature operating constrains. For instance, Sumitomo Electric Industries and Kyoto University developed a low temperature sodium related battery – the novel sodium-containing material can be melted at 330 K [84]. The inventor claimed that the new battery can achieve an energy density as high as 290 W h/L [84]. In addition, a part of the outcome to the "Wind to Battery" project led by Xcel Energy was recently presented in [83], mainly on the field results and analyses quantifying the ability and the value of NaS battery EES toward wind generation integration support [15] and [83].

Table 5. Selected sodium-sulfur battery energy storage facilities [77], [81], [82] and [83].

Name/locations	Rated power/capacity	Application area
Kawasaki EES test facility, Japan	0.05 MW	The 1st large-scale, proof principle, operated in 1992
Long Island Bus's BES System, New York, US	1 MW/7 MW h	Refueling the fixed route vehicles
Rokkasho Wind Farm ES project, Japan	34 MW/244.8 M W h	Wind power fluctuation mitigation
Saint Andre, La reunion, France	1 MW	Wind power on an island
Graciosa Island, Younicos, Germany	3 MW/18 MW h	Wind & solar power EES for islands, commissioning 2013
Abu Dhabi Island, UAE	40 MW	Load levelling

Nickel–cadmium (NiCd) Batteries

A NiCd battery uses nickel hydroxide and metallic cadmium as the two electrodes and an aqueous alkali solution as the electrolyte. It normally has relatively high robust reliabilities and low maintenance requirements. The weaknesses of NiCd batteries are: cadmium and nickel are toxic heavy metals, resulting in environmental hazards [9] and [85]; the battery suffers from the memory effect – the maximum capacity can be dramatically decreased if the battery is repeatedly recharged after being only partially discharged [86].

To date there have been very few commercial successes using NiCd batteries for utility-scale EES applications. One example is at Golden Valley, Alaska, in the US [77]. This NiCd facility was officially put into operation in 2003 by Golden Valley Electric Association. It offers services in spinning reserve, power supply and compensation to an "electrical island system" due to the geographic restrictions, i.e. remote areas[77] and [87]. The system has the ability to deliver the rated power at ~27 MW for 15 min or 40 MW for 7 min, and the efficiency is in the range of 72–78% with the operating temperature at 233–323 K [77], [87] and [88]. The local cold temperature was the primary driving force behind the choice of the NiCd battery. It was reported that the NiCd technology for utility-scale EES applications was not pursued further after the Golden Valley installation [8] and [77]. It seems unlikely that NiCd batteries will be heavily used for future large-scale EES projects.

Other Candidates of Battery Energy Storage

The Nickel–metal Hydride (NiMH) battery is similar to the NiCd battery except that a hydrogen-absorbing alloy is used as the electrode instead of cadmium. It has moderate specific energy (~70–100 W h/kg) and relatively high energy density (~170–420 W h/L), significantly better than those of the NiCd battery [89], [90], [91] and [92]. Other advantages of NiMH batteries over NiCd batteries include a reduced "memory effect" and they are more environmental friendly. NiMH batteries have the longer cycle life in comparison with Li-ion batteries [89]. The NiMH battery has a wealth of applications from portable products to HEVs & EVs and potential industrial standby applications, such as UPS devices [89], [90] and [93]. However, the significant barrier for EES applications is the high rate of self-discharge, losing ~5–20% of its capacity within the first 24 h after fully charging [89], [90] and [91]. It is also sensitive to deep cycling – the performance decreases after a few hundreds full cycles [89], [90], [91] and [92].

The technology of sodium nickel chloride battery (also known as ZEBRA battery) is similar to that of the NaS battery. The ZEBRA battery has moderate specific energy (~94–120 W h/kg), energy density

(~150 W h/L), specific power (150–170 W/kg), and a high operating temperature (~523–623 K) [4], [94], [95], [96], [97] and [98]. The advantages consist of good pulse power capability, cell maintenance free, very little self-discharge and relatively high cycle life. This battery technology has been applied in EV demonstrations and Rolls Royce has used it to replace lead–acid in surface ships applications [97]. GE launched Durathon sodium–metal halide battery for UPS and utility market, which can be considered as a continuing improvement on the ZEBRA technology [98]. A GE Durathon battery manufacturing facility in New York was officially opened in 2012 [98]. Recently, a new venture, FIAMM Energy Storage Solutions, also started to produce such batteries (named SoNick batteries) for stationary storage applications [99]. With regard to its drawback, the battery takes 12–15 h to heat up after it has been solidified (frozen) [94]. In addition, only a few companies have been involved in the development of this technology and have produced this type of battery, which may limit its potential [4], [94] and [98].

Flow Battery Energy Storage (FBES)

A flow battery stores energy in two soluble redox couples contained in external liquid electrolyte tanks. These electrolytes can be pumped from the tanks to the cell stack which consists of two electrolyte flow compartments separated by ion selective membranes. The operation is based on reduction-oxidation reactions of the electrolyte solutions. During the charging phase, one electrolyte is oxidized at the anode and another electrolyte is reduced at the cathode, and the electrical energy is converted to the electrolyte chemical energy. The above process is reversed during the discharging phase.

Flow batteries can be classified into the categories of redox flow batteries and hybrid flow batteries, depending on whether all electroactive components can be dissolved in the electrolyte. Fig. 8 shows a schematic diagram of a vanadium redox flow battery system. A crucial advantage of FBES is that the power of a FBES system is independent of its storage capacity. The power of the FBES system is determined by the size of the electrodes and the number of cells in the stack; whereas the storage capacity is determined by the concentration and the amount of electrolyte [26], [100] and [101]. Also, the very small self-discharge is an inherent strength of the FBES system due to the electrolytes being stored in separate sealed tanks [4] and [13]. Drawbacks of flow batteries include low performance resulting from non-uniform pressure drops and the reactant mass transfer limitation, relatively high manufacturing costs and more complicated system requirements compared to traditional batteries [102] and [103].

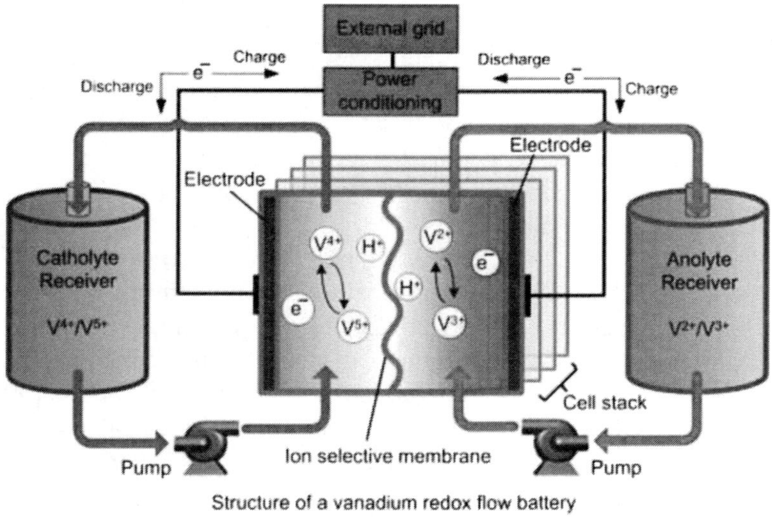

Figure 8. Schematic diagram of a vanadium redox flow battery system.

FBES facilities have been demonstrated at a few hundred kW and even multi-MW levels, and there are not many commercially available FBES systems at present [4],[19] and [104]. The current research activities undertaken cover: low-cost, efficient and reliable electrodes; highly permselective and durable membranes; power and energy management of large-scale FBES systems, etc. Some types of flow battery technologies have been used or can potentially be used for utility EES applications, including vanadium redox, zinc bromine and polysulfide bromine, which are described in the following three subsections.

Vanadium Redox Flow Battery (VRB)

The VRB is one of the most mature flow battery systems [4] and [87]. The VRB stores energy by using vanadium redox couples (V^{2+}/V^{3+} and V^{4+}/V^{5+}) in two electrolyte tanks (Fig. 8). VRBs exploit the vanadium in these four oxidation states which makes the flow battery have only one active element in both anolyte and catholyte [100]. During the charge/discharge cycles, H^+ ions are exchanged through the ion selective membrane. The chemical reaction is: $V^{4+} \leftrightarrow V^{5+} + e^-$ and $V^{3+} + e^- \leftrightarrow V^{2+}$; the cell voltage is ~1.4 V[100] and [102].

VRBs have quick responses (faster than 0.001 s) and can operate for 10,000–16,000+ cycles [18] and [105]. They have relatively high efficiencies, up to ~85%[100] and [105]. Manufacturers can design VRBs to provide continuous power (discharge duration time

24+ hours) [4] and [106]. Although VRBs now tends to expand their range of applications by enhancing the physical scale, there are some technical challenges that need to be solved, for instance, low electrolyte stability and solubility leading to low quality of energy density [107] and [108]. Also, the relatively high operating cost needs to be further reduced [103].

VRBs can be used in a large number of applications, mainly including enhancing power quality used for stationary applications and UPS devices, improving load levelling and power security, supporting the intermittent nature of renewable energy-based power generation. Some VRB facilities worldwide are introduced in Table 6. Currently, two projects on VRBs have been funded with a combined cost of £1.2 million in the UK. One project has been developed by Scottish Power, the University of Southampton and others, which planned to test a 100 kW redox flow battery for utility EES [6]. Another VRB energy storage system project has been developed by C-Tech Innovation Ltd, E.ON UK plc. and other institutes, which is especially for storing surplus energy from renewable energy sources [108]. Both of these two projects intend to be developed to a larger scale after the successes of initial small-scale trials [6] and [109].

Table 6. Selected vanadium redox flow battery energy storage facilities [67], [105], [107], [110] and [111].

Name/locations	Power/capacity	Application area
Edison VRB EES facility, Italy	5 kW, 25 kW h	Telecommunications back-up application
Wind power EES facility King Island, Australia	200 kW, 800 kW h	Integrated wind power, foil fuel energy with EES
Wind Farm EES project, Ireland	2 MW, 12 MW h	Wind power fluctuation mitigation, grid integration
VRB EES facility installed by SEI, Japan	1.5 MW, 3 MW h	Power quality application
VRB facility by PacifiCorp, Utah, U.S.	250 kW, 2 MW h	Peak power, voltage support, load shifting
VRB EES system build by SEI, Japan	500 kW, 5 MW h	Peak shaving, voltage support

Zinc Bromine (ZnBr) Flow Battery

ZnBr flow batteries belong to the hybrid flow batteries category. In a ZnBr battery, two aqueous electrolyte solutions contain the reactive components, which are based on zinc and bromine elements, stored in two external tanks. During the charging/discharging phases, these two electrolyte solutions flow through the cell stack consisting of carbon-plastic composite electrodes with compartments. Thus the reversible electrochemical reactions occur in these electrolytic cells. The corresponding chemical reactions are: $2Br^- \leftrightarrow Br_2 + 2e^-$ and $Zn^{2+} + 2e^- \leftrightarrow Zn$ [4] and [102].

The ZnBr flow battery has relatively high energy density (~30–65 W h/L) and cell voltage (1.8 V) [4] and [26]. It also has deep discharge capability and good reversibility[19] and [102]. Module sizes vary from 3 kW to 500 kW, with estimated lifetimes of 10-20 years and discharge durations of up to ~10 h [4], [112] and [113]. The disadvantages of the ZnBr battery are: material corrosion, dendrite formation and relatively low cycle efficiencies (around 65–75%) compared to traditional batteries, which can limit its applications [4], [114] and [115]. Furthermore, ZnBr batteries normally operate in a narrow temperature range [102] and [116]. Utility EES applications using ZnBr batteries are in the early stage of demonstration/commercialization. ZBB Energy Corporation and Premium Power Corporation have developed this technology for commercial purposes (50 kW h, recently tested up to ~2 MW) [14]. The firm RedFlow in Australia successfully commercialized a fully functional ZnBr module product, named ZBM, which delivers up to 3 kW of continuous power (5 kW peak) and up to 8 kW h of energy; the company claimed that it can achieve up to 80% DC-DC max net energy efficiency [113]. In 2011, U.S. electric utilities conducted early trials of 0.5 MW/2.8 MW h transportable ZnBr systems for grid support and reliability [117]. In the same year, Sacramento Municipal Utility District (SMUD) planned to demonstrate a 1 MW ZnBr flow battery system for multi EES applications [100] and [118]. A relevant project, named "flow batteries for grid scale energy storage", has been implemented by Lawrence Berkeley National Laboratory in Berkeley, U.S. [100].

Polysulfide Bromine (PSB) Flow Battery

A PSB system uses sodium bromide and sodium polysulphide as salt solution electrolytes. The chemical reactions are: $3Br^- \leftrightarrow Br_3^- + 2e^-$ and $2S_2^{2-} \leftrightarrow S_4^- + 2e^-$[4] and [19]. The significant advantages of PSB systems are: the materials of two electrolytes are abundant and highly soluble in aqueous electrolytes, and they are also cost-effective [19]. The voltage generated across the membrane is ~1.5 V; the PSB system has a fast response time, reacting within 20 ms [4] and [105]. PSBs have a wide range of potential application areas, especially for power system frequency control and voltage control due to their fast response characteristic. Because bromine and sodium sulfate crystals are produced during the chemical reactions, this may result in environmental issues.

Several PSB systems have been demonstrated at multi-kW scales. For instance, a 100 kW stack using PSB technology had been built by the UK Company Innogy, with a net efficiency of ~75% [104]. Concerning large-scale PSB facility deployment, Regenesys Technologies had tried to build a 15 MW/120 MW h energy storage plant at a power station in the UK; another demonstration plant to be located at Tennessee Valley in the U.S. was designed with a 12 MW/120 MW h capacity for EES to support a wind power plant operation [4]. However, due to engineering difficulties and financial constraints, the construction of these two large storage plants was ceased and the demonstration plants were uncompleted [77], [104] and [119]. Thus the PSB technology for large-scale EES applications still needs practical experience.

Capacitor and Supercapacitor

A capacitor is composed of at least two electrical conductors (normally made of metal foils) separated by a thin layer of insulator (normally made of ceramic, glass or a plastic film). When a capacitor is charged, energy is stored in the dielectric material in an electrostatic field [4], [120] and [121]. Its maximum operating voltage is dependent on the breakdown characteristics of the dielectric material. Capacitors are appropriate for storing small quantities of electrical energy and conducting a varying voltage; they have a higher power density and shorter charging time compared to conventional batteries [70]. However, they have limited capacity, relatively low energy density and high energy dissipation due to the high self-discharge losses [4], [120], [121] and [122]. According to these characteristics, capacitors can be used for some power quality applications, such as high voltage power correction, smoothing the output of power supplies, bridging and energy recovery in mass transit systems.

Supercapacitors, also named electric double-layer capacitors or ultracapacitors, contain two conductor electrodes, an electrolyte and a porous membrane separator (refer to Fig. 9) [13]. Due to their structures, supercapacitors can have both the characteristics of traditional capacitors and electrochemical batteries. The energy is stored in the form of static charge on the surfaces between the electrolyte and the two conductor electrodes. The supercapacitors with high-performance are based on nano materials to increase electrode surface area for enhancing the capacitance.

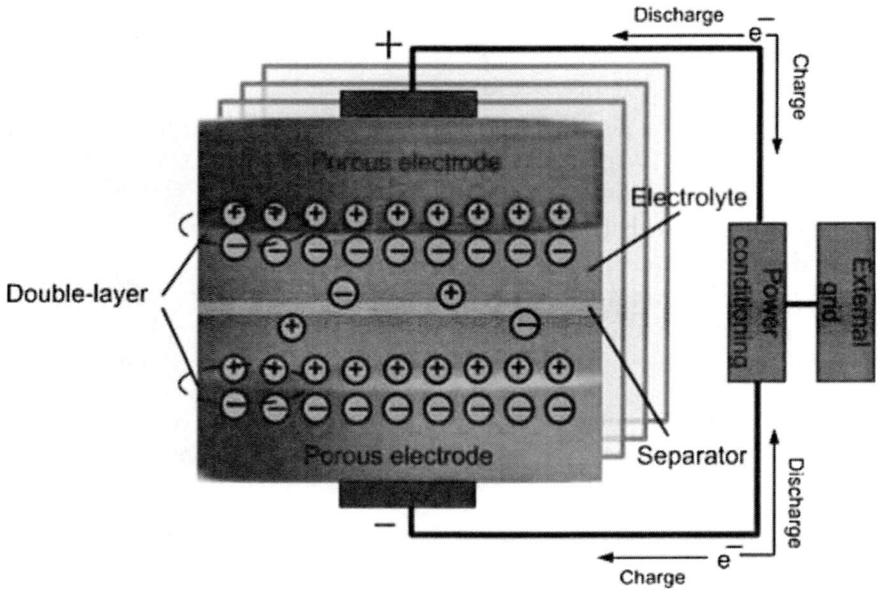

Figure 9. Schematic diagram of a supercapacitor system.

The power and energy densities of supercapacitors are between those of rechargeable batteries and traditional capacitors [20], [123] and [124]. The most important features of supercapacitors are their long cycling times, more than 1×10^5 cycles, and high cycle efficiency, ~84–97% [4] and [66]. However, the daily self-discharge rate of supercapacitors is high, ~5–40%, and the capital cost is also high, in excess of 6000 $/kW h [4], [10], [13] and [125]. Thus supercapacitors are well suited for short-term storage applications but not for large-scale and long-term EES. Typical applications in power quality consist of pulse power, hold-up/bridging power to equipment, solenoid and valve actuation in factories, UPS devices, etc. There are a number of manufacturers producing supercapacitors worldwide (refer to Table 7).

Table 7. Selected manufacturers of supercapacitors for utility applications [20], [67], [126] and [127].

Device/Company name	Country	Technical information
Super capacitor, CAP-XX	Australia	Single cell 2.3–2.9 V, up to ~2.4 F, 233-358 K
Gold capacitor, Panasonic	Japan	Single cell 2.3–5.5 V, 0.1–2000 F
Ultracapacitor/Boostcap, Maxwell	U.S.	Single cell 2.2–2.7 V, 1–3000 F, UPS, pulse, transportation
Supercapacitor, NEC	Japan	3.5–12 V, 0.01–6.5 F, power quality application
Supercapacitor, Siemens	Germany	21 MJ/5.7 W h, 2600 F, metro distribution net application
Supercapacitor, TVA company	U.S.	200 kW, supporting the start of high power dc machines

Research and development in supercapacitors has been very active in recent years. Some recent good quality reviews have focused on the recent development of materials for chemical capacitive energy storage, such as an overview of carbon materials for super-capacitors is given in [24] and an overview of graphene-based electrodes can be found in [25]. To be more specific, a new composite material formed by dispersing ultra-small silicon nanoparticles in polyaniline was developed as the electrode material for supercapacitors [128]. The integration of a short-term supercapacitor EES device in a doubly fed induction generator has been studied in order to smooth the fast wind-induced power variations [129]. One UK EPSRC funded project aiming to develop high-performance supercapacitors with enhanced energy density had been implemented. The prototype had been tested for designing an effective and sustainable power system. Some achievements of this project were published in 2013 [130].

Superconducting Magnetic Energy Storage (SMES)

A typical SMES system is composed of three main components which include: a superconducting coil unit, a power conditioning subsystem, and a refrigeration and vacuum subsystem [13], [109] and [131]. The SMES system stores electrical energy in the magnetic field generated by the Direct Current (DC) in the superconducting coil which has been cryogenically cooled to a temperature below its superconducting critical temperature. In general, when current passes through a coil, the electrical energy will be dissipated as heat due to the resistance of the wire; however, if the coil is made from a superconducting material, such as mercury or vanadium, under its superconducting state (normally at a very low temperature), zero resistance occurs and the electrical energy can be stored with almost no losses. One commonly used superconducting material is Niobium–Titanium which has a superconducting critical temperature of 9.2 K[4] and [132]. In the discharging phase, the SMES system can release the stored electrical energy back to the Alternating Current (AC) system, by a connected power converter module. The magnitude of stored energy is determined by the self-inductance of coil and the current flowing through it [133]. A simplified structure of a SMES system is illustrated in Fig. 10.

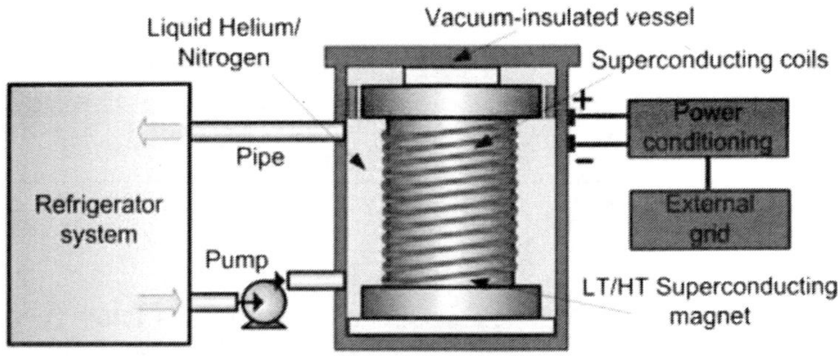

Figure 10. Schematic diagram of a SMES system.

Superconducting coils can be classified into two groups: Low Temperature Superconducting (LTS) coils, working at ~5 K, and High Temperature Superconducting (HTS) coils, working at ~70 K [13] and [131]. The LTS-SMES technology is more mature and commercially available while the HTS-SMES is currently in the development stage. SMES devices in the range of 0.1–10 MW have been used commercially; while SMES systems with 100 MW h could be available in the next decade.

The features of SMES include relatively high power density (up to ~4000 W/L), fast response time (millisecond level), very quick full discharge time (less than 1 min), high cycle efficiency (~95–98%) and long lifetime (up to ~30 years) [4], [66], [114] and [134]. In contrast to rechargeable batteries, SMES devices are capable of discharging near to the totality of the stored energy with little degradation after thousands of full cycles. The drawbacks are that they have high capital cost (up to 10,000 $/kW h, 7200 $/kW), high daily self-discharge (10–15%) and a negative environmental impact due to the strong magnetic field [4], [14] and [114]. Moreover, the coil is sensitive to small temperature variations which can cause the loss of energy. From the above, SMES is suitable for short-term storage in power and energy system applications and it is expected to have an important role in the increased use of intermittent renewable energy [131]. Table 8 shows selected SMES facilities with their application fields.

Table 8. Some projects of superconducting magnetic energy storage [67], [131], [133] and [138].

Locations/organizations	Technical data	Features/applications
Proof principle, tested in a grid in Germany	5 KJ, 2 s to max 100 A at 25 K	World first significant HTS-SMES, by ASC
Nosoo power station in Japan	10 MW	Improve system stability and power quality
Upper Wisconsin by American Transmission	3 MW/0.83 kW h, each 8 MV A	Power quality application reactive power support
Bruker EST in Germany	2 MJ	High temperature superconductors
Korea Electric Power Corporation, Hyundai	3 MJ, 750 kV A	Improving power supply quality for sensitive loads
Chubu Electric Power Co. in Japan	7.3 MJ/5 MW and 1 MJ	Provide comparison to transient voltage
University of Houston, Super Power & others	20 kW, up to 2 MJ class	UHF-SMES, voltage distribution

Recently, considerable research and development effort has been made: (1) to reduce the costs of superconducting coils and related refrigeration systems; (2) to develop HTS coil materials which are less cryogenically sensitive [14], [109] and [131]. Since 2011, SuperPower Inc., in partnership with ABB Inc., Brookhaven national laboratory and the Texas center for superconductivity at the University of Houston has been developing an advanced SMES demonstrator 20 kW ultra-high field SMES system with a capacity up to 2 MJ [135]. This demonstration project aims to pave the way of grid-scale SMES technology for the U.S. electric grid operation and for the integration of renewable sources. The University of Bath in the UK continuously focus on SMES technology development and a relevant on-going project funded by the UK EPSRC aims to investigate HTS-SMES as part of hybrid EES systems for renewable energy micro-grids[136] and [137].

Solar Fuels

Solar fuel is a relatively new technology to EES. Approaches to produce solar fuels include: (1) natural photosynthesis; (2) artificial photosynthesis; (3) thermochemical approaches [4], [139] and [140]. A number of fuels can be produced by solar energy, such as solar hydrogen, carbon-based fuels, and solar chemical heat pipe [4], [140],[141], [142] and [143]. These fuels can be stored and subsequently provide the basis for later electricity generation.

For the first two approaches to produce solar fuels, solar energy is captured via photosynthesis and then stored in chemical bonds, i.e., the sunlight is used to convert water and/or carbon dioxide into oxygen and other materials [144]. Fig. 11 shows a comparison of natural and artificial photosynthesis. The artificial system for water-splitting catalysts generally relies on scarce elements, e.g., Ruthenium (*Ru*), Palladium (*Pd*) and Rhenium (*Re*) [140] and [145]. For example, sunlight can be captured by Ruthenium (*Ru*) as a catalyst, and electrons moves from the donor (marked as "D") to the acceptor (Fig. 11) [140] and [145].

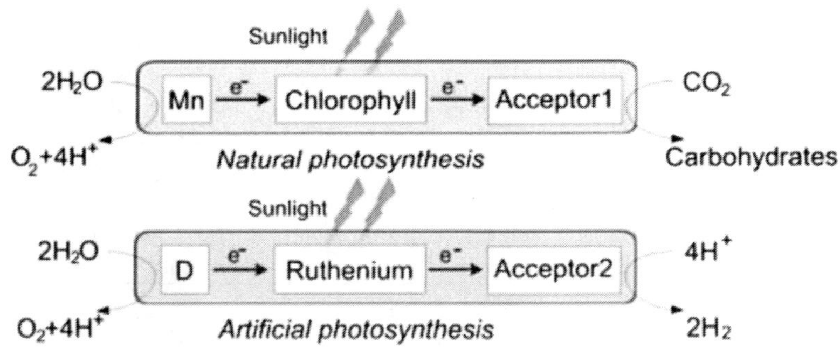

Figure 11. Comparison of natural and artificial photosynthesis [140].

The thermochemical approach uses thermal processes for solar fuels production, which involve the generation of very high temperatures in a closed environment to split water into its constituent parts [4]. Thus this method is more dependent on strong sunlight compared to the other two [144] and [146]. After the solar radiant energy is concentrated by heliostats, an endothermic chemical transformation is carried out in a reaction vessel. The reaction produces hydrogen and/or carbon monoxide and/or other materials [142],[143] and [147].

Solar fuel technology is currently at the development stage. The power rating of solar fuels is potentially up to 20 MW and the specific energy estimate is from 800 W h/kg to 100,000 W h/kg [4] and [148]. The storage duration can range from a few hours to several months [4]. One drawback of artificial photosynthesis is that the water-splitting catalysts normally depend on scarce, expensive elements [145]. Another disadvantage is that solar fuel facilities need a large area to place devices to concentrate sunlight, especially when using the thermochemical approach to produce solar fuels.

Research in solar fuels has recently undergone substantial advances, making it possible for it to become cost-effective for utility EES applications in the near future. There are on-going research projects in the U.S., the Netherlands, South Korea, Singapore, Japan and China. In the US, there are several organizations focusing on this area, such as Energy Innovation Hub at DoE, the MIT spin-out Sun Catalytix and the Princeton University spin-out Liquid Light. The "Towards BioSolar Cells" research programme has focused on increasing the photosynthetic efficiency and creating solar collectors [140] and [149]. Concerning the issue of the reliance on scarce and expensive elements, one important breakthrough in the development of using earth-abundant, relatively cheap catalysts (e.g. cobalt and phosphate) and silicon-based semiconductors for the water-splitting process has been recently reported by the Nocera's team from MIT [150] and [151]. Asia's pioneering solar fuel research laboratory at Nanyang Technological University of Singapore has also made effort on the investigation of affordable approaches to extract large amounts of hydrogen from water using sunlight for engineering applications [151] and [152].

Hydrogen Storage and Fuel Cell

Hydrogen energy storage systems use two separate processes for storing energy and producing electricity (refer to Fig. 12). The use of a water electrolysis unit is a common way to produce hydrogen which can be stored in high pressure containers and/or transmitted by pipelines for later use (Fig. 12) [8] and [13]. When using the stored hydrogen for electricity generation, the fuel cell (also known as regenerative fuel cell) is adopted, which is the key technology in hydrogen EES.

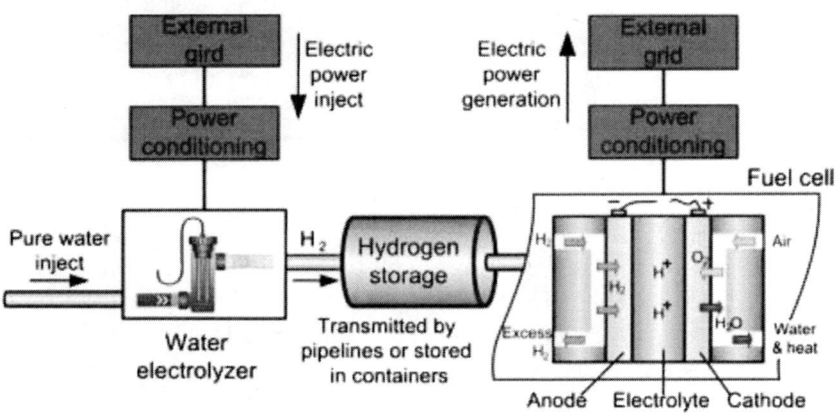

Figure 12. Topology of hydrogen storage and fuel cell.

Fuel cells can convert chemical energy in hydrogen (or hydrogen-rich fuel) and oxygen (from air) to electricity [8], [13] and [153]. The overall reaction is: $2H_2 + O_2 \rightarrow 2H_2O$ + energy [154]. Electrical and heat energy are released during the process (Fig. 12). Depending on the choice of fuel and electrolyte, there are six major groups of fuel cells, which are: Alkaline Fuel Cell (AFC), Phosphoric Acid Fuel Cell (PAFC), Solid Oxide Fuel Cell (SOFC), Molten Carbonate Fuel Cell (MCFC), Proton Exchange Membrane Fuel Cell (PEMFC) and Direct Methanol Fuel Cell (DMFC) [154]. Their chemical reactions and application fields are listed in Table 9.

Table 9. Chemical reactions of main fuel cells [154] and [155].

Fuel cell Type	Chemical reactions at anodes and cathodes	Applications
AFC	$2H_2 + 4OH^- \rightarrow 4H_2O + 4e^-$ $O_2 + 2H_2O + 4e^- \rightarrow 4OH^-$	Military, space applications
PAFC	$2H_2 \rightarrow 4H^+ + 4e^-$ $O_2 + 4H^+ + 4e^- \rightarrow H_2O$	Distributed generation
SOFC	$O^{2-}(S) + H_2(g) \rightarrow H_2O(g) + 2e^-$ $1/2O_2(g) + 2e^- \rightarrow O^{2-}(s)$	Utility EES, distributed generation
MCFC	$H_2O + CO_3^{2-} \rightarrow H_2O + CO_2 + 2e^-$ $2H_2 + 4OH^- \rightarrow 4H_2O + 4e^-$	Electric utility EES, distributed generation
PEMFC	$H_2(g) \rightarrow 2H^+ + 2e^-$ $1/2O_2(g) + 2H^+ + 2e^- \rightarrow H_2O$	Backup power, small distributed generation
DMFC	$CH_3OH + H_2O \rightarrow CO_2 + 6H^+ + 6e^-$ $3/2O_2 + 6e^- + 6H^+ \rightarrow 3H_2O$	Transportation, portable devices

In general, the electricity generation by using fuel cells is quieter, produces less pollution and is more efficient than the fossil fuel combustion approach [156]. Other features include easy scaling (potential from 1 kW to hundreds of MW) and compact design[153] and [157]. Fuel cell systems combined with hydrogen production and storage can provide stationary or distributed power (primary electrical power, heating/cooling or backup power) and transportation power (potentially replacing fossil fuels for vehicles)[153] and [154]. Such hydrogen EES systems can offer capacity and power independence in energy production, storage and usage, due to the separate processes. It should be noted that the disposal of exhaust fuel cells must consider degradation and recycling while toxic metals are used as electrodes or catalysts. Many of the relevant aspects and approaches have been under investigation [158] and [159]. For instance, palladium in catalysts of fuel cells can be reprocessed into other products in theory [159].

Currently, hydrogen EES with fuel cell technology is in the development and demonstration stage. Stationary power applications are relatively mature. In 2012 nearly 80% of total investment in the global fuel cell industry was made by the U.S. companies[153]. Cost reduction and durability verification/improvement are essential to deploy this technology in large-scale EES applications [109]. Some research or demonstration projects are in place and on-going across the world. The world's first utility-scale test of a stand-alone renewable energy system integrated with hydrogen storage and fuel cells was installed in Norway, which delivered power with required quality and high reliability[157]. One of the world's largest biogas fuel cell power plants was launched in 2012 in California (2.8 MW), which converts biogas into electricity and usable high-quality heat[160]. In 2013, the US Naval Air Warfare Center Weapons Division in California successfully tested a novel 5 kW trailer-mounted regenerative fuel cell system to use solar power to produce hydrogen with fuel cells [161]. Since 2013, McPhy and Enertrag AG in Germany have worked jointly to develop economic wind-hydrogen solutions for EES and for transportation fuel cell applications [162]. Currently, the on-going hydrogen storage and fuel cell relevant projects include IdealHy (the Netherlands), RE4CELL (Spain), Sapphire (Norway), SmartCat (France), etc.

Thermal Energy Storage (TES)

TES encompasses a variety of technologies that store available heat energy using different approaches in insulated repositories [6] and [26]. A TES system normally consists of a storage medium in a reservoir/tank, a packaged chiller or built-up refrigeration system, piping, pump(s), and controls. Based on the range of operating temperature, TES can be classified into two groups: low-temperature TES (consisting of aquiferous low-temperature TES and cryogenic energy storage) and high- temperature TES (including latent (fusion) heat TES, sensible heat TES and concrete thermal storage) [4], [163], [164], [165] and [166]. Aquiferous low-temperature TES normally uses water cooled/iced and reheating processes, which is more suitable for peak shaving and industrial cooling loads [4]. Cryogenic energy storage employs a cryogen (such as liquid nitrogen or liquid air) to achieve the electrical and thermal energy conversion. For instance, Liquid Air Energy Storage (LAES) is attracting attention due to the high expansion ratio from the liquid state to the gaseous state and the high power densities of liquid air compared to that of gaseous state of air. Latent heat TES employs Phase Change Materials (PCMs) as the storage media and uses the energy absorption or emission in liquid-solid transition of these PCMs at constant temperature. Concrete thermal storage utilizes concrete or castable ceramics to store heat energy, normally supported by synthetic oil as a heat transfer fluid. The above TES technologies have different features with various applications. For instance, latent heat storage can provide a relatively high storage density with a small dimension reservoir, thus the use of this technology in buildings receives attention [21]. In addition, cryogenic energy storage is expected to be used for future grid power management.

The TES system can store large quantities of energy without any major hazards and its daily self-discharge loss is small (~0.05–1%); the reservoir offers good energy density and specific energy (80–500 W h/L, 80–250 W h/kg) and the system is economically viable with relatively low capital cost (3–60 $/kW h) [4], [10], [166], [167] and [168]. However, the cycle efficiency of TES systems is normally low (~30–60%) [4]. TES has been used in a wide spectrum of applications, such as load shifting and electricity generation for heat engine cycles.

With particular focus on using TES for power system and grid applications, there are many active research projects worldwide and, in addition, numerous demonstration projects are built, under construction or planned. The UK based company Highview Power Storage designed and assembled a pilot LAES facility (300 kW/2.5 MW h storage capacity) which has been in operation at Scottish and Southern Energy's 80 MW biomass plant since 2010 [168] and [169]. In February 2014, this firm has been awarded £8 million funding from the UK government for a 5 MW/15 MW h demonstration LAES project; the designed LAES system will be alongside one landfill gas generation plant in the UK [169]. A TES system in an office building was built by a joint U.S. and China demonstration project in Beijing, which can reduce peak electric energy consumption of 6100 kW h per month [170]. A new central energy plant including an ice-based TES system is being built in South Florida. The completely built plant will have a total capacity of 11,500 tons of chilled water with 68,000 ton-hour of TES [170]. A 15 MW commercial power plant, named "Solar Tres Power Tower", is being built in Spain by Torresol Energy, and it uses molten salt as the working fluid to store heat energy [165]. A wind power generation system combined with a sensible heat storage facility had been proposed (Fig. 13) [165]. The electrical energy from wind power is used to heat a bulk storage material; the heat energy is recovered to produce water vapor which in turn drives a turbo-alternator to generate electricity. A detailed study of load shifting of nuclear power plants by using cryogenic energy storage technology was recently reported in [171]. A UKERC funded project, "the future role of TES in the UK energy system", has investigated the potential for, and limitations of, the role of TES in the transition to a sustainable low carbon energy supply system; the project has also studied the suitability of TES in managing energy generation and distribution systems with large-scale penetration [6].

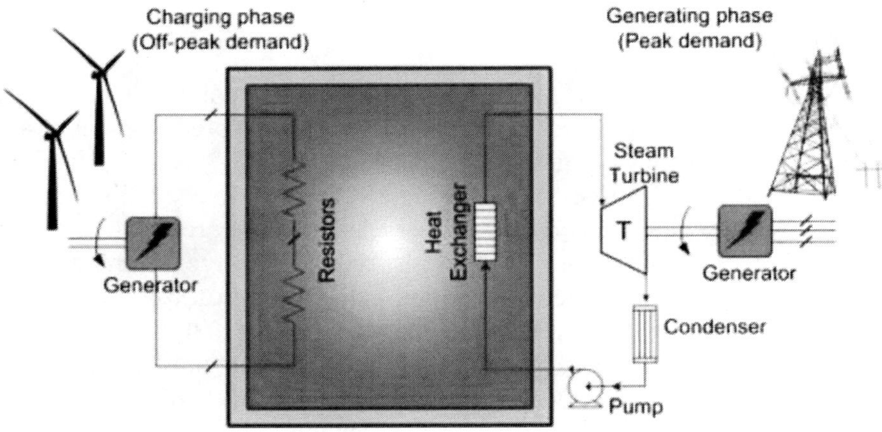

Figure 13. A sensible heat storage system for wind power generation.

Hybrid Electrical Energy Storage

Hybrid EES refers to the integration of at least two different EES technologies into one system or application. Advantages of each EES technology can be utilized to achieve specific requirements, meet harsh working environments, optimize the whole system performance or improve the cycle efficiency. For example, the first large pilot power plant (ADELE) utilizes CAES and TES technologies for enhancing the efficiency and avoiding fossil fuel consumption [42] and [43]. The combination of supercapacitor and battery technologies can offer relatively large storage capacity and very fast charge/discharge rates. One project in the theme of a hybrid supercapacitor-battery system, funded by E.ON, was completed in UK and a corresponding demonstration system was developed[6]. Another project funded by UK EPSRC, named "Ultra battery feasibility - investigation into the combined battery- supercapacitor for hybrid electric vehicle applications", was completed in 2012, and some achievements were recently published in [172]. Researchers in Japan High

Energy Accelerator Research Organization, Tohoku University and others have designed a back-up system for renewable energy power generation, which combines a liquid hydrogen refrigeration-based SMES system with a hydrogen-fuel cell system [109] and [173].

COMPARISON AND EVALUATION OF ELECTRICAL ENERGY STORAGE TECHNOLOGIES

It is well recognized that no single EES technology can meet the requirements for all power system applications. Comprehensive analysis of different EES technologies is conducted and Table 10, Table 11 and Table 12 provide the matrices to clearly show the positions of different EES performance and characteristics. The selection of representative matrices in Table 10, Table 11 and Table 12 is derived with consideration of the focuses in both academic research and industry application areas, which is drawn from the comprehensive literature review in this paper. The selection of EES indices also relies on the assessment in the characteristics of different EES options against the requirements of power system applications, which will be discussed in Section 5.

Table 10. Technical characteristics of electrical energy storage technologies.

Techn ology	Energy density (W h/L)	Power density (W/L)	Specific energy (W h/kg)	Specific power (W/kg)	Power rating (MW)	Rated energy capacity (MW h)
PHS	0.5–1.5 [4], 1–2 [26]	0.5–1.5[4], ~1[26],	0.5–1.5 [4]	–	100–5000 [4], 30 [34], < 4000 [114]	500–8000 [4], 180 Okinawa PHS[34] and [77]
Large-scale CAES	3–6 [4], 2–6[26]	0.5–2 [4], ~1 [26]	30–60 [4]	–	Up to 300 [4], 110 & 290[39], 1000[70]	~ < 1000[10], 580 & 2860[38] and [42]
Overground small CAES	Higher than large-scale CAES	Higher than large-scale	140 at 300 bar[174]	–	0.003–3 [51]Potential ~10[175]	~0.01[10], ~0.002–0.0083[51]

		CAES				
Flywheel	20–80 [4],[26] and [123]	1000–2000 [4], ~5000[26]	10–30 [4], 5–100 [57], 5–80[176]	400–1500[4]	<0.25 [4], 3.6[60], 0.1–20[13] and [177]	0.0052 [60], 0.75[70], up to 5[177]
Lead–acid	50–80 [4], 50–90 [70]	10–400[4]	30–50 [4], 25–50[178]	75–300[4], 250[70], 180[57]	0–20 [4], 0–40 [14], 0.05–10 [179]	0.001–40 [179]More than 0.0005[180]
Li-ion	200–500 [4], 200–400 [26], 150 [70]	1500–10,000[26]	75–200 [4], 90[70], 120–200[181]	150–315[4], 300[70], 500–2000 [57]	0–0.1 [4], 1–100 [73], 0.005–50[182]	0.024 [79], ~0.004–10 [182]
NaS	150–250 [4], 150–300 [26]	~140–180 [26]	150–240 [4], 100 [183], 174[184]	150–230[4], 90–230 [9], 115 [13],	<8 [4], <34[14]	0.4–244.8 [81], 0.4 [185]
NiCd	60–150 [4], 15–80 [26], 80 [70]	80–600[26]	50–75 [4], 50[70], 45–80[71]	150–300[4], 160[13], 150[70],	0–40 [4], 27[88], 40 [186]	6.75[57] and [88]
VRB	16–33 [4], 25–35 [19]	~ < 2[26]	10–30 [4]	166 [187]	~0.03–3 [4], 2 [188]possible 50[5]	<60 [13], 2 [88], 3.6 [189]
ZnBr	30–60 [4], ~55–65 [26]	~ < 25[26]	30–50 [4], 80[190], 75 [191]	100 [190], 45 [191]	0.05–2 [4], 1–10 [73]	0.1–3 [13], 4[14], 0.05 & 0.5[192]
PSB	~20–30 [123]	~ < 2[26]	~15–30 [123]	–	1–15 [4], 1[193], 0.004[194]	Potential up to 120 [193], 0.06[194]
Capacitor	2–10 [4], ~0.05 [124]	100,000+[4],	0.05–5 [4], <~0.05[121] and [124]	~100,000[4], >~3000–10^7[124]	0–0.05 [4]	–
Super-capacitor	10–30 [4], ~10–30 [123]	100,000+[4],	2.5–15 [4], ~0.05–15[124]	500–5000[4], ~10,000[124]	0–0.3 [4], ~0.3+[26] ~0.001–0.1[70]	0.0005 [70]

SMES	0.2–2.5 [4], ~6 [26]	1000–4000 [4], ~2500[26]	0.5–5 [4], 10–75 [195]	500–2000[4]	0.1–10[4] and [14], ~1–10 [70]	0.0008 [70], 0.015 [138], 0.001 [196]
Solar fuel	500–10,000[4]	–	800–100,000[4]	–	0–10 [4], 6 and developing 20[197]	–
Hydrogen Fuel cell	500–3000 [4]	500+ [4]	800–10,000[4], ~150–1500 [124]	500+ [4], ~5–800[124]	<50 [4], <10[26], 58.8[199]	0.312 [198], developing 39[200]
TES	80–120, 120–200, 200–500[4]	–	80–120, 80–200 [4], 150–250 [4]	10–30 [4]	0.1–300 [4], 15 [165], 10[201]	–
Liquid air Storage	4–6 times than CAES at 200 bar [202]	–	214 [174]	–	10–200 [8], 0.3 [168]	2.5 [168]

Table 11. Additional technical characteristics of electrical energy storage technologies.

Technology	Daily self-discharge (%)	Lifetime (years)	Cycling times (cycles)	Discharge efficiency (%)	Cycle efficiency (%)	Response time
PHS	Very small[4] and [192]	40–60 [4], 40+[69], 30+[175]	10,000–30,000[14]	~87 [114]	70–85 [4], 70–80 [175]87 [33], 75–85 [203]	Minutes[114], not rapid discharge[203]
Large-scale CAES	Small [4], Almost zero[192]	20–40 [4], 30[70], 20+[69] and [203]	8000–12,000 [14]	~70–79[114]	42,54[4] and [42]AA-CAES 70[43] and [203]	Minutes[114]
Over-ground small CAES	Very small [51]	23+[51]	Test 30,000stop/starts[51]	~75–90[51]	–	Seconds–minutes[114]
Flywheel	100 [4], ⩾20% per hour [57]	~15 [4], 15+[69], 20 [114]	20,000+ [4], 21,000+[69]	90–93[114]	~90–95 [4], 90 & 95 [70]	<1 cycle[114], seconds[203]
Lead–acid	0.1–0.3 [4], <0.1 [57], 0.2[69]	5–15 [4] and [57], 13 [69]	500–1000 [4], 200–1800 [13]	85 [114]	70–80[4], 63–90 [14], 75–80 [204]	<1/4 cycle[114]milliseconds

Li-ion	0.1–0.3 [4], 1 & 5 [13]	5– 15 [4], 14–16[205]	1000–10,000 [4], up to 20,000 [9]	85 [114]	~90–97 [4], 75–90 [73]	Milliseconds, <1/4 cycle[14]
NaS	Almost zero[13] and [185]	10–15 [4], 15[69], 12–20 [192]	2500 [4], 3000[206] 2500–4500 [14]	85 [114]	~75–90 [4], 75 [206], 75–85 [204]	–
NiCd	0.2–0.6 [4],0.3[57], 0.03–0.6[14]	10–20 [4], 3–20[13], 15–20 [57]	2000–2500 [4], 3500 [179]	85 [114]	~60–70 [4], 60–83 [14]	Milliseconds, <1/4 cycle[14]
VRB	Small [4], very low [13]	5–10 [4], 20 [193]	12,000+ [4], 13,342 [69]	~75–82[207]	75–85[4] and [62], 65–75 [73]	<1/4 cycle[14]
ZnBr	Small[4] and [100]	5–10 [4], 10 [69], 8–10 [205]	2000+ [4], 1500[69]	~60–70[208]	~65–75 [4], 66–80 [14], 66 [114]	<1/4cycle[114]
PSB	Small [4]Almost zero[193]	10–15 [4], 15[209]	–	–	~60–75 [4], 60–75 [209]	20 ms [116]
Capacitor	40 [4], ~50 in about 15 minutes [122]	~5 [4], ~1–10[122]	50,000+ [4], 5000 (100% DoD)[210]	~75–90[127]	~60–70 [4], 70+[210]	Milliseconds, <1/4 cycle[14]
Super-capacitor	20–40 [4], 5[10], 10–20[211]	10–30 [4], 10–12[66]	100,000+ [4], 50,000+[69]	95 [114]Up to ~98 [127]	~90–97 [4], 84–95 [66]	Milliseconds, ¼ cycle[114]
SMES	10–15 [4]	20+[4], 30 [114]	100,000+4], 20,000+ [14]	95 [114]	~95–97 [4], 95–98 [66], 95 [70]	Milliseconds, <1/4 cycle[114]
Solar fuel	Almost zero[4]	–	–	–	~20–30 [4], planned eff.>54 [197]	–
Hydrogen Fuel cell	Almost zero[4] and [192]	5–15 [4], 20 [119]20+[212]	1000+ [4], 20,000+[212]	59 [114]	~20–50 [4], 32 [106], 45–66 [213]	Seconds, <1/4 cycle[114]
TES	0.05–1 [4]	10–20 [4], 5–15[4], 30 [203]	–	–	~30–60 [4]	Not for rapid response[203]
Liquid air Storage	Small[169] and [214]	25+[214]	–	–	55–80+[214]	Minutes[215]

Table 12. Other technical and economical characteristics of electrical energy storage technologies.

Technology	Suitable storage duration	Discharge time at power rating	Power capital cost ($/kW)	Energy capital cost ($/kW h)	Operating and maintenance cost	Maturity
PHS	Hours–months [4], long-term [27]	1–24 h+[4], 6–10 h [73] 10 h[175]	2500–4300[73], 2000–4000[175]	5–100[4], 10–12 [114]	0.004 $/kW h [70], ~3 $/kW/year [72]	Mature
Large-scale CAES	Hours–months [4], long-term [27]	1–24 h+ [4], 8–20 h [73]	400–800[4], 800–1000[175]	2–50 [4], 2–120[8], 2 [70]	0.003 $/kW h [70], 19–25 $/kW/year[72]	CAES commercialized, AA-CAES developing
Over-ground small CAES	Hours–months, long-term [27]	30 s–40 min[51], 3 h [216]	517[114], 1300–1550[216]	1MVA from £296 k[51], 200–250[216]	Very low [51]	Early commercialized
Flywheel	Seconds–minutes [4]short-term(<1 h) [27]	Up to 8 s [4], 15 s–15 min[175]	250–350[4]	1000–5000 [4], 1000–14,000[8]	~0.004 $/kW h[70], ~20 $/kW/year [72]	Early commercialized
Lead–acid	Minutes–days[4], short-to-med. term	Seconds–hours [4], up to 10 h [14]	300–600[4], 200–300[114], 400 [206]	200–400[4], 50–100 [57], 330 [206]	~50 $/kW/year [72]	Mature
Li-ion	Minutes–days[4], short-to-med. term	Minutes–hours[4], ~1–8 h[209]	1200–4000[4], 900–1300[57], 1590[73]	600–2500[4], 2770–3800 [73]	–	Demonstration
NaS	Long term[82]	Seconds–hours [4], ~1 h[209]	1000–3000 [4], 350–3000 [8]	300–500[4], 350[206], 450 [217]	~80 $/kW/year [72]	Commercialized
NiCd	Minutes–days[4], Short and	Seconds–hours [4], ~1–8 h [209]	500–1500 [4]	800–1500[4], 400–	~20 $/kW/year [72]	Commercialized

	long term			2400 [57]		
VRB	Hours–months [4], Long term[27]	Seconds–24 h+ [4], 2–12 h [106]	600–1500 [4]	150–1000[4], 600[217]	~70 $/kW/year [72]	Demo/early commercialized
ZnBr	Hours–months [4] long term [27]	Seconds–10 h+ [4], ~10 h [209]	700–2500 [4], 400 [87], 200 [114]	150–1000[4], 500[71]	–	Demonstration
PSB	Hours–months [4] long term [27]	Seconds–10 h+ [4], ~10 h [209]	700–2500 [4]	150–1000[4], 450[217]	–	Developing
Capacitor	Seconds–hours [4], ~5 h [210]	Milliseconds –1 h [4]	200–400[4],	500–1000[4],	13 $/kW/year [72], <0.05 $/kWh [210]	Commercialized
Super-capacitor	Seconds–hours [4]short-term(<1 h) [27]	Milliseconds –1 h [4], 1 min[209], 10 s[216]	100–300[4], 250–450 [216]	300–2000[4]	0.005 $/kWh [70], ~6 $/kW-year [114]	Developing /demo.
SMES	Minutes–hours [4]short-term (<1 h)[27]	Milliseconds –8 s [4], up to 30 min [209]	200–300[4], 300[114], 380–489[216]	1000–10,000[4], 500–72,000[114]	0.001 $/kWh [70], 18.5 $/kW/year [72]	Demo/early commercialized
Solar fuel	Hours–months [4]	1–24 h+ [4]	–	–	–	Developing
Hydrogen Fuel cell	Hours–months [4]	Seconds–24 h+ [4]	500[114], 1500–3000[154]	15 [114], 2–15€/kWh[204]	0.0019–0.0153 $/kW [154]	Developing /demo.
TES	Minutes–days[4], minutes–months [4]	1–8 h [4], 1–24 h+ [4], 4–13 h [203]	200–300[4], 250[203], 100–400[203]	20–50[4], 30–60 [4], 3–30 [4]	–	Demo/early commercialized
Liquid air Storage	Long-term[214]	Several hours[168] and [214]	900–1900[214]	260–530[214]	–	Developing /demo.

Size of storage devices is an important factor for many applications. Fig. 14 shows the comparison of power density and energy density of different EES technologies. For a given amount of energy, the higher the power and energy densities are, the smaller the volume of the required energy storage system will be. In Fig. 14, the highly compact technologies suitable for volume-limited applications can be found at the top right corner and the large volume consuming storage systems are located at the bottom left corner. It can be seen that most batteries, flywheel and fuel cells have relatively moderate energy densities and power densities. PHS and CAES have lower densities, thus they are mainly used in stationary EES and require large reservoirs for grid scale applications. Supercapacitors and capacitors have very high power densities but low energy densities. The densities of flow batteries are commonly lower than those of conventional batteries. The Li-ion battery has both a high energy density and a high power density, which leads to widespread uses in portable devices and promising potential in transportation and other small-scale EES applications.

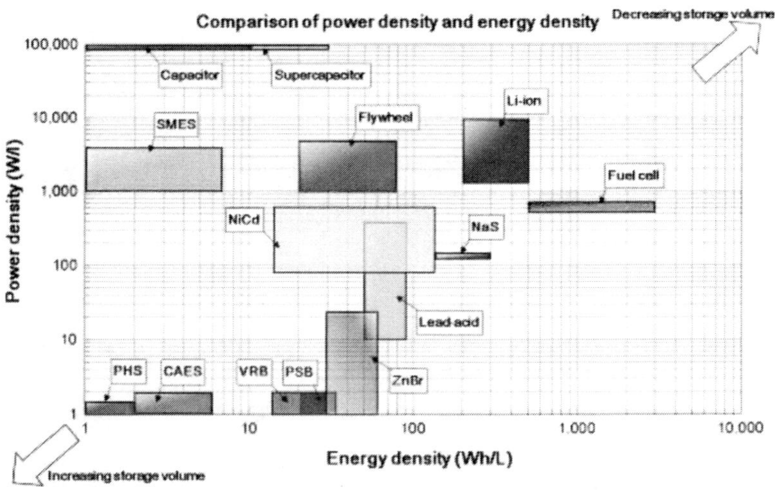

Figure 14. Comparison of energy density and power density.

Specific energy and specific power are important indices which represent the total energy and power per unit weight. Fig. 15 presents the comparison of the specific energy and specific power of EES technologies. For acquiring a certain amount of energy, the higher the specific power and the specific energy are, the lighter the weight of the EES system will be. The EES technologies suitable for light weight applications can be found at the top right corner of the figure. It can be seen that SMES capacitors and supercapacitors have high specific power but low specific energy; because of their fast response time (Table 11), they are more suitable for power quality applications for electric power (current) delivery. Fuel cells and TES have high specific energy with low specific power. Flywheel, flow batteries and most conventional batteries are located at the middle levels in terms of specific power and specific energy, which may serve for different EES application domains. Li-ion batteries are outstanding due to both high specific energy and specific power, which offers a reasonable explanation to the current broad range of development and applications for Li-ion batteries.

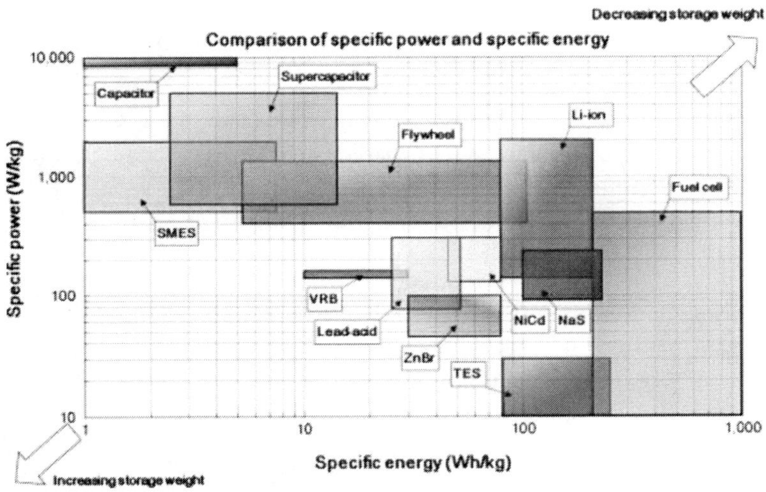

Figure 15. Comparison of specific energy and specific power.

Fig. 16 shows a comparison of power ratings and rated energy capacities of EES technologies. The nominal discharge time duration at the rated power is also shown within the range from seconds to months. The data of practical EES facilities/plants in Section 2 are marked in Fig. 16 to highlight the positions of their characteristics. This figure indicates the general application areas of current EES systems and also provides a guiding range for potential future applications. From Table 12 and Fig. 16, EES technologies can be categorized by the nominal discharge time at rated power: (1) discharge time less than 1 hour: flywheel, supercapacitor and SMES; (2) discharge time up to around 10 hours: over ground small-scale CAES, Lead–acid, Li-ion, NiCd, ZnBr and PSB; (3) discharge time longer than 10 h: PHS, underground large-scale CAES, liquid air energy storage, VRB, solar fuel, fuel cell and TES.

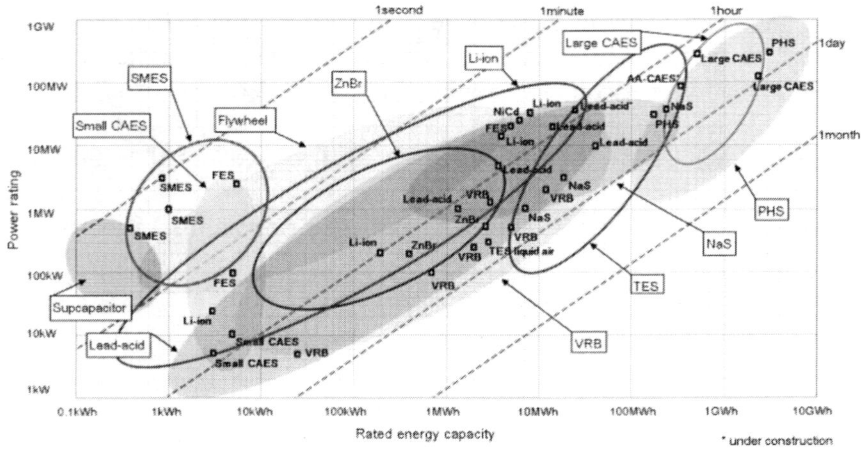

Figure 16. Comparison of power rating and rated energy capacity with discharge time duration at power rating. The marked data of EES facilities from the cited references in Section 2 of the paper).

Cycle efficiency, also named the round-trip efficiency, is the ratio of the whole system electricity output to the electricity input. Derived from the data shown in Table 11, Fig. 17shows the comparison of cycle efficiencies of EES technologies. Most commercialized (including early commercialized) techniques have medium-to-high cycle efficiencies (above 60% from all cited references), such as PHS, flywheels, conventional batteries, flow batteries, capacitors, supercapacitors and SMES. CAES, TES, solar fuels and fuel cells have low cycle efficiency (below 60%) reported in some published literatures. In general, the efficiency has been continuously improved with time through dedicated research and development efforts. For instance, the cycle efficiency for CAES has increased from 42% (in 1978), ~54% (in 1991) to the expected 70% (AA-CAES, announced by RWE power in 2011, ADELE project) [38], [39], [42] and [43]. Table 11 lists the discharge efficiencies of EES technologies. Discharge efficiency represents the energy transmission ability from the energy-storing phase to the energy-releasing phase, which contributes to the overall cycle efficiency achieved. For example, the compressed air UPS products from Flow battery have relatively high discharge efficiency (75–90%), which improved the related cycle efficiency and became a key factor for the company to launch the product successfully [51].

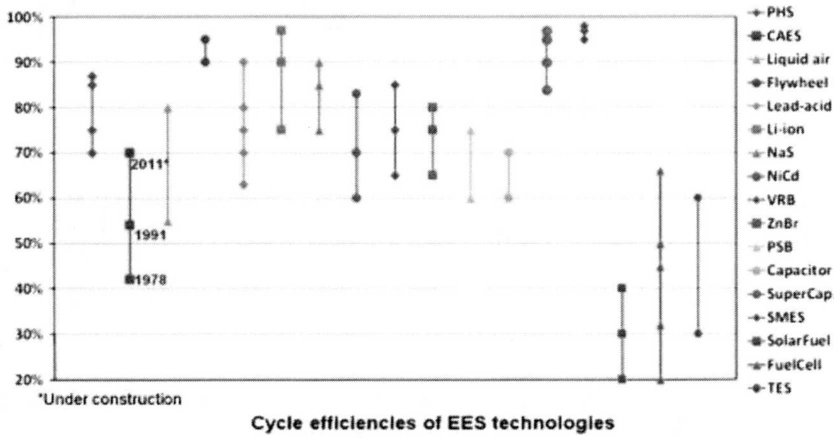

Figure 17. Comparison of cycle efficiencies of EES technologies.

Self-discharge is related to energy dissipation, in the forms of heat transfer losses in thermal storage, air leakage losses in compressed air storage, electrochemical losses in batteries, etc. The level of self-discharge of an EES system is one of the major factors in deciding the associated suitable storage duration. From Table 11 and Table 12, PHS, CAES, NaS batteries, flow batteries, fuel cells and solar fuels have very small daily self-discharge ratios so it is technically possible for the energy to be stored in long-term durations (up to months); most conventional batteries (except NaS batteries) have daily self-discharge ratios, from 0.03% to 5%, which can be used for medium-term storage durations (up to days); SMES, flywheel, capacitors and supercapacitors have very high daily self-charge ratios, from 10% to 100%, that is, they could completely release their stored energy after a few hours or even shorter. Hence, they can only be utilized for short-term storage durations (up to hours). TES encompasses a variety of technologies and thus it may be suitable for medium-term and/or long-term storage durations.

Lifetime and cycling times are two factors which affect the overall investment cost. Low lifetime and low cycling times will increase the cost of maintenance and replacement. Table 11 shows the comparison of lifetime and cycling times of different EES technologies discussed in the paper. It can be seen that these two indices more or less associate with the EES technology's category – in the form of the type of energy stored in the system. Mechanical energy storage systems, including PHS, CAES and flywheels, normally have high cycling times (around 10,000 or more) which mainly depend on their mechanical components. The cycle times for EES with energy stored in electrical energy, such as SMES, capacitors and supercapacitors, are normally higher than 20,000. FromTable 11, it can be seen that the cycle abilities of conventional batteries are not as high as other EES systems mainly due to chemical deterioration with the accumulated operating time.

The maturities of EES technologies are linked to the level of commercialization, the technical risk and the related economic benefits. Table 12 compares the levels of technical maturity of the EES technologies reviewed in the paper. The technology maturity level for utility EES applications can be classified into five categories: (1) Developing (AA-CAES, PSB and solar fuel); (2) Demonstration (liquid air storage, Li-ion, VRB, ZnBr, supercapacitor, SMES, fuel cell and TES); (3) Early Commercialized (over-ground small CAES and flywheel); (4) Commercialized (conventional CAES, NaS, NiCd and capacitor); (5) Mature (PHS & Lead–acid). It can be seen that several technologies are undergoing breakthroughs from one category evolving to another (Table 12). The technologies in the developing stage are technically possible and have great potential for future EES projects.

Table 12 lists the EES cost in terms of energy capacity and the cost for Operating and Maintenance (O&M). Fig. 18 presents the comparison of energy capital cost and annual O&M cost. A complete economic analysis of EES technologies needs to consider not only the capital cost but also the O&M cost and the impact of the equipment lifetime. For instance, although the energy capital cost of lead–acid battery is relatively low, it may not be the best option for large-scale EES applications due to its relatively high O&M cost and short lifetime. The cost of EES is tending to decreases with the continuous effort in research and development, and some key technology breakthroughs can lead to dramatic changes in cost. From Fig. 18, among the mature and commercialized techniques, PHS and CAES have lowest energy capital costs compared to all other technologies; NaS, VRB and Lead–acid battery have relatively high O&M cost. As shown in Table 12, TES is in the low range in terms of energy capital cost; SMES and flywheel are suitable for high power and small-scale applications as they are cheap in terms of the power capital cost but expensive in terms of the energy capital cost. It should be noted that the capital cost of a specific EES system varies in terms of the timescale of EES construction, the location of the plant/facility and the size of the system. The economic analyses of different EES technologies in various application scenarios are attracting attention due to their great application prospects, which can be found in [40], [171], [218],[219] and [220].

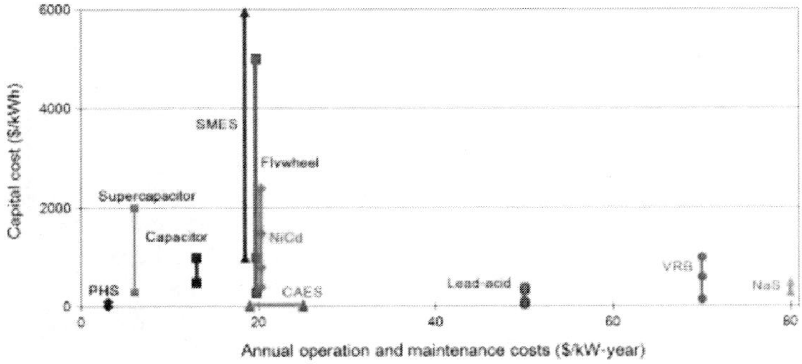

Figure18. Comparison of energy capital cost and annual operating & maintenance cost.

ANALYSIS AND RECOMMENDATIONS OF EES TECHNOLOGIES FOR VARIOUS APPLICATIONS

The application outlooks and potentials of EES in power system operations have been widely reported in recent years [6], [13], [14], [26], [114] and [123]. Table 13 summarizes the current and promising EES options for various applications with corresponding specifications. The applications of EES technologies cover a wide range including the maintenance of power quality, power system protections and energy management.

Table 13. Overview of current and potential electrical energy storage options for various applications with their specifications.

Application area	Application characteristics & specifications (refer to[6], [10], [11], [12], [13],[14], [15], [16], [26], [114],[123], [209] and [224])	Experienced and promising EES technology options	Related references
Power quality	~ < 1 MW, response time (~milliseconds, <1/4 cycle), discharge duration (milliseconds to seconds)	Experienced: flywheels, batteries, SMES, capacitors, supercapacitors. Promising: flow batteries	[26,80,129,221, 222]
Ride-through	~100 kW–10 MW, response time (up to ~1 s), discharge	Experienced: batteries and flow	[4,60,87,126,223]

capability (bridging power)	duration (seconds to minutes and even hours)	batteries; Promising: fuel cells, flywheels and supercapacitors	
Energy managem ent	Large (>100 MW), medium/small (~1–100 MW), response time (minutes), discharge duration (hours–days)	Experienced: Large (PHS, CAES, TES); small (batteries, flow batteries, TES) Promising: flywheels, fuel cells	[34,39,56,88,17 0]
More specific applications			
Integratio n renewable smoothing intermitte nt	Up to ~20 MW, response time (normally up to 1 s, <1 cycle), discharge duration (minutes to hours)	Experienced: flywheels, batteries and supercapacitors; Promising: flow batteries, SMES and fuel cells	[56,157,173,217 ,226]
Integratio n renewable for back-up	~100 kW–40 MW, response time (seconds to minutes), discharge duration (up to days)	Experienced: batteries and flow batteries; Promising: PHS, CAES, solar fuels, and fuel cells	[13,26,40,76,10 8]
Emergenc y back-up power	Up to ~1 MW, response time (milliseconds to minutes), discharge duration (up to ~24 h)	Experienced: batteries, flywheels, flow batteries; Promising: small-scale CAES and fuel cells	[13,227,228,229]
Telecomm unications back-up	Up to a few of kW, response time (milliseconds), discharge duration (minutes to hours)	Experienced: batteries; Promising: fuel cells, supercapacitors and flywheels	[114,228,230,23 1]
Ramping and load following	MW level (up to hundreds of MW), response time (up to ~1 second), duration (minutes to a few hours)	Experienced: batteries, flow batteries and SMES; Promising: fuel cells	[13,88,232,233]
Time shifting	~1–100 MW and even more, response time (minutes), discharge duration (~3–12 h)	Experienced: PHS, CAES and batteries; Promising: flow batteries, solar fuels, fuel cells and TES	[26,234,235,236]
Peak shaving	~100 kW–100 MW and even more, response time (minutes),	Experienced: PHS, CAES and batteries;	[114,237,238]

	discharge duration (hour level, ~ < 10 h)	Promising: flow batteries, solar fuels, fuel cells and TES	
Load levelling	MW level (up to several hundreds of MW), response time (minutes), discharge duration (~12 h and even more)	Experienced: PHS, CAES and batteries; Promising: flow batteries, fuel cells and TES	[38,159,218,239]
Seasonal energy storage	Energy management, 30–500 MW, quite long term storagedischarge duration (up to weeks), response time (minutes)	Promising: PHS, TES and fuel cells; Possible: large-scale CAES and solar fuels	[26,240,241,242]
Low voltage ride-through	Normally lower than 10 MW, response time (~milliseconds), discharge duration (up to minutes)	Experienced: Flywheels, batteries; Promising: flow batteries, SMES and supercapacitors	[13,129,244]
Transmission and distribution stab.	Up to 100 MW, response time (~milliseconds, <1/4 cycle), discharge duration (milliseconds to seconds)	Experienced: batteries and SMES; Promising: flow batteries, flywheels and supercapacitors	[26,114,224,245,246]
Black-start	Up to ~40 MW, response time (~minutes), discharge duration (seconds to hours)	Experienced: small-scale CAES, batteries, flow batteries; Promising: fuel cells and TES	[15,39,87]
Voltage regulation and control	Up to a few of MW, response time (milliseconds), discharge duration (up to minutes)	Experienced: batteries and flow batteries; Promising: SMES, flywheels and supercapacitors	[247,248,249]
Grid/network fluctuation suppression	Up to MW level, response time (milliseconds), duration (up to ~minutes)	Experienced: batteries, flywheels, flow batteries, SMES, capacitors and supercapacitors,	[114,137,225,250]
Spinning reserve	Up to MW level, response time (up to a few seconds), discharge duration (30 minutes to a few hours)	Experienced: batteries; Promising: small-scale CAES, flywheels, flow batteries, SMES and	[251,252,254,255]

		fuel cells	
Transportation applications	Up to ~50 kW, response time (milliseconds–seconds), discharge duration (seconds to hours)	Experienced: batteries, fuel cells and supercapacitors; Promising: flywheels, liquid air storage and solar fuels	[256,257,258,259]
End-user electricity service reliability	~ up to 1 MW, response time (milliseconds, <1/4 cycle), storage time at rated capacity (0.08–5 hours)	Experienced: batteries; Promising: flow batteries, flywheels, SMES and supercapacitors	[6,13,114,170]
Motor starting	Up to ~1 MW, response time (milliseconds–seconds), discharge duration (seconds to minutes)	Experienced: batteries and supercapacitors; Promising: flywheels, SMES, flow batteries and fuel cells	[51,114,198]
Uninterruptible power supply	Up to ~5 MW, response time (normally up to seconds), discharge duration (~10 min to 2 h)	Experienced: Flywheels, supercapacitors, batteries; Promising: SMES, small CAES, fuel cells, flow batteries	[57,262,263,264]
Transmission upgrade deferral	~10–100 + MW, response time (~minutes), storage time at rated capacity (1–6 h)	Experienced: PHS and batteries; Promising: CAES, flow batteries, TES and fuel cells	[4,101,220]

The EES technologies used for maintaining power quality will need to have very fast response time (at the millisecond level). Flywheels, conventional batteries, SMES, capacitors and supercapacitors are well suited for this service. Some flow batteries, such as VRB and PSB, are also technically suitable for this service. For studies of EES systems for power quality applications refer to [63], [80], [129], [221] and [222]. The papers [129] and [221] present the power quality enhancement of renewable energy source power (wind and photovoltaic power generation systems) by using supercapacitors and SMES respectively. In [231], the applications of VRB systems to maintain power quality have been studied. One example from the study is the Sumitomo Electric Industries (SEI) VRB facility (rated power 170 kW) which works in combination with a wind turbine (275 kW) for stabilization of wind turbine output fluctuations; the system demonstrated good performance [222].

When EES is adopted for bridging power, they are required to have moderate power rating (100 kW–10 MW) and response time (up to around 1 s), in order to provide the continuity of power supply at energy gap periods (up to several hours), such as the time interval for switching the system from one source of power generation to another. Conventional batteries and flow batteries are suitable for this application. Flywheels, supercapacitors and fuel cells are also reported for such types of applications [60],[126] and [223]. Piller Power Systems Ltd. has practical experiences in using flywheels as ride-through power sources [60]. The article [223] presents a decoupled *P–Q* control strategy of a supercapacitor energy storage system for low voltage ride-through as well as damping enhancement of the doubly fed induction generator system. The fault ride-through capability of the generator has been investigated for a severe symmetrical three-phase to ground fault on the grid bus.

EES plays an important role in energy management for optimizing energy uses and decoupling the timing of generation and consumption of electric energy. Time shifting and peak shaving are typical applications in energy management. Energy management application can be further classified based on the application power rating: large-scale (above 100 MW) and medium/small-scale (~1–100 MW). PHS, large-scale CAES and TES can be used for large-scale energy management. Flow batteries, large-scale conventional batteries, fuel cells and solar fuels are more suitable for small/medium-scale energy management. Practical examples are described in [34], [39], [88] and [170]. For instance, both the Huntorf and the McIntosh large-scale CAES plants provide the functions of energy management.

Table 13 also summarizes an overview of EES options for more specific applications with their specifications. A brief explanation of most of these applications is given below:

- Integration of renewable power generation: The inherent intermittent renewable generation can be backed up, stabilized or smoothed through integration with EES facilities. The recent research, development and demonstrations in this area are described in [76], [108], [157] and [173].
- Emergency and telecommunications back-up power: In the case of power failure, EES systems can be operated as an emergency power supply to provide adequate power to important users including telecommunication systems until the main supply is restored, or to ensure the system enabling orderly shutdown. For emergency back-up power, instant-to-medium response time and relatively long duration of discharge time are required. The suitable EES technologies for this

application are given in Table 13, and relevant reports and studies can be found in [227],[228] and [229]. For example, one of the world's first utility (hybrid) CAES back-up systems was recently installed at a Co-op Bank data center to provide an emergency supply of electricity [227]. For telecommunications back-up, the instant response time is essential. Several technologies are suited to this application (see Table 13) and related publications include [114], [230] and [231].

- Ramping and load following: EES facilities can provide support in following load changes to electricity demand. The relevant research and demonstration projects can be found in [77], [232] and [233]. One EES trial project, named Irvine Smart Grid Demonstration, using advanced batteries (25 kW) in California offers services in load following and voltage support [77]. A flexible load following operation mode for operating a NaS battery EES system with its control method has been studied in[233].
- Time shifting: Time shifting can be achieved by storing electrical energy when it is less expensive and then using or selling the stored energy during peak demand periods. EES technologies are required to provide power ratings in the range of around 1-100 MW. PHS, CAES and conventional batteries have experience in this service; flow batteries, solar fuels and TES have demonstration plants or are potentially available for this application. Some associated discussions can be found in [234], [235] and [236].
- Peak shaving and load levelling: Peak shaving means using energy stored at off-peak periods to compensate electrical power generation during periods of maximum power demand. This function of EES can provide economic benefits by mitigating the need to use expensive peak electricity generation.
- Load levelling is a method of balancing the large fluctuations associated with electricity demand. Conventional batteries and flow batteries in peak shaving applications, as well as in load following and time shifting, need a reduction in overall cost and an increase in the cycling times to enhance their competitiveness. Economic and technical studies and related demonstrations of these two applications are shown in [159], [218], [237], [238], [239] and [243]. For instance, an optimization model of the weekly economic operation of isolated systems was developed and then applied to two Spanish isolated power systems in the Canary Islands [237].
- Seasonal energy storage: Storing energy in the time frame of months, for community seasonal space heating and the energy networks with large seasonal variation in power generation and consumption. EES technologies which have a very large energy capacity and almost zero self-discharge are required. At present, there are no commercialized

EES technologies for this application and storing fossil fuels is still a practical solution. PHS, hydrogen-based fuel cells, CAES, TES and solar fuels have potential to serve this application. Some relevant research and development is introduced in [26], [240], [241] and [242].

- Low voltage ride-through: It is crucial to some electrical devices, especially to renewable generation systems. It is a capability associated with voltage control operating through the periods of external grid voltage dips. High power ability and instant response are essential for this application. In order to smooth the fast wind-induced power variations and to reinforce the low voltage ride-through capability, the integration of a supercapacitor EES device in a doubly fed induction generator design was studied in [129]. A VRB-based EES system was simulated, to improve low voltage ride through capability and to stabilize output power of direct-drive permanent magnet wind power system [244].
- Transmission and distribution stabilization: EES systems can be used to support the synchronous operation of components on a power transmission line or a distribution unit to regulate power quality, to reduce congestion and/or to ensure the system operating under normal working conditions. Instant response and relatively large power capacity with grid demand are essential for such applications. Studies in this area can be found in [26], [114], [245] and [246].
- Black-start: EES can provide capability to a system for its start-up from a shutdown condition without taking power from the grid. A typical example the Huntorf CAES plant that provides black-start power to nuclear units located near to the North Sea[39] and [42].
- Voltage regulation and control: Electric power systems react dynamically to changes in active and reactive power, thus influencing the magnitude and profile of the voltage in networks [193]. With the functions of EES facilities, the control of voltage dynamic behaviors can be improved. Several EES technologies can be used or potentially used for voltage control solutions (Table 13). A preliminary analysis of a pilot project for the exploitation of EES devices for distribution network voltage regulation was given in [247]. The modeling and measurement of a high-speed composite flywheel system to regulate a specific DC voltage on a metro network has been studied in [248].
- Grid/network fluctuation suppression: Some power electronic, information and communication systems in the grid/network are highly sensitive to power related fluctuation. EES facilities can provide the function to protect these systems, which requires the capabilities of high ramp power rates and high cycling times with fast

response time. Some recent research and development is given in [114], [137],[225] and [250].

- Spinning reserve: In the case of a fast increase in generation (or a decrease in load) to result in a contingency, EES systems can feature the function of spinning reserve. The EES units must respond immediately and have the ability of maintaining the outputs for up to a few hours. This function is described and studied in [13], [251],[252], [253], [254] and [255].
- Transportation applications: Providing power to transportation, such as HEVs and EVs. High energy density, small dimension, light weight and fast response are necessary for implemented EES units. The research and development of EES applications on transportation can be found in [256], [257], [258], [259] and [260]. For instance, a hybrid powertrain using fuel cell, battery, and supercapacitor technologies for the tramway was simulated based on commercially available devices, and a predicative control strategy was implemented for performance requirements [256].
- Uninterruptible Power Supply (UPS): EES systems can feature the function of UPS to maintain electrical load power in the event of the power interruption or to provide protection from a power surge. A typical UPS device offers instantaneous (or near to instantaneous) reaction, by supplying energy mostly stored in batteries, flywheels or supercapacitors. Some research papers regarding to UPS by using different EES technologies can be found in [9], [57], [261], [262], [263] and [264].
- Standing reserve: In order to balance the supply and demand of electricity on a certain timescale, EES facilities/plants can provide service as temporary extra generating units to the middle-to-large scale grid. Standing reserve can be used to deal with actual demand being greater than forecast demand and/or plant breakdowns. The descriptions related to this application were introduced in [6],[114], [265] and [266].

CONCLUDING REMARKS

This paper provides an overview of the current development of various types of EES technologies, from the recent achievements in both the academic research community and industrial sectors. A comprehensive analysis is carried out based on the relevant technical and economic data, which leads to a number of tables and figures showing a detailed comparison of various EES technologies from different perspectives.

Further discussion on EES power system application potentials is given based on the current characteristics of EES and the relevant application specifications. The overview has shown a synthesis of the state-of-the-art in important EES technologies, which can be used for supporting further research and development in this area and for assessing EES technologies for deployment.

The review identified that PHS plants have been deployed worldwide, mainly due to its technological maturity. Since PHS has relatively low power/energy densities it is mainly used in stationary large-scale EES. The Li-ion battery has relatively high power/energy densities and specific power/energy, which has resulted in the current broad range of development, particularly in small-scale EES applications. The cycle efficiencies of EES technologies have been continuously improved with time through development efforts leading to technology breakthroughs, and most commercialized techniques normally have medium-to-high cycle efficiencies. The energy capacity and the self-discharge of EES systems are the major factors in deciding the associated suitable storage duration. From the overview, it is clear that there is no suitable commercialized technology for seasonal energy storage at present. Several EES technologies, such as PHS, fuel cells and TES, have the potential to be applied in this area. On the whole, the various applications with different network sizes will have different decision-making factors to consider when choosing suitable EES options for deployment. For the national regulator, the level of technological maturity, reliability and potential environmental impacts (such as the toxic chemical materials used in batteries described in Section 3) may be considered as the main decision-making factors and the cost-effectiveness may not be particularly important; for the end-user customers or local (private) networks, in addition to the above factors, the investment cost and the economic gain will also be dominant factors.

From this overview, it can be seen that the current technologies have wide ranging technological characteristics. With a suitable combination of different technologies, EES can meet most technical requirements for different power system and network operations. However, apart from PHS, most EES technologies are not cost-effective or mature enough for widespread implementation within the current large network operation regulation and energy market frame. On the other hand, the benefits brought to power system operation by utilizing EES technologies need further exploration. The capital and the maintenance cost of an EES system varies with the timescale of construction, the location of the facility, the size of the system, the material chosen for storing energy (such as PCMs in TES) and many other factors. Although a number of demonstration projects or EES trial stations were completed, the corresponding detailed techno-economic analysis, which can enhance the relevant database to practical EES experience, is still not sufficient. The widespread deployment of EES will depend on advances in relevant technologies, but it also relies on progress in further quantification and analysis of the benefits brought by EES.

ACKNOWLEDGMENTS

The authors would like to thank EPSRC for funding support to the IMAGES research project (EP/K002228/1), Advantage West Midlands and the European Regional Development Fund for Birmingham Science City Energy Efficiency and Demand Reduction project.

REFERENCES

1. Electricity information. International energy agency. Organisation for Economic Cooperation and Development (OECD) Publishing. Published 13th August 2013.
2. 2012 key world energy statistics. IEA. Printed in France by Soregraph. n.d. accessed 15.05.14..
3. Skea J, Nishioka S, Policies and practices for a low-carbon society. Modelling long-term scenarios for low carbon societies. In: Strachan N, Foxon T, Fujino TJ, editors. Climate Policy, vol. 8; 2008. p. 5–16.
4. Chen H, Cong TN, Yang W, Tan C, Li Y, Ding Y. Progress in electrical energy storage system: a critical review. Prog Nat Sci 2009;19:291–312.
5. Status of electrical energy storage systems. Dti report. DG/DTI/00050/00/00, urn no. 04/1878. UK Department of Trade and Industry; 2004. p. 1–24.
6. The future role for energy storage in the UK main report. Energy Research Partnership (ERP) technology report published June 2011.
7. Strbac G, Aunedi M, Pudjianto D, et al. Strategic assessment of the role and value of energy storage systems in the UK low carbon energy future. Report for Carbon Trust. Imperial College London. Published June 2012. accessed 26.05.14..
8. Taylor P, Bolton R, Stone D, Zhang XP, Martin C, Upham P, et al. Pathways for energy storage in the UK. Technical report. Centre for low carbon futures. Published 27th March 2012. accessed 11.06.14..
9. Review of electrical energy storage technologies and systems and of their potential for the UK. Dti Report. DG/DTI/00055/00/00. UK Department of Trade and Industry; 2004.
10. Ibrahim H, Ilinca A, Perron J. Energy storage systems—characteristics and comparisons. Renew Sust Energy Rev 2008;12:1221–50.

11. Hall PJ, Bain EJ. Energy-storage technologies and electricity generation. Energy Policy 2008;36:4352–5.
12. Liu C, Li F, Ma L-P, Cheng H-M. Advanced materials for energy storage. Adv Mater 2010;22:E28–62.
13. Díaz-González F, Sumper A, Gomis-Bellmunt O, Villafáfila-Robles R. A review of energy storage technologies for wind power applications. Renew Sust Energy Rev 2012;16:2154–71.
14. Beaudin M, Zareipour H, Schellenberglabe A, Rosehart W. Energy storage for mitigating the variability of renewable electricity sources: an updated review. Energy Sust Dev 2010;14:302–14.
15. Zhao H, Wu Q, Hu S, Xu H, Rasmussen CN. Review of energy storage system for wind power integration support. Appl Energy 2015;137:545–53. http:// dx.doi.org/10.1016/j.apenergy.2014.04.103 accessed 26.08.14..
16. Evans A, Strezov V, Evans TJ. Assessment of utility energy storage options for increased renewable energy penetration. Renew Sust Energy Rev 2012;16:4141–7.
17. Connolly D, Lund H, Mathiesen BV, Leahy M. A review of computer tools for analysing the integration of renewable energy into various energy systems. Appl Energy 2010;87:1059–82.
18. Zhou L, Qi ZP. A review of flywheel energy storage system. Proceedings of ISES World Congress 2007 (Vol. I – Vol. V). Berlin, Heidelberg: Springer, Berlin Heidelberg; 2009. p. 2815–19.
19. Weber AZ, Mench MM, Meyers JP, Ross PN, Gostick JT, Liu Q. Redox flow batteries: a review. J Appl Electrochem 2011;41:1137–64.
20. Sharma P, Bhatti TS. A review on electrochemical double-layer capacitors. Energy Convers Manage 2010;51:2901–12.
21. Zhou D, Zhao CY, Tian Y. Review on thermal energy storage with phase change materials (PCMs) in building applications. Appl Energy 2012;92:593–605.
22. Dunn B, Kamath H, Tarascon J-M. Electrical energy storage for the grid: a battery of choices. Science 2011;334:928–35.
23. Whittingham MS. Materials challenges facing electrical energy storage. MRS Bull 2008;33:411–9.
24. Zhai Y, Dou Y, Zhao D, Fulvio PF, Mayes RT, Dai S. Carbon materials for chemical capacitive energy storage. Adv Mater 2011;23:4828–50.
25. Xu C, Xu B, Gu Y, Xiong Z, Sun J, Zhao XS. Graphene-based electrodes for electrochemical energy storage. Energy Environ Sci 2013;6:1388.
26. Electrical energy storage: white paper. Technical report. Prepared by electrical energy storage project team. International Electrotechnical Commission (IEC), Published December 2011. accessed 15.05.14..

27. Brito AV, editor. Dynamic modelling. Molina MG. Chapter 4: Dynamic modelling and control design of advanced energy storage for power system applications. InTech; 2010. M.G.
28. Abele A, Elkind E, Intrator J, Washom B, et al. 2020 Strategic Analysis of Energy Storage in California. University of California, Berkeley School of Law; University of California, Los Angeles; and University of California, San Diego. California Energy Commission; 2011.
29. Energy storage – packing some power. The Economist. Published 3rd March 2012. accessed 08.06.14..
30. Figueiredo FC, Flynn PC. Using diurnal power price to configure pumped storage. IEEE Trans Energy Convers 2006;21:804–9.
31. Bath County Pumped Storage Station. n.d. accessed 05.05.14..
32. Pumped-hydro energy storage: potential for transformation from single dams. JRC IET scientific and technical report.EUR 25239 EN.Institute for Energy and Transport. Published on 24th February 2012.
33. Hawaiian Electric Company. Clean Energy – Issues and Challenges – Energy Storage. n.d. accessed 24.05.14..
34. Fujihara T, Imano H, Oshima K. Development of pump turbine for seawater pumped-storage power plant. Hitachi Rev 1998;47:199–202.
35. Pickard WF. The history, present state, and future prospects of underground pumped hydro for massive energy storage. Proc IEEE 2012;100:473–83.
36. Papaefthymiou SV, Karamanou PG, Papathanassiou SA, Papadopoulos MP. A wind-hydro-pumped storage station leading to high RES penetration in the autonomous island system of Ikaria. IEEE Trans Sust Energy 2010;1:163–72.
37. Bueno C, Carta JA. Wind powered pumped hydro storage systems, a means of increasing the penetration of renewable energy in the Canary Islands. Renew Sust Energy Rev 2006;10:312–40.
38. Samir S. Large energy storage systems handbook. In: Levine JG, editor. CRC Press; 2011. p. 112–52.
39. Succar S, Williams RH. Compressed air energy storage: theory, resources, and applications for wind power. Princeton Environmental Institute. Energy Anal Group; 2008.
40. Madlener R, Latz J. Economics of centralized and decentralized compressed air energy storage for enhanced grid integration of wind power. Appl Energy 2013;101:299–309.
41. Raju M, Kumar Khaitan S. Modeling and simulation of compressed air storage in caverns: a case study of the Huntorf plant. Appl Energy 2012;89:474–81.
42. Finkenrath M, Pazzi S, D'Ercole M. Status and technical challenges of advanced Compressed Air Energy Storage (CAES) technology. In: Int. workshop on environment and alternative energy; 2009.

43. RWE power. ADELE – Adiabatic compressed-air energy storage (CAES) for electricity supply. n.d. accessed 07.06.14..
44. John JS. Texas to host 317 MW of compressed air energy storage – After a long wait, a second big CAES project gets underway in the U.S. Published 9th July 2013. accessed 25.05.14..
45. Succar S, Denkenberger DC, Williams RH. Optimization of specific rating for wind turbine arrays coupled to compressed air energy storage. Appl Energy 2012;96:222–34.
46. LightSail Energy Ltd. We store energy in compressed air. n.d. accessed 12.06.14..
47. 47. Hartmann N, Vöhringer O, Kruck C, Eltrop L. Simulation and analysis of different adiabatic compressed air energy storage plant configurations. Appl Energy 2012;93:541–8.
48. Rutqvist J, Kim H-M, Ryu D-W, Synn J-H, Song W-K. Modeling of coupled thermodynamic and geomechanical performance of underground compressed air energy storage in lined rock caverns. Int J Rock Mech Min Sci 2012;52:71–81.
49. Compressed-air energy storage using hard-rock geology: test facility and results. USA Electr. Power Res. Inst. (EPRI); 2005. p. 1–2.
50. Schulte RH, Critelli N, Jr., Holst K, Huff G. Lessons from Iowa: development of a 270 megawatt compressed air energy storage project in Midwest independent system operator. Technical report. Sandia National Laboratories. Published January 2012. accessed 23.05.14..
51. Flowbattery (formerly Pnu Power). Compressed air batteries. n.d. accessed 08.10.13..
52. Wang J, Luo X, Yang L, Shpanin LM, Jia N, Mangan S, et al. Mathematical modeling study of scroll air motors and energy efficiency analysis—part II. IEEE/ASME Trans Mechatronics 2011;16:122–32.
53. Proczka JJ, Muralidharan K, Villela D, Simmons JH, Frantziskonis G. Guidelines for the pressure and efficient sizing of pressure vessels for compressed air energy storage. Energy Convers Manage 2013;65:597–605.
54. Sun H, Luo X, Wang J. Management and control strategy study for a new hybrid wind turbine system. In: IEEE Conf. Decis. Control Eur. Control Conf., IEEE; 2011. p. 3671–6.
55. Pena-Alzola R, Sebastian R, Quesada J, Colmenar A. Review of flywheel based energy storage systems. In: 2011 Int. Conf. Power Eng. Energy Electr. Drives, IEEE; 2011. p. 1–6.
56. Díaz-González F, Sumper A, Gomis-Bellmunt O, Bianchi FD. Energy management of flywheel-based energy storage device for wind power smoothing. Appl Energy 2013;110:207–19.

57. Hadjipaschalis I, Poullikkas A, Efthimiou V. Overview of current and future energy storage technologies for electric power applications. Renew Sust Energy Rev 2009;13:1513–22.
58. Sebastián R, Peña Alzola R. Flywheel energy storage systems: review and simulation for an isolated wind power system. Renew Sust Energy Rev 2012;16:6803–13.
59. Source from Beacon Power Corporation. Beacon Power inaugurates 20MW flywheel plant in NY. Published 21st July 2011. accessed 23.05.14..
60. Piller power systems products. Energy storage. n.d. accessed 12.06.14..
61. Mulcahy TM, Hull JR, Uherka KL, Abboud RG, Juna JJ. Test results of 2-kW h flywheel using passive PM and HTS bearings. IEEE Trans Appl Supercond 2001;11:1729–32.
62. Amodeo SJ, Chiacchiarini HG, Solsona JA, Busada CA. High-performance sensorless nonlinear power control of a flywheel energy storage system. Energy Convers Manage 2009;50:1722–9.
63. Suzuki Y, Koyanagi A, Kobayashi M, Shimada R. Novel applications of the flywheel energy storage system. Energy 2005;30:2128–43.
64. Earle Waghorne W. Viscosities of electrolyte solutions. Philos Trans Roy Soc A Math Phys Eng Sci 2001;359:1529–43.
65. Song JY, Wang YY, Wan CC. Review of gel-type polymer electrolytes for lithium-ion batteries. J Power Sources 1999;77:183–97.
66. Smith SC, Sen PK, Kroposki B. Advancement of energy storage devices and applications in electrical power system. In: 2008 IEEE Power Energy Soc. Gen. Meet. – Convers. Deliv. Electr. Energy 21st Century, IEEE; 2008. p. 1–8.
67. Zhang W, Qiu M, Lai X. Application of energy storage technologies in power grids. Power Syst Technol 2008:32.
68. 68. James P, Dunlop PE. Batteries and charge control in stand-alone photovoltaic systems – fundamentals and application. Technical report. Florida Solar Energy Center. Sandia National Laboratories; 1997.
69. Kondoh J, Ishii I, Yamaguchi H, Murata A, Otani K, Sakuta K, et al. Electrical energy storage systems for energy networks. Energy Convers Manage 2000;41:1863–74.
70. Farret FA, Simões MG. Integration of alternative sources of energy. John Wiley & Sons Inc.; 2006. p. 262–300.
71. Baker J. New technology and possible advances in energy storage. Energy Policy 2008;36:4368–73.
72. 72. Paul B. The future of electrical energy storage: the economics and potential of new technologies. Report. Business Insights (Energy). 2009. Table of Contents and brochure. accessed 28.05.14..

73. Rastler D. Electricity energy storage technology options: a white paper primer on applications, costs, and options. Electric Power Research Institute (EPRI). Technical report. Published December 2010.
74. Ecoult. UltraBattery: the new dimension in lead–acid battery technology. n.d. accessed 12.06.14..
75. Wald ML. Wind drives growing use of batteries. New York Times. Published 28th July 2010. accessed 19.05.14..
76. Hodson H. Texas-sized battery aims to green up the grid. New Sci 2013;217(February):20.
77. US Development of Energy (DOE) global energy storage database. n.d. accessed 16..06.14..
78. Subburaj AS, Kondur P, Bayne SB, Giesselmann MG, Harral MA. Analysis and review of grid connected battery in wind applications; 2014. p. 1–6.
79. Tweed K. UK launches Europe's largest energy storage trial. IEEE Spectrum. Published 1st August 2013. accessed 04.05.14..
80. Daly J. Japan looks at recycling vehicle batteries for renewable power. Published on 20th February 2014. accessed 26.02.14..
81. Kawakami N, Iijima Y, Fukuhara M, Bando M, Sakanaka Y, Ogawa K, et al. Development and field experiences of stabilization system using 34MW NAS batteries for a 51MW wind farm. In: 2010 IEEE Int. Symp. Ind. Electron., IEEE; 2010, p. 2371–6.
82. Kamibayashi M, Nichols DK, Oshima T. Development update of the NAS battery. IEEE/PES Transm. Distrib. Conf. Exhib., vol. 3, IEEE; n.d., p. 1664–8.
83. Tewari S, Mohan N. Value of NAS energy storage toward integrating wind: results from the wind to battery project. IEEE Trans Power Syst 2013;28:532–41.
84. Sumitomo Electric annual report 2011 (year-end 31st March 2011). Sumitomo Electric ingenious dynamic; 2011. p. 16-7.
85. Lacerda VG, Mageste AB, Santos IJB, da Silva LHM, Da Silva MdoCH. Separation of Cd and Ni from Ni–Cd batteries by an environmentally safe methodology employing aqueous two-phase systems. J Power Sources 2009;193:908–13.
86. Linden D, Reddy TB, editors. Handbook of batteries. McGraw-Hill; 2002. p. 18–28.
87. Divya KC, Østergaard J. Battery energy storage technology for power systems—an overview. Electr Power Syst Res 2009;79:511–20.
88. Poullikkas A. A comparative overview of large-scale battery systems for electricity storage. Renew Sust Energy Rev 2013;27:778–88.
89. Zhu WH, Zhu Y, Davis Z, Tatarchuk BJ. Energy efficiency and capacity retention of Ni–MH batteries for storage applications. Appl Energy 2013;106:307–13.

90. Fetcenko MA, Ovshinsky SR, Reichman B, Young K, Fierro C, Koch J, et al. Recent advances in NiMH battery technology. J Power Sources 2007;165:544–51.
91. Nickel-based Batteries. n.d. accessed 15.05.14..
92. NiMH Nickel–Metal Hydride: a ''Sensitive Battery''. n.d. accessed 16.05.14..
93. Ying T, Gao X, Hu W, Wu F, Noreus D. Studies on rechargeable NiMH batteries. Int J Hydrogen Energy 2006;31:525–30.
94. Peak oil - the electric vehicle imperative. Battery technologies. Meridian International Research. Project report. Published 2nd December 2005. p. 104–12 (chapter 7).
95. Dustmann C-H. Advances in ZEBRA batteries. J Power Sources 2004;127:85–92.
96. O'Sullivan TM, Bingham CM, Clark RE. Zebra battery technologies for all electric smart car. In: Int. Symp. Power Electron. Electr. Drives, Autom. Motion, 2006. SPEEDAM 2006, IEEE; n.d., p. 243.
97. Energy storage – ZEBRA batteries. Rolls-Royce. n.d. accessed 16.05.14..
98. Battery Power, Reimagined. GE Energy Storage. n.d. accessed 26.05.14..
99. FIAMM Sodium Nickel (SoNick) batteries. FIAMM Energy Storage Solutions. n.d. accessed 10.06.14..
100. Yang Z, Zhang J, Kintner-Meyer MCW, Lu X, Choi D, Lemmon JP, et al. Electrochemical energy storage for green grid. Chem Rev 2011;111:3577–613.
101. Eyer J, Iannucci J, Butler PC. Estimating electricity storage power rating and discharge duration for utility transmission and distribution deferral: a study for the DOE energy storage program. Sandia National Laboratories. Technical report. Printed November 2005.
102. Nguyen T, Savinell RF. Flow batteries. Electrochem Soc Interface 2010;19:54–6.
103. Taylor EJ. Flow battery structures to improve performance and reduce manufacturing cost. The 2012 DOE Energy Storage Program Peer Review and Update Meeting; 2012.
104. Modiba P. Electrolytes for redox flow battery systems. Ph.D. thesis. Department of Chemistry and Polymer Science. the University of Stellenbosch, December 2010.
105. Gonzalez A, Ó'Gallachóir B, McKeogh E, Lynch K. Study of electricity storage technologies and their potential to address wind energy intermittency in Ireland. Final report. Sustainable energy research group and Rockmount capital partners Cork. RE/HC/03/001. Published May 2004.
106. Barton JP, Infield DG. Energy storage and its use with intermittent renewable energy. IEEE Trans Energy Convers 2004;19:441–8.

107.Leung P, Li X, Ponce de León C, Berlouis L, Low CTJ, Walsh FC. Progress in redox flow batteries, remaining challenges and their applications in energy storage. RSC Adv 2012;2:10125.
108.Walsh F. Progress & challenges in the development of flow battery Technology. In: The 1st Int. Flow Battery Forum (IFBF); 2010, p. 1–8.
109.AEA Technology plc. Energy storage and management study. Scottish Government. Published 2nd October 2010. accessed 11.06.14..
110.Holzman DC. The vanadium advantage: flow batteries put wind energy in the bank. Environ Health Perspect 2007;115:A358–61.
111.Peters R, O'Malley L. Storing renewable power. The 2nd primer in the series: making renewable energy a priority. The Pembina Institute. Published June 2008.
112.Brekken TKA, Yokochi A, von Jouanne A, Yen ZZ, Hapke HM, Halamay DA. Optimal energy storage sizing and control for wind power applications. IEEE Trans Sust Energy 2010;2:69–77.
113.RedFlow ZBM zinc-bromide flow battery. RedFlow advanced energy storage limited. October 2012. accessed 31.05.14..
114.Shoenung SM. Characteristics and technologies for long- vs. short-term energy storage: a study by the DOE energy storage systems program. Technical report. SAND2001-0765. Sandia National Laboratories. United States Department of Energy. Published March 2001.
115.Arai J, Iba K, Funabashi T, Nakanishi Y, Koyanagi K, Yokoyama R. Power electronics and its applications to renewable energy in Japan. IEEE Circ Syst Mag 2008;8:52–66.
116.Tong W. Wind power generation and wind turbine design. WIT Press; 2010. p. 678–82.
117.Rastler D. Electric energy storage applications in the electric enterprise: overview of applications and U.S. demonstration activities. Electric Power Research Institute (EPRI). EDF energy storage work shop. October 2011.
118.Energy commission awards 227,000 USD to SMUD for energy storage research. The California Energy Commission. Published 30th November 2011. accessed 28.05.14..
119.Baxter R. Energy storage: a nontechnical guide. PennWell Books; 2006.
120.Lin FC, Dai X, Xu ZA, Li J, Zhao ZG. High density capacitors. High Power Laser Part Beams 2003;1:94–6.
121.Arepalli S, Fireman H, Huffman C, Moloney P, Nikolaev P, Yowell L, et al. Carbon-nanotube-based electrochemical double-layer capacitor technologies for spaceflight applications. JOM 2005;57:26–31.
122.Capacitors age and capacitors have an end of life: a white paper from the experts in business-critical continuity. Emerson network power. n.d. accessed 27.05.14..

123. Electricity storage technology comparison. Electricity Storage Association (ESA). n.d. accessed 22.11.13..
124. Winter M, Brodd RJ. What are batteries, fuel cells, and supercapacitors? Chem Rev 2004;104:4245–70.
125. Kusko A, DeDad J. Short-term, long-term, energy storage methods for standby electric power systems. In: Fourtieth IAS Annu. Meet. Conf. Rec. 2005 Ind. Appl. Conf. 2005, vol. 4, IEEE; n.d., p. 2672–8.
126. Maxwell ultracapacitors. Maxwell technologies. n.d. accessed 27.05.14..
127. CAP-XX supercapacitors product guide 2013. CAP-XX Ltd. Online.. Available at: http://www.cap-xx.com/products/products.php. Accessed: 14-May- 2014..
128. Liu Q, Nayfeh MH, Yau S-T. Supercapacitor electrodes based on polyaniline– silicon nanoparticle composite. J Power Sources 2010;195:3956–9.
129. Abbey C, Joos G. Supercapacitor energy storage for wind energy applications. IEEE Trans Ind Appl 2007;43:769–76.
130. Markoulidis F, Lei C, Lekakou C. Fabrication of high-performance supercapacitors based on transversely oriented carbon nanotubes. Appl Phys A 2013;111:227–36.
131. Ali MH, Dougal RA. An overview of SMES applications in power and energy Systems. IEEE Trans Sust Energy 2010;1:38–47.
132. Superconducting magnets for fusion. Magnets and magnetic materials. Clefs CEA. Winter 2007–2008. 56. accessed 07.06.14..
133. Yuan W. Second-generation high-temperature superconducting coils and their applications for energy storage. Springer theses. Ph.D. thesis by the University of Cambridge 2011.
134. Schainker RB. Executive overview: energy storage options for a sustainable energy future. In: IEEE Power Eng. Soc. Gen. Meet. 2004, vol. 2, IEEE; n.d., p. 2310–5.
135. Superconducting Magnetic Energy Storage (SMES) project overview, SuperPower Inc. n.d. accessed 05.06.14..
136. Nie Z, Xiao X, Kang Q, Aggarwal R, Zhang H, Yuan W. SMES-battery energy storage system for conditioning outputs from direct drive linear wave energy converters. IEEE Trans Appl Supercond 2013;23:5000705.
137. Zhu J, Qiu M, Wei B, Zhang H, Lai X, Yuan W. Design, dynamic simulation and construction of a hybrid HTS SMES (high-temperature superconducting magnetic energy storage systems) for Chinese power grid. Energy 2013;51:184–92.
138. Hassenzahl WV, Hazelton DW, Johnson BK, Komarek P, Noe M, Reis CT. Electric power applications of superconductivity. Proc IEEE 2004;92:1655–74.
139. Swedish consortium for artificial photosynthesis. n.d. accessed 26.05.14..

140. The Royal Society of Chemistry. Solar fuels and artificial photosynthesis: science and innovation to change our future energy options. Report at the 1st UK Solar to Fuels Symposium. Published January 2012. accessed 12.06.14..
141. Amao Y, Shuto N, Furuno K, Obata A, Fuchino Y, Uemura K, et al. Artificial leaf device for solar fuel production. Faraday Discuss 2012;155:289.
142. Steinfeld A. Solar hydrogen production via a two-step water-splitting thermochemical cycle based on Zn/ZnO redox reactions. Int J Hydrogen Energy 2002;27:611–9.
143. Steinfeld A, Kuhn P, Reller A, Palumbo R, Murray J, Tamaura Y. Solarprocessed metals as clean energy carriers and water-splitters. Int J Hydrogen Energy 1998;23:767–74.
144. Styring S. Solar fuels: vision and concepts. Ambio Royal Swedish Acad Sci 2012;41(Suppl 2):156–62.
145. Styring S. Artificial photosynthesis for solar fuels. Faraday Discuss 2012;155:357.
146. Pagliaro M, Konstandopoulos AG, Ciriminna R, Palmisano G. Solar hydrogen: fuel of the near future. Energy Environ Sci 2010;3:279–87.
147. Lovegrove K, Luzzi A, McCann M, Freitag O. Exergy analysis of ammoniabased thermochemical solar power systems. Sol Energy 1999;66:103–15.
148. Fehrenbacher K. Apple reveals big solar, fuel cell plans for data center. Published 20th February 2012. accessed 28.05.14..
149. Solar fuels and artificial photosynthesis. BioSolar Cells. Published June 2013. accessed 29.05.14..
150. Reece SY, Hamel JA, Sung K, Jarvi TD, Esswein AJ, Pijpers JJH, et al. Wireless solar water splitting using silicon-based semiconductors and earth-abundant catalysts. Science 2011;334:645–8.
151. Tran PD, Wong LH, Barber J, Loo JSC. Recent advances in hybrid photocatalysts for solar fuel production. Energy Environ Sci 2012;5:5902.
152. Faunce TA, Lubitz W, Rutherford AW (Bill), MacFarlane D, Moore GF, Yang P, et al. Energy and environment policy case for a global project on artificial photosynthesis. Energy Environ Sci 2013;6:695.
153. Fuel Cells. Energy Efficiency & Renewable Energy. U.S. Department of Energy. n.d. accessed 28.03.14..
154. Mekhilef S, Saidur R, Safari A. Comparative study of different fuel cell technologies. Renew Sust Energy Rev 2012;16:981–9.
155. Kirubakaran A, Jain S, Nema RK. A review on fuel cell technologies and power electronic interface. Renew Sust Energy Rev 2009;13:2430–40.
156. Ozarslan A. Large-scale hydrogen energy storage in salt caverns. Int J Hydrogen Energy 2012;37:14265–77.

157. Nakken T, Strand L, Frantzen E, Rohden R, Eide PO, The Utsira wind-hydrogen system operational experience. In: Eur. Wind Energy Conf; 2006. p. 1–9.
158. Borup R, Meyers J, Pivovar B, Kim YS, Mukundan R, Garland N, et al. Scientific aspects of polymer electrolyte fuel cell durability and degradation. Chem Rev 2007;107:3904–51.
159. Gleason W. Fuel cell technician's guide. Cengage Learn 2012:285–94.
160. Largest renewable biogas fuel cell installation operating in the USA. FuelCell Energy Inc. n.d. accessed 16.05.14..
161. US Navy testing mobile regenerative fuel cell/solar system. Fuel Cells Bull 2013; 2013: 6.
162. Dauerthal, Fanjas LM. McPhy to continue ENERTRAG's HyTec electrolyzer activities. McPhy Energy. Published 27th September 2013. accessed 25.05.14..
163. Demirbas MF. Thermal energy storage and phase change materials: an overview. Energy Sources, Part B Econ Plan, Policy 2006;1:85–95.
164. Sharma A, Tyagi VV, Chen CR, Buddhi D. Review on thermal energy storage with phase change materials and applications. Renew Sust Energy Rev 2009;13:318–45.
165. Energy storage technologies for wind power integration. Report. Université Libre de Bruxelles Faculté des Sciences Appliquées, Service BEAMS groupe Energie. ULB. Published March 2010. accessed 14.06.14..
166. Thermal Energy Storage. Cristopia Energy Systens. accessed 18.06.14..
167. Aziz MBA, Zain ZM, Baki SRMS, Muslam MN. Review on performance of thermal energy storage system at S & T Complex, UiTM Shah Alam, Selangor. 2010 IEEE Control Syst. Grad. Res. Colloq. (ICSGRC 2010), IEEE; 2010, p. 49–54.
168. Gent E. Liquid air energy storage could become £1bn industry. The Institution of Engineering and Technology (IET) Engineering and Technology (E&T) Magazine. Published 9th May 2013.
169. Darby M. Liquid air plant gets £8 million for energy storage trial. Utility Week (UW). Published 14th February 2014. accessed 04.03.14..
170. Clark L. The benefits of ice-based thermal energy storage. Published 6th May 2010. accessed 16.12.13..
171. DeForest N, Mendes G, Stadler M, Feng W, Lai J, Marnay C. Optimal deployment of thermal energy storage under diverse economic and climate conditions. Appl Energy 2014;119:488–96.
172. Fairweather AJ, Stone DA, Foster MP. Evaluation of UltraBatteryTM performance in comparison with a battery-supercapacitor parallel network. J Power Sources 2013;226:191–201.
173. Hamajima T, Amata H, Iwasaki T, Atomura N, Tsuda M, Miyagi D, et al. Application of SMES and fuel cell system combined with liquid hydrogen

vehicle station to renewable energy control. IEEE Trans Appl Supercond 2012;22:5701704.

174. Liquid air energy network forms in UK: focus on transportation and energy storage. Published May 2013. accessed 30.05.14..

175. Electricity storage: technology brief. Technology Policy Brief, Energy Technology Systems Analysis Programme. International Renewable Energy Agency. Published April 2012. accessed 28.05.14..

176. Ries G, Neumueller H-W. Comparison of energy storage in flywheels and SMES. Phys C Supercond 2001;357–360:1306–10.

177. Beacon power technology. Beacon Power Corporation. n.d. accessed 24.05.14..

178. San Martín JI, Zamora I, San Martín JJ, Aperribay V, Eguía P. Energy storage technologies for electric applications. In: Int. Conf. on Renewable Energies and Power Quality (ICREPQ'11) 2011.

179. McDowall J. Integrating energy storage with wind power in weak electricity grids. J Power Sources 2006;162:959–64.

180. Emerging energy storage technologies in Europe. Rapport Frost & Sullivan; 2003.

181. Du Pasquier A, Plitz I, Menocal S, Amatucci G. A comparative study of Li-ion battery, supercapacitor and nonaqueous asymmetric hybrid devices for automotive applications. J Power Sources 2003;115:171–8.

182. Energy storage & renewables overview, solutions and customer case studies. Saft Batteries. n.d. accessed 29.05.14..

183. Beck F, Rüetschi P. Rechargeable batteries with aqueous electrolytes. Electrochim Acta 2000;45:2467–82.

184. Yamamura T, Wu X, Ohta S, Shirasaki K, Sakuraba H, Satoh I, et al. Vanadium solid-salt battery: solid state with two redox couples. J Power Sources 2011;196:4003–11.

185. Wen Z, Cao J, Gu Z, Xu X, Zhang F, Lin Z. Research on sodium sulfur battery for energy storage. Solid State Ionics 2008;179:1697–701.

186. McDowall J. High power batteries for utilities - the world's most powerful battery and other developments. In: IEEE Power Eng. Soc. Gen. Meet. 2004, vol. 2, IEEE; n.d. p. 2034–7.

187. Barote L, Weissbach R, Teodorescu R, Marinescu C, Cirstea M. Stand-alone wind system with vanadium redox battery energy storage. In: 2008 11th Int. Conf. Optim. Electr. Electron. Equip., IEEE; 2008. p. 407–12.

188. Gonzalez A, Ó'Gallachóir B, McKeogh E, Lynch K. Study of electricity storage technologies and their potential to address wind energy intermittency in Ireland. Final report. Sustainable energy research group and Rockmount capital partners Cork. RE/HC/03/001. Published May 2004.

189. Gills Onions Project–Fact Sheet: 600kW/3.6MWh VRB energy storage system. Prudent Energy VRB Systems. n.d. accessed 23.05.14..

190.Pistoia G. Electric and hybrid vehicles: power sources, models, sustainability, infrastructure and the market. Elsevier; 2010. p. 373.
191.Li P. Energy storage is the core of renewable technologies. IEEE Nanotechnol Mag 2008;2:13–8.
192.Moghaddam MP, Haghifam MR, Yousefi GR. Electric energy storage systems in a market-based economy: comparison of emerging and traditional technologies. Renew Energy 2009;34:2630–9.
193.Ponce de León C, Frías-Ferrer A, González-García J, Szánto DA, Walsh FC. Redox flow cells for energy conversion. J Power Sources 2006;160:716–32.
194.The kWh sodium polysulfide/bromine redox flow battery energy storage for experimental prototype has been developed. Energy storage division of Dalian National Laboratory for Clean Energy. Published 31th March 2011. accessed 24.05.14..
195.Ribeiro PF, Johnson BK, Crow ML, Arsoy A, Liu Y. Energy storage systems for advanced power applications. Proc IEEE 2001;89:1744–56.
196.Sannomiya T, Hayashi H, Ishii T, Ikeda R. Test results of compensation for load fluctuation under a fuzzy control by a 1 kW h/1 MW SMES. IEEE Trans Appl Supercond 2001;11:1908–11.
197.The beta plant is currently being planned: testing in the megawatt range. ETOGAS GmbH. n.d. accessed 26.05.14..
198.Electric fuel introduces practical, zero-emission transportation.Electric Fuel Ltd. accessed 16.06.14..
199.Construction of world's largest fuel cell park expected to commence in 2012. Fuel Cell Today. Source from Fuel Cell Energy Inc. Published 17th October 2012. accessed 18.06.14..
200.INGRID 39 MWh grid-connected renewable energy storage project. Fuel Cells Bull 2012; 2012: 9.
201.Solar Two demonstrates clean power for the future. Sun Lab produced U.S. Development of Energy. n.d. accessed 17.06.14..
202.Liquid Air Energy Storage (LAES). U.S. Energy Storage Association. n.d. accessed 30.05.14..
203.Andrepont SJ. Energy storage – thermal energy storage coupled with turbine inlet cooling. In: 14th annual electric power conf. & exhibition. n.d. accessed 17.06.14..
204.Kaldellis JK, Zafirakis D. Optimum energy storage techniques for the improvement of renewable energy sources-based electricity generation economic efficiency. Energy 2007;32:2295–305.
205.Rydh CJ, Sanden BA. Energy analysis of batteries in photovoltaic systems – part II: energy return factors and overall battery efficiencies. Energy Convers Manage 2005;46:1980–2000.

206.Schoenung S. Energy storage systems cost update: a study for the DOE energy storage systems program. Technical report. Sandia National Laboratories; 2011.
207.Tokuda N, Kanno T, Hara T, Shigematsu T, et al. Development of a redox flow battery system. Sei Tech Rev 2000:50.
208.Gevorkian P. Large-scale solar power systems: construction and economics, vol. 08. Cambridge University Press; 2012.
209.Boer PD, Raadschelders J. Flow batteries. Briefing paper. Leonardo ENERGY; June 2007. accessed 27.05.14..
210.Miller RJ. Capacitors for power grid storage - multi-hour bulk energy storage using capacitors. JME, Inc. and Case Western Reserve University. In: TransAtlantic Workshop on Storage Technologies for Power Grids; 2010.
211.Pickard WF, Shen AQ, Hansing NJ. Parking the power: strategies and physical limitations for bulk energy storage in supply–demand matching on a grid whose input power is provided by intermittent sources. Renew Sust Energy Rev 2009;13:1934–45.
212.Smith W. The role of fuel cells in energy storage. J Power Sources 2000;86:74–83.
213.Schaber C, Mazza P, Hammerschlag R. Utility-scale storage of renewable energy. Electr J 2004;17:21–9.
214.Highview power storage: secure, clean power. Highview Power; 2011. accessed 24.12.13..
215.Liquid air in the energy and transport systems: opportunities for industry and innovation in the UK full report. The Centre for Low Carbon Futures. Released 9th May 2013. accessed 19.06.14..
216.Dan R. Overview of energy storage options for the electric enterprise. Electric Power Research Institute (EPRI); 2009. accessed 29.06.14..
217.Barin A, NevesCanha L, Magnago K, da Rosa Abaide A. Fuzzy multi-sets and multi-rules: analysis of hybrid systems concerning renewable sources with conventional and flow batteries. In: 2009 15th Int. Conf. Intell. Syst. Appl. to Power Syst., IEEE; 2009, p. 1–6.
218.Kerestes RJ, Reed GF, Sparacino AR. Economic analysis of grid level energy storage for the application of load leveling. In: 2012 IEEE Power Energy Soc. Gen. Meet., IEEE; 2012. p. 1–9.
219.Bradbury K, Pratson L, Patiño-Echeverri D. Economic viability of energy storage systems based on price arbitrage potential in real-time U.S. electricity markets. Appl Energy 2014;114:512–9.
220.Heymans C, Walker SB, Young SB, Fowler M. Economic analysis of second use electric vehicle batteries for residential energy storage and load-levelling. Energy Policy 2014;71:22–30.

221.Nakao K, Ishiyama A, Seo HR, Kim AR, Park M, Yu IK. Power quality enhancement of renewable energy source power network using SMES system. Phys C Supercond 2011;471:1409–12.
222.T, Kumamoto T, Deguchi H, Hara T. Applications of a vanadium redox-flow battery to maintain power quality. In: IEEE/PES Transm. Distrib. Conf. Exhib., vol. 2, IEEE; 2002. p. 1065–70.
223.Rahim AHMA, Nowicki EP. Supercapacitor energy storage system for fault ride-through of a DFIG wind generation system. Energy Convers Manage 2012;59:96–102.
224.Keeping the electricity transmission system in balance. National Grid. n.d. accessed 19.06.14..
225.Ni B, Sourkounis C. Control strategies for energy storage to smooth power fluctuations of wind parks. In: Melecon 2010–15th IEEE Mediterr. Electrotech. Conf., IEEE; 2010. p. 973–8.
226.YekiniSuberu M, Wazir Mustafa M, Bashir N. Energy storage systems for renewable energy power sector integration and mitigation of intermittency. Renew Sust Energy Rev 2014;35:499–514.
227.World first – environmentally friendly air-powered generator system installed at Stockport Pyramid. Flowbattery. accessed 31.10.14..
228.Ribeiro E, Cardoso AJM, Boccaletti C. Fuel cell-supercapacitor system for telecommunications. In: 5th IET Int. Conf. Power Electron. Mach. Drives (PEMD 2010), IET; 2010. p. 1–6.
229.Kyriakopoulos A, O'Sullivan D, Hayes JG, Griffiths J, Egan MG. Kinetic energy storage for high reliability power supply back-up. APEC 07 – Twenty-Second Annu. In: IEEE Appl. Power Electron. Conf. Expo., IEEE; 2007. p. 1158–63.
230.Horner RE, Proud NJ. The key factors in the design and construction of advanced flywheel energy storage systems and their application to improve telecommunication power back-up. In: Proc. Intelec'96 – Int. Telecommun. Energy Conf., IEEE; 1996. p. 668–75.
231.Krishnan KJ, Kalam A, Zayegh A. H2 optimisation and fuel cell as a back-up power for telecommunication sites. In: 2013 Int. Conf. Circuits, Power Comput. Technol., IEEE; 2013. p. 1274–7.
232.Li S, Tomsovic K, Hiyama T. Load following functions using distributed energy resources. In: 2000 Power Eng. Soc. Summer Meet. (Cat. No.00CH37134), vol. 3, IEEE; 2000. p. 1756–61.
233.Hida Y, Yokoyama R, Shimizukawa J, Iba K, Tanaka K, Seki T. Load following operation of NAS battery by setting statistic margins to avoid risks. In: IEEE PES Gen. Meet., IEEE; 2010. p. 1–5.
234.Yang M, Dixon RK, Taylor P, Nyamdash B, Denny E, O'Malley M. The viability of balancing wind generation with large scale energy storage. Energy Policy 2010;38:7200–8.

235.Rudolf V, Papastergiou KD. Financial analysis of utility scale photovoltaic plants with battery energy storage. Energy Policy 2013;63:139–46.
236.Purvins A, Sumner M. Optimal management of stationary lithium-ion battery system in electricity distribution grids. J Power Sources 2013;242:742–55.
237.Lobato E, Sigrist L, Rouco L. Use of energy storage systems for peak shaving in the Spanish Canary Islands. In: 2013 IEEE Power Energy Soc. Gen. Meet., IEEE; 2013. p. 1–5.
238.Oudalov A, Cherkaoui R, Beguin A. Sizing and optimal operation of battery energy storage system for peak shaving application. In: 2007 IEEE Lausanne Power Tech, IEEE; 2007. p. 621–5.
239.Kloess M, Zach K. Bulk electricity storage technologies for load-leveling operation – an economic assessment for the Austrian and German power market. Int J Electr Power Energy Syst 2014;59:111–22.
240.Converse AO. Seasonal energy storage in a renewable energy system. Proc IEEE 2012;100:401–9.
241.Peerce-Landers P. The role of hydrogen in seasonal energy storage. Published 14th November 2013. accessed 17.12.13..
242.Worthington MA. Seasonal thermal energy storage applications for district energy systems. IDEA 101ST Annual Conference & Trade Show; 2010.
243.Thomas PR, Walker TJ. Demonstration of Community Energy Storage fleet for load leveling, reactive power compensation, and reliability improvement. In: 2012 IEEE Power Energy Soc. Gen. Meet., IEEE; 2012. p. 1–4.
244.Wang W, Ge B, Bi D, Qin M, Liu W. Energy storage based LVRT and stabilizing power control for direct-drive wind power system. In: 2010 Int. Conf. Power Syst. Technol., IEEE; 2010. p. 1–6.
245.Rogers JD, Schermer RI, Miller BL, Hauer JF. 30-MJ superconducting magnetic energy storage system for electric utility transmission stabilization. Proc IEEE 1983;71:1099–107.
246.Irie F, Takeo M, Sato S, Katahira O, Fukui F, Okada H, et al. A field experiment on power line stabilization by a SMES system. IEEE Trans Magn 1992;28:426–9.
247.Celli G, Pilo F, Soma GG, Dal Canto D, Pasca E, Quadrelli A. Benefit assessment of energy storage for distribution network voltage regulation. In: CIRED 2012 Work. Integr. Renewables into Distrib. Grid, IET; 2012. p. 1–4.
248.Wheeler N. Voltage regulation application of a kinetic energy storage system. In: Second IEE int. conf. power electron. mach. drives, vol. 2004. IEE; 2004, p. 605–8.
249.Quesada J, Sebastián R, Castro M, Sainz JA. Control of inverters in a low voltage microgrid with distributed battery energy storage. Part I: primary control. Electr Power Syst Res 2014;114:126–35.
250.Energy storage: added stability for electrical grids. European Voice. Published 26th September 2013. accessed 17.12.13..

251. Kazempour SJ, Hosseinpour M, Moghaddam MP. Self-scheduling of a joint hydro and pumped-storage plants in energy, spinning reserve and regulation markets. In: 2009 IEEE Power Energy Soc. Gen. Meet., IEEE; 2009, p. 1–8.
252. Dursun B, Alboyaci B. The contribution of wind-hydro pumped storage systems in meeting Turkey's electric energy demand. Renew Sust Energy Rev 2010;14:1979–88.
253. Mercier P, Cherkaoui R, Oudalov A. Optimizing a Battery energy storage system for frequency control application in an isolated power system. IEEE Trans Power Syst 2009;24:1469–77.
254. Campbell C. Business case for advanced energy storage: spinning reserve. The Daily Energy Report. Published 4th April 2011. accessed 17.12.13..
255. Habibi MS. Model for impact of storage on spinning reserve requirements and distributed generation. In: 33rd Southeast. Symp. Syst. Theory, IEEE; 2001. p. 161–5.
256. Torreglosa J, Garcia P, Fernandez L, Jurado F. Predictive control for the energy management of a fuel cell-battery-supercapacitor tramway. IEEE Trans Ind Inf 2013;10:276–85.
257. Thounthong P, Chunkag V, Sethakul P, Davat B, Hinaje M. Comparative study of fuel-cell vehicle hybridization with battery or supercapacitor storage device. IEEE Trans Veh Technol 2009;58:3892–904.
258. Steiner M, Klohr M, Pagiela S. Energy storage system with ultracaps on board of railway vehicles. In: 2007 Eur. Conf. Power Electron. Appl., IEEE; 2007, p. 1–10.
259. Lukic SM, Bansal RC, Rodriguez F, Emadi A. Energy storage systems for automotive applications. IEEE Trans Ind Electron 2008;55:2258–67.
260. Amjadi Z, Williamson SS. Review of alternate energy storage systems for hybrid electric vehicles. In: 2009 IEEE Electr. Power Energy Conf., IEEE; 2009. p. 1–7.
261. Olivo T. Analysis of ultra capacitors as UPS energy storage devices. In: Proc. IEEE SoutheastCon 2010, IEEE; 2010. p. 398–401.
262. Zhou L, Qi ZP. Modeling and control of a flywheel energy storage system for uninterruptible power supply. In: 2009 Int. Conf. Sustain. Power Gener. Supply, IEEE; 2009. p. 1–6.
263. Racine MS, Parham JD, Rashid MH. An overview of uninterruptible power supplies. In: Proc. 37th Annu. North Am. Power Symp. 2005, IEEE; 2005. p. 159–64.
264. Lahyani A, Venet P, Guermazi A, Troudi A. Battery/supercapacitors combination in Uninterruptible Power Supply (UPS). IEEE Trans Power Electron 2013;28:1509–22.
265. The future value of storage in the UK with generator intermittency. Dti report. DG/ DTI/00040/00/00. Published 2004. accessed 17.12.13..

266.Black M, Strbac G. Value of bulk energy storage for managing wind power fluctuations. IEEE Trans Energy Convers 2007;22:197–205.

CITATION

Xing Luo, Jihong Wang, Mark Dooner, Jonathan Clarke, Overview of current development in electrical energy storage technologies and the application potential in power system operation, Applied Energy, Volume 137, 1 January 2015, Pages 511-536, ISSN 0306-2619, http://dx.doi.org/10.1016/j.apenergy.2014.09.081.

Index